Global Trends in Smog and Air Pollution Control

Qamar Ali
https://orcid.org/0000-0002-8178-7591
Virtual University of Pakistan, Pakistan

Muhammad Ashfaq
The University of Faisalabad, Pakistan

IGI Global
Scientific Publishing
Publishing Tomorrow's Research Today

Vice President of Editorial	Melissa Wagner
Director of Acquisitions	Mikaela Felty
Director of Book Development	Jocelynn Hessler
Production Manager	Mike Brehm
Cover Design	Jose Rosado

Published in the United States of America by
IGI Global Scientific Publishing
701 East Chocolate Avenue
Hershey, PA, 17033, USA
Tel: 717-533-8845 | Fax: 717-533-7115
Website: https://www.igi-global.com E-mail: cust@igi-global.com

Library of Congress Cataloging-in-Publication Data

(CIP Data Pending)
ISBN13: 9798337324692
Isbn13Softcover: 9798337324708
EISBN13: 9798337324715
British Cataloguing in Publication Data
A Cataloguing in Publication record for this book is available from the British Library.

Editorial Advisory Board

Table of Contents

Detailed Table of Contents

Air pollution in South Asia and China reached critical levels during 2000–2023, with annual $PM_{2.5}$ averaging 58.21 μg/m^3, far above the 10 μg/m^3 safety guideline. Peaks rose to 135.60 μg/m^3, while the lowest levels stood at 12.50 μg/m^3. Agriculture contributed most, with ammonia emissions averaging 456.30 kt and reaching as high as 5,812 kt. Transport demand was intense, averaging 785.45 thousand barrels per day and peaking above 8,500. Industry added steady pressure, averaging 27.18% of GDP and reaching 46.20% in some nations. Urbanization averaged 38.75% but at times exceeded 65%, with mixed effects. Renewable energy provided consistent relief, averaging 36.52% of supply and cutting $PM_{2.5}$ by up to –0.72. Econometric results confirmed that agriculture, transport, and industry increased pollution, while renewables reduced it. Threshold analysis showed that risks doubled when income exceeded USD 9,000, but renewable benefits also became stronger. Long-run tests confirmed convergence, with error correction near –0.47. Policy analysis of 158 documents revealed strong negative sentiment toward burning (–0.87), urban sprawl (–0.91), and industry gaps (–0.85), while renewables (+0.82) and community adaptation (+0.65) received strong support. The evidence highlights farming and transport as the most critical threats and renewable adoption as the strongest solution for clean air and sustainable growth.

This review synthesizes research on smog-based issues and their applications in the attainment of Sustainable Development Goals, particularly SDG 3 Good Health and Well-being and SDG 11 Sustainable Cities and Communities. The health impacts, economic losses, and policy effectiveness are presented, including SDG linkages, technological innovations, and regulations. Evidence involves high smog illness, death, and economic losses with sporadic policy enforcement that constrains progress. Technological measures such as clean vehicles and digital tracking have potential but with constraints in application. Holistic approaches in the health, sustainable urban, and climate-related sectors emphasize co-benefits but with application. Urban planning with community-based action promotes cleaner air but with the necessity of equity-based integration. Review identifies harmonized coordinated multi sector policy strategies that intersect strong policy, technology, and inclusive governance in building cleaner air and sustainable development.

Smog, which is a combination of high particulate matter and gaseous pollutants, is one of the major environmental issues with immense implications for sustainable development. This chapter applies an integrated assessment framework to understand the multi-faceted, complex relationships that exist between smog and the United Nations' Sustainable Development Goals (SDGs). It reviews the direct and indirect effects of atmospheric pollution on the key SDGs such as Good Health and Well-being (SDG 3), Affordable and Clean Energy (SDG 7), Sustainable Cities (SDG 11), Climate Action (SDG 13) and Decent Work and Economic Growth (SDG 8). The analysis shows both serious threats to achieving the goals from smog and significant opportunities for synergistic policy co-benefits. The chapter concludes by arguing that the target of smog can only be achieved by going beyond sectoral boundaries in a mainstreaming approach to air quality into energy, transportation, industrial, and urban planning policy. Drawing on empirical evidence and a case study carried out on air quality management in China, the chapter concludes that multi-pollutant, integrated solutions are not only an environmental necessity but an essential foundation for achieving inclusive, environmentally sustainable economic development and protecting public health for present and future generations.

Smog, a mixture of air pollutants, dust, and smoke coupled with fog under direct sunlight, significantly jeopardizes human health and the environment by diminishing visibility. Smog has emerged as an escalating global concern with significant repercussions for human health, ecosystems, and economies all around the world. It has also emerged has major issue for environmental sustainability, health and overall economic performance in Pakistan. The study's goal is to critically analyzed the interdependent issues of urban smog, energy transition, and sustainable development in Pakistan, with an aim to suggest an effective policy framework for climate-resilient urbanization. Hence, this paper proposes an integrated policy framework that aligns smog mitigation, energy transition, and sustainable urban development to ensure climate-resilient cities in Pakistan. This policy paper rigorously analyzes the systemic causes of smog, associating it with industrial emissions, automobile pollution, agricultural produce incineration, garbage combustion and inadequate urban design. Notwithstanding established legal frameworks, inadequate enforcement and bureaucratic inefficiencies intensify the situation. This report, informed by expert perspectives from the Pakistan Economics Frontier's World Energy Day event, advocates a comprehensive policy framework to alleviate smog via sustainable urbanization, energy transition, and regional cooperation. The suggested framework underscores legal reforms, technology advancements, and cohesive governance to tackle smog at its source. Essential interventions comprise the implementation of Euro 6 gasoline requirements, the transition to electric vehicles, the modernization of public transportation, and the enhancement of climate diplomacy. Investments in green infrastructure, such as sustainable urban planning, vertical gardens, and afforestation, are crucial for the effective long-term management of this issue. This study is consistent with Sustainable Development Goals (SDG 7, 11, and 13) and promotes a policy approach that is grounded in evidence and data to improve climate resilience and reduce emissions. By Adopting renewable energy, green investment, stimulating multispectral. Reciprocity Pakistan can achieve environmental and economical sustainability.

Pakistan's air quality crisis is persistent, severe, and costly. Fine particulate matter (PM$_{2.5}$) concentrations routinely exceed the World Health Organization's (WHO) 2021 guideline of 5 $\mu g/m^3$ annual average; in 2023 Pakistan's national average was 73.7 $\mu g/m^3$—about 15× the WHO guideline—placing the country among the world's most polluted according to IQ Air's global assessment. These levels drive large health losses, productivity declines, and wide social disruptions, including repeated school closures in Punjab during the winter smog season. The World Bank estimates the economic cost of ambient air pollution in Pakistan at roughly 6.5% of GDP annually, with major contributions from residential solid-fuel use, transport, industry (including brick kilns), agriculture (residue burning), and waste burning. Policy momentum has accelerated (Euro-5 fuel adoption, a National Clean Air Policy, Punjab's Clean Air Plan, kiln modernization), yet implementation gaps, limited monitoring, and the regional "airshed" nature of the problem keep Pakistan highly exposed.

Türkiye and Pakistan, despite differing levels of economic development, face similar macroeconomic challenges such as income inequality, inflation and debt. Both countries also experience environmental pressures from industrialization and rapid urbanization, with air pollution emerging as a critical concern affecting economic productivity and sustainable development. This study conducts a comparative analysis of air pollution in Türkiye and Pakistan from 1980 to 2023, using the Autoregressive Distributed Lag (ARDL) bounds testing approach to examine the short and long run relationships between air pollution, urbanization, industrialization, energy consumption and macroeconomic policies. The analysis tests the Environmental Kuznets Curve (EKC) for Türkiye and the pollution-led growth hypothesis for Pakistan. Findings reject the EKC for Türkiye but confirm pollution-led growth in Pakistan, offering insights for effective environmental regulation and sustainable development strategies.

This chapter will investigate sustainable practices that can effectively address these intertwined challenges. By emphasizing green infrastructure—such as urban forests, green roofs, permeable pavements, and community gardens—we can identify strategies that improve urban environments. Green infrastructure not only enhances air and water quality but also creates vital spaces for recreation and social interaction, thereby fostering community well-being. Moreover, this chapter will examine the necessity of integrating environmental policies with public health initiatives. Sustainable urban planning can lead to the creation of resilient communities, where health considerations are embedded in urban design. By promoting equitable access to green spaces, enhancing public transportation, and investing in clean energy solutions, cities can cultivate healthier living environments that benefit all residents.

the article adopts an interdisciplinary legal analysis that integrates environmental science, human rights theory, and public international law. The inquiry begins with a definitional framework of the environment as a composite of biotic and abiotic elements, followed by an exploration of its socio-legal dimensions. Ultimately, the article advances a prescriptive methodology, arguing for the recognition of an enforceable universal right to a healthy environment. This methodological design combines descriptive legal mapping with evaluative critique, proposing that sustainable governance requires a holistic integration of environmental law into the human rights system. The study thus positions itself at the intersection of environmental protection and fundamental rights, offering a framework for both academic discourse and policy-making.

Air pollution and smog remain critical global health challenges, contributing significantly to respiratory illnesses, cardiovascular diseases, and premature mortality. Despite advancements in air quality monitoring and control strategies, urbanization, industrialization, and vehicular emissions continue to exacerbate pollution levels in many regions. This paper examines the health implications of global trends in smog and air pollution, highlighting the interconnectedness between environmental degradation and public health outcomes. It further evaluates contemporary strategies for cleaner air, including technological innovations, regulatory frameworks, and sustainable urban planning, as well as their effectiveness in reducing exposure-related health risks. The study underscores the necessity for integrated approaches that combine policy intervention, public awareness, and cross-sectoral collaboration to achieve long-term improvements in air quality and public well-being.

South Africa is arguably one of the major emitters experiencing high levels of air pollution. The blessing of natural ground resources come with a price as major mining activities contribute to seasonal air pollution and smog issues. The primary objective of the study is to investigate if corporate sustainability performance influences debt pricing. The study employs quantitative research and adopts a positivist perspective. Archival data were employed as extracted from companies constituted in the FTSE/JSE Responsible Investment Index selected primarily for their environmental impact.

Preface

INTRODUCTION

Smog and air pollution have become one of the most topical environmental and public-health concerns of the twenty-first century. The high rate of industrialization, growth of urban settlements, unregulated emission, and the increased use of energy has aggravated the air quality in both developed and developing world. Now almost every continent is polluted by air and poses threats to human health, destruction of the ecosystems, and is an economic burden to the countries. Such problems underscore an important world-wide fact that clean air is no longer an implicit environmental advantage but a basic human need, and requires joint scientific, technological, and policy interventions.

The literature of smog formation, atmospheric chemistry, urban pollution source and mitigation technology has expanded exponentially in two decades since the onset of the past century. Different strategies are being taken to countries, some of them are stricter emissions rules, renewable energy sources, and modern transport systems, others are sophisticated air quality monitoring, behavior interventions, and pollution control methods, horticulture. Concurrently, cross-disciplinary research in the field of environmental science, policy, engineering, health, planning, and behavioural research is now imperative in the development of solutions that are effective, fair and scalable.

The edited volume gathers the works of different scholars in the global arena who analyze the issue of smog and air pollution through various perspectives: scientific, socioeconomic, technological, legal, and behavioural. The book tries to come up with a multifaceted, systemic, and progressive view of how societies around the world can learn more about smog, react more efficiently, and create the long-term ways through which people can breathe cleaner air and build sustainable cities.

This book aims to achieve a synthesis of the causes, effects, and control measures of smog through the combination of scientific knowledge, practical study, policy formulation, and novel approaches. It is supposed to be a single source to be used by researchers, engineers, policymakers, business leaders, and environmental scientists, as well as students who would be dedicated to bringing cleaner, healthier, and more resilient communities across the globe.

CHAPTER OVERVIEW

Chapter 1 frames scientific and empirical backgrounds of smog in South Asia and China, providing specific evidence on the movements of PM 2.5, agricultural, transport-related, industry-related, renewable energy interactions and econometric analysis which plots drivers and thresholds of pollution.

Chapter 2 assesses the connections between smog and the Sustainable Development Goals (SDGs): health, urban sustainability, technological adoption, and multi-sector governance as pathways to cleaner air are considered to be most important.

Chapter 3 builds upon this evaluation by offering a framework that both links smog with the SDGs including health, clean energy, sustainable city, climate action and economic growth and also identifies both systemic risks and the potential co-benefits.

Chapter 4 offers a Pakistan-centered approach to the smog, including the role of energy transitions, urban governance, climate resilience, and policy reforms that should be implemented to address the increasing pollution rates.

Chapter 5 focuses on the air quality issue in Pakistan, which is a severe crisis with economic, environmental, and social facets such as the health-related losses, the loss of productivity, and the gaps in the current policies.

In chapter 6, the econometric models comparing Türkiye and Pakistan are used to examine the Environmental Kuznets Curve (EKC), pollution based growth, industrialization, and macroeconomic factors regarding air quality.

Chapter 7 focuses on understanding how pollution exposure and community resilience can be achieved by integrated environmental and public-health planning using green infrastructure and urban health strategies.

Chapter 8 addresses the horticulture-based solutions to mitigating smog globally which includes urban forests, vertical gardens, green roofs, and plants based systems in the capture of pollutants.

Chapter 9 uses biblimetric analysis to research green entrepreneurship and determines the trends in global research on the subject, clusters concerning the innovation, and the contribution that sustainable enterprises offer to fight air pollution.

Chapter 10 is a synthesis of the world evidence on the causes, health impacts, pollutants and mitigation framework relating to smog, with case studies of Asia, Europe and other parts of the world.

Chapter 11 examines the ways to reduce smog in the world by technology, governance, community action, and SDG-oriented reform of policy.

Chapter 12 focuses on the discussion of smog in the larger framework of environmental rights, advocating the globalization of a human right to a healthy environment.

Chapter 13 relates smog to the health of the people by examining the respiratory and cardiovascular effects and evaluating technological and policy solutions to clean air.

Chapter 14 examines the connection between performance of corporate sustainability and cost of debt, including price of environmental risks in the South African pollution-intensive industries.

CONCLUSION

Smog and air pollution are international issues which transcend geographic, socioeconomic, and institutional frontiers. All the contributions to research in this volume help to prove that it is possible to get cleaner air by incorporating the elements of science, technology, policy innovation, interaction with the population, and cross-sectoral cooperation. It is impossible to establish effective air quality management by isolated interventions; this will require co-ordinated governance, fair investments and commitment by the society in a long-term basis.

The book is also intended to educate and motivate scholars, practitioners and policymakers who want to find practical solutions towards cleaner air and healthier communities. The volume has added to the world discussion on air quality improvement by offering a wide range of case studies, interdisciplinary and future-oriented approaches to it, and given a platform upon which to build evidence-based, equitable, and long-term solutions.

It is our great pleasure to thank all the participating authors, reviewers, and institutions and their knowledge, which have enriched this volume. We hope that this book will help make informed decisions, become innovative and build strong communities and nations in the future where all people can access clean air.

Chapter 1
Understanding Smog

Qamar Ali
https://orcid.org/0000-0002-8178-7591
Virtual University of Pakistan, Pakistan

Muhammad Aamir Mahmood
https://orcid.org/0009-0007-4452-2627
Virtual University of Pakistan, Pakistan

Muhammad Ashfaq
The University of Faisalabad, Pakistan

ABSTRACT

Air pollution in South Asia and China reached critical levels during 2000–2023, with annual $PM_{2.5}$ averaging 58.21 µg/m^3, far above the 10 µg/m^3 safety guideline. Peaks rose to 135.60 µg/m^3, while the lowest levels stood at 12.50 µg/m^3. Agriculture contributed most, with ammonia emissions averaging 456.30 kt and reaching as high as 5,812 kt. Transport demand was intense, averaging 785.45 thousand barrels per day and peaking above 8,500. Industry added steady pressure, averaging 27.18% of GDP and reaching 46.20% in some nations. Urbanization averaged 38.75% but at times exceeded 65%, with mixed effects. Renewable energy provided consistent relief, averaging 36.52% of supply and cutting $PM_{2.5}$ by up to –0.72. Econometric results confirmed that agriculture, transport, and industry increased pollution, while renewables reduced it. Threshold analysis showed that risks doubled when income exceeded USD 9,000, but renewable benefits also became stronger. Long-run tests confirmed convergence, with error correction near –0.47. Policy analysis of 158 documents revealed strong negative sentiment toward burning (–0.87), urban sprawl (–0.91), and industry gaps (–0.85), while renewables (+0.82) and commu-

DOI: 10.4018/979-8-3373-2469-2.ch001

nity adaptation (+0.65) received strong support. The evidence highlights farming and transport as the most critical threats and renewable adoption as the strongest solution for clean air and sustainable growth.

1. UNDERSTANDING SMOG: FROM HISTORICAL ORIGINS TO FUTURE CHALLENGES

1.1. Introduction to Smog: The Evolving Environmental Crisis

Figure 1. Key pollution drivers and renewable energy responses in South Asia and China.

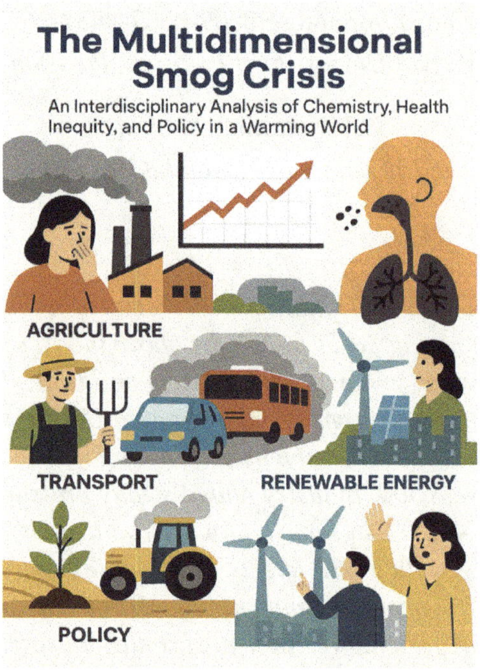

Figure 2. Smog impacts on air quality, health inequity, and policy responses.

Smog is one of the world's most widespread environmental problems, an atmospheric phenomenon with several interacting mechanisms, and with serious consequences for human health, ecosystems and climate systems(Benis et al., 2021; Mancini et al., 2017). Although the term itself, a portmanteau of "smoke" and "fog," originated during London's industrial revolution, it has since come to mean many types of atmospheric pollution across the world. Today smog is a serious global health problem, and about 46 percent of Americans (156 million people) live in areas where air pollution exceeds healthy levels, either of ozone or particulate. The geographical spread of smog has moved on from its industrial roots, with mega-cities on continents coming under the smog's effects and showcasing the transnationality of air quality issues(Richelle, 2025).

The history of smog is informative because it is a problem that has changed significantly since its discovery. London's notorious 1952 Great Smog, which led to thousands of deaths, was caused by coal burning and meteorological conditions that led to sulfur dioxide and particulate matter being trapped. Today, however, the smog takes a different form in different regions: photochemical smog plagues cities such as Los Angeles, where nitrogen oxides and volatile organic compounds combine in sunlight to create ground-level ozone, while smog in rapidly developing economies continues to stem from coal-burning. This has been evolving in response to changing energy systems, patterns of economic development, and technological change that have altered the chemical makeup and seasonal patterns of smog events around the world(Bilal et al., 2022).

Smog is a multidimensional phenomenon - it's both a local health threat and a regional environmental issue, as well as a global climate problem. The modern smog is complicated by its multiple sources of emissions, complex atmospheric chemistry and the varying impacts to different sectors of society. As explained in "Breathless: The Exhausting Reality of Air Pollution," a few days spent in cities such as Paris, London, or Rome subjects residents to as much pollution as smoking two or three cigarettes, and in Delhi, the toxic smog is as health threatening as smoking 50 cigarettes a day. Given that this public health emergency requires a broad understanding and innovative solutions, the study of smog is progressively important for urban planners, policymakers and health experts around the world(Giles-Corti et al., 2022).

1.2. The Chemical Makeup and Formation of Smog

1.2.1. Major Pollutants: the Raw Materials for Smog

Smog is formed by the emission of primary pollutants which are directly released into the atmosphere from known sources. Among the most important are the nitrogen oxides (NOx), especially nitrogen dioxide (NO2), which is entirely released by high-temperature combustion processes in vehicles and power plants. The transportation sector is the source of about 45% of NOx emissions in the United States. Volatile Organic Compounds (VOCs) are another important class of primary contaminants, volatile organic compounds in the form of vapors that come from fuels, solvents, paints, natural sources such as vegetation. These compounds range in chemical complexity and reactivity from simple to very complex and some VOCs are precursors for hazardous air toxins such as benzene and formaldehyde, which are known or suspected carcinogens(Protano et al., 2021).

Particulate matter (PM) is the third major category of primary contaminants and is made up of microscopic solid or liquid particles in air. These particles are distinguished by size rather than chemical makeup, and the particles most frequently monitored for health impacts are PM2.5 (particles less than 2.5 micrometers in size) and PM10 (particles less than 10 micrometers in size). Primary particles are emitted directly from sources such as diesel engines, construction sites, industrial processes and biomass burning. Importantly, particulate matter varies tremendously in its chemical composition, and may contain sulfates, nitrates, ammonia, sodium chloride, black carbon, mineral dust, and water. Due to their heterogeneity, such particles have the potential to transport adsorbed toxic species deep into the human respiratory system and are therefore of particular concern to human health(Mazziotta et al., 2023).

1.3. Secondary Pollutants: Chemical Transformations

Smog formation is a complex set of reactions in atmospheric chemistry, which convert primary pollutants into secondary pollutants, which are often more toxic than their original precursors. Ground-level ozone (O3), which is the main component of photochemical smog, is produced by photochemical reactions involving nitrogen oxides (NOx) and volatile organic compounds (VOCs), in the presence of solar energy. This secondary pollutant is distinctively different from the protective ozone layer in the stratosphere; on ground level, ozone is a strong respiratory irritant whose effects have been described as "a sunburn of the lungs." The non-linear nature of the ozone formation process in which precursor emissions and pollution levels are related means that control strategies are complex(Zhao et al., 2023).

Secondary particulate matter is a second important class of transformed pollutants formed by atmospheric reactions of gaseous precursors. For example, sulfur dioxide (SO2) can oxidize to produce sulfate particles and nitrogen oxides can change to nitrate particles. Secondary particles may constitute a major portion of PM2.5 levels, particularly in the region downwind of large sources. The atmospheric lifetime of these pollutants is sufficient to cause regional-scale transport, so smog may form one place due to emissions many hundreds of kilometers away. This conversion process poses important policy problems since pollution control must deal with not only local emissions but regional precursor emissions as well(Olivier & Berdowski, 2021).

1.4. Emerging Problems and Non-conventional Pollutants

Emerging contaminants in smog have recently been identified by scientists, making controls and monitoring of smog even more difficult. For example, polycyclic aromatic hydrocarbons (PAHs) are a group of carbon and hydrogen-containing organic compounds that are created when fossil fuels, wood, garbage, or other organic material are burned in an incomplete manner. Of more than 100 PAHs known to be prevalent in the environment, 15 are listed in the Report on Carcinogens, evidencing their potent toxicity. These compounds tend to attach to particulate matter and enter deep into the respiratory system and may even enter the blood stream(Beachey, 2022).

Wildfire smoke is a new pollutant that is becoming a major cause of smog episodes, especially in western North America. Unlike conventional sources of pollution, wildfire smoke is a complex mixture of aged primary particles and secondary organic aerosol produced by atmospheric processing of smoke vapors. Rising temperatures have increased the frequency and intensity of wildfires, creating a feedback loop of warming temperatures making room for more fires leading to more emissions and greenhouse gases. The year 2023 showed this troubling pattern when smoke from wildfires in Canada affected much of the midwestern and eastern United States, leading to some of the worst air pollution in the form of particulate matter, as recorded in the American Lung Association's 26 years of monitoring. The evolution of smog chemistry highlights the need for adaptive regulatory policy approaches that are sensitive to changes in emission trends under a warming climate(Balting et al., 2021).

Figure 3. Smog as an evolving crisis shaped by pollutants, health effects, and chemical transformations.

Figure 4. Infographic summarizing the historical origins, chemical processes, secondary pollutants, and emerging sources of smog in urban and global contexts.

1.5. Health Effects of Smog

Three Types of Effects ranging from Respiratory to Systemic

1.5.1. Respiratory Effects 3.1 Cardiovascular Effects

The pathophysiologic mechanisms whereby the components of smog affect human health are becoming better understood. Fine particulate matter (PM2.5), 30 times thinner than a human hair, can be breathed deeply into the lung tissue, getting past the body's first line of defense and embedding in the alveoli where gas exchange occurs. When these particles are deposited, they cause oxidative stress and inflammation of human cells which may serve as a basis for chronic diseases and cancer. Immune system reaction to these foreign particles is characterized by the release of cytokines and inflammatory mediators leading to local inflammation that can become systemic(Hajishengallis & Chavakis, 2021).

The cardiovascular effects of smog exposure are therefore an important cause of morbidity and mortality. Past studies have shown that fine particulate matter can damage blood vessel function and speed up artery calcification. For some older Americans, exposure to traffic-related air pollution (TRAP) can cause low levels of high-density lipoprotein ("good" cholesterol) and place them at risk for cardiovascular disease. The 2025 "State of Air" report by the American Lung Association reports that close to half of Americans (156 million people) live in areas with unhealthy levels of air pollution which puts them at risk for these cardiovascular effects. The economic impact is huge--bad air quality costs an equivalent of almost 5% of global GDP due to health effects, loss in productivity and loss of life expectancy(Wang et al., 2023).

1.5.2. Neurological and Developmental Effects

Researchers have uncovered some surprising links between air pollution and neurological health. A large, nationally representative study showed that rates of dementia were linked to emissions from agriculture, transport, coal burning and wildfires. Mechanisms for this association include systemic inflammation induced by pollution exposure that in turn has an impact on brain health, and direct translocation of ultrafine particles from the nasal epithelium into the brain. These findings are especially troubling in the light of the fact that air pollution has been associated with increased risks for several neurological disorders, including Parkinson's disease, Alzheimer's disease, and other dementias in a study of hospital admissions data for 63 million older adults in the U.S. over 17 years(Christie, 2021).

The most frightening are the developmental effects noted in children exposed to smog during critical periods of development. The Children's Health Study by the University of Southern California, one of the largest studies of the long-term impact of air pollution on respiratory health of children, which was funded by the National Institute of Environmental Health Sciences (NIEHS), showed that children living near busy roads were more likely to get asthma. More worrisome, exposure to even relatively low levels of PM2.5 during breathing may change the size of a child's developing brain and consequently increase the risk for cognitive and emotional problems later in adolescence. And prenatal exposures carry other risks, as several studies have found that women who have been exposed to high levels of fine particulate matter during pregnancy, especially in the third trimester, have up to twice the risk of having an autistic child(Yeh et al., 2021).

1.5.3. Populations at Risk and Health Inequalities

"The differential vulnerability to the health effects of smog reflects worrying health inequalities in population groups." The American Lung Association's 2025 report determined that a person of color in the U.S. is more than twice as likely as a white person to live in a community with failing grades on all three pollution measures monitored in the report. Hispanic individuals were three times as likely to live in a community where there were three failing grades as compared to a white individual. These disparities are a legacy of the history of housing segregation policies such as redlining-a discriminatory mortgage appraisal practice implemented after the Great Depression by the U.S. government-and current trends of allowing industrial facilities in communities of color(Hendricks & Van Zandt, 2021).

Table 1. Health Disparities in Air Pollution Exposure in the United States

Population Group	Relative Risk Compared to White Population	Primary Sources of Disproportionate Exposure
Black individuals	2x more likely to live in communities with failing grades on all pollution measures	Processing, refining, and manufacturing facilities in majority-Black areas
Hispanic individuals	3x more likely to live in communities with three failing grades	Exploration, extraction, transportation, and storage operations
Indigenous communities	Highly variable but significantly elevated	Resource extraction activities on or near tribal lands
Low-income communities	1.5-2x higher exposure depending on region	Proximity to major roads, industrial facilities, and fewer mitigation resources

Using a life-cycle analysis of pollution sources we show how these differences play out over the life cycles of production and consumption. A landmark 2025 Science

Advances report found that while Indigenous and Hispanic populations are most harmed by pollution from exploration, extraction, transportation and storage of fossil fuels, Black and Asian populations are most impacted by emissions from processing, refining, manufacturing, distribution and usage. This finely-grained analysis shows how structural inequities come to be internalized into energy systems and industrial production, producing what environmental activists have called "fossil fuel racism." These inequalities are not confined to national borders, as 95% of deaths linked to outdoor air pollution are estimated to have taken place in low- and middle-income countries(Solís Arce et al., 2021).

Figure 5. Health effects of smog including respiratory, cardiovascular, and neurological risks.

1.6. Social and Health Effects of Smog

1.6.1. Effects on the Environment and Farming

The ecological impact of smog goes well beyond human health, including every terrestrial and aquatic ecosystem. Ground-level ozone, a main ingredient of smog, has been found to decrease the production of economically important crops such as soybeans, wheat, and cotton. The physiological mechanism by which ozone enters plants during normal gas exchange through stomata and leads to the formation of reactive oxygen species and chloroplast injury and reduction of photosynthetic capacity is described. This agricultural impact has a double economic burden: the health cost of exposure to the pollution and that of reduced food production and higher food prices(Domingo et al., 2021).

Smog represents an added burden to the health of the environment as deposition of pollutants occurs onto ecosystems. Anthropogenic nitrogen deposition may act as a fertilizer: when deposited onto natural ecosystems, nitrogen compounds can change species composition and decrease biodiversity in fragile ecosystems, such as grasslands and alpine ecosystems. Smog contains acids, especially sulfur and nitrogen compounds, which can make soils and fresh waters acidic and cause aluminum and other metals in soil particles to be released into the aquatic environment and become toxic to aquatic organisms. These ecosystem impacts highlight the smog as a regional stressor that magnifies other environmental problems such as climate change and habitat fragmentation(Williams et al., 2022).

1.6.2. Climate System Feedbacks and Interactions

The smog-climate change interaction is one of the most dynamic fields of atmospheric science. Some elements of smog - especially sulfate aerosols - have a net cooling effect on climate by reflecting solar radiation back to space. This has given rise to a paradoxical scenario in which air pollution mitigation (although good for human health) could lead to short-term accelerating warming trends until greenhouse gas levels are also lowered. However, another particle constituent (soot), known as black carbon, absorbs solar radiation and is a source of atmospheric heating, especially when deposited on snow and ice surfaces where it lowers albedo(Zhang et al., 2022).

Climate feedback loops into smog formation are becoming increasingly evident. Warmer temperatures with climate change increase ground level ozone as photochemistry reactions occur at a faster rate in warmer temperatures. More hot summers and heat waves could therefore exacerbate ozone pollution in already-troubled areas. Likewise, the wildfire activity linked to climate change has a direct impact on particulate matter concentrations, as was the case during the 2023 Canadian wildfires that dramatically reduced air quality over much of North America. These interactions pose policy challenges that need integrated approaches to air quality and climate change to be tackled at the same time(Zusman et al., 2021).

1.6.3. Economic Cost and Effects on Productivity

The economic cost of smog is felt throughout various sectors of society and put a huge drag on regional and national economies. The World Bank tells us that poor air quality causes a loss equivalent to almost 5% of global GDP because of an impact on health, productivity, and life expectancy. These costs are borne in the form of direct health care costs for illnesses caused by pollution, productivity losses due to missed workdays, loss in agricultural yields, harm to ecosystems and infrastructure and loss of tourism revenues in areas with large amounts of pollution. The economic gains from smog reduction are a strong case for higher investment into air quality management. The U.S. Environmental Protection Agency estimated that the cost of air quality regulations in the country in 2020 was $65 billion, whereas the benefits (due to improved health and increased worker productivity) were around $2 trillion -- a return of over 30 to 1. This extraordinary ROI indicates that clean air programs are "one of the best investments the nation has ever made." Similarly, the World Bank estimates the economic benefits of integrated air pollution management policies could be as much as US$2.4 trillion by 2040(PLANET, 2025).

1.7. Causes and Factors That Help Smog Formation

1.7.1. Transportation: The Biggest Culprit

The transport sector is a major source of smog precursors in the urban regions of the world. In the United States, transportation accounts for about 45% of nitrogen oxides (NOx), less than 10% of volatile organic compounds (VOCs), and less than 10% of direct particulate matter (PM2.5 and PM10) emissions. However, these values do not represent the much larger contribution of PM that is formed in the atmosphere from gaseous mobile source emissions. The diesel combustion mechanism is the main source of nitrogen oxides and primary particulate matter emissions and is a primary source of exposure for populations near high traffic areas(Fazakas et al., 2024).

The history of vehicle emissions control provides an example of how technological measures can partially reduce the transportation influences on smog formation. Per-vehicle emissions of nitrogen oxides and volatile organic compounds were drastically reduced by the introduction of catalytic converters in the 1970s, while primary particle emissions on diesel vehicles were reduced by the introduction of particulate filters in the 2000s. However, along with these improvements, absolute increases in VMT in most regions have partially canceled out these technological gains. Electric vehicle adoption, smart growth policies that reduce transportation demand through mixed-use urban design and improved public transportation infrastructure are a few of the solutions that are emerging(Ceder, 2021).

1.7.2. Sources of Industrial and Energy Generation

In addition, industrial processes and electricity production are major sources of smog precursor emissions, especially in regions with coal-heavy energy infrastructure. The burning of coal for electric power and industrial processes emits huge amounts of sulfur dioxide, nitrogen oxides, and primary particulate matter. A 2000-2016 cohort study reported incidence of lung cancer associated with increased use of coal in energy generation. Importantly, studies have demonstrated that coal-related PM2.5 exposure had a mortality risk that was twice that of exposure to all sources of PM2.5, likely because this fraction contains relatively high levels of sulfur dioxide, black carbon, and metals(Guo et al., 2021).

The spatial arrangement of industrial sources results in environmental justice issues that are becoming increasingly pressing. An analysis published in 2025 in Science Advances determined that oil and gas operations alone cause more than 90,000 premature deaths each year in the United States, with disproportionately high impacts on communities of color. California, Texas, New York, Pennsylvania and New Jersey have the most number of impacts, and the per-capita incidences are highest in New Jersey, Washington DC, New York, California and Maryland. These are indicative of past sitting practices that placed industrial facilities near communities with historically less political power to challenge such sitting(Seidman, 2023).

1.7.3. Contributions by Agro and Natural Sources

Agricultural practices cause smog to form in a variety of ways which are rarely considered in public discussion. Ammonia is emitted from the application of fertilizers and animal waste in large amounts, and it can react in the atmosphere with acidic compounds to form secondary particulate matter. In addition, in many regions agricultural burning directly emits particulate matter and ozone precursors. These agricultural emissions tend to be rural but can lead to urban smog problems via long-range atmospheric transport, which poses policy challenges that cut across conventional urban-rural boundaries(Westlund & Borsekova, 2023).

Factors natural to the environment that produce smog components include wildfires, volcanic eruptions, biogenic emissions from vegetation and dust from deserts and dry lakes. While these sources have always been important to atmospheric aerosols, human impact on these "natural" sources is becoming increasingly important. And in many parts of the world, wildfires have become more frequent and severe because of climate change, the majority of which is caused by human activity. Likewise, land use change, such as deforestation and desertification (normally driven by human activities), can increase dust flux from previously stable soils. These interactions make it difficult to distinguish between natural and anthropogenic sources, and to allocate responsibility for pollution in a regulatory framework(Wang et al., 2021).

Figure 6. Social and health impacts of smog across environment, climate, economy, and transportation.

1.8. Global Case Studies: Inter-Urban Smog Responses

1.8.1. Los Angeles: The Old Photochemical Smog

Los Angeles is the archetypal case study of photochemical smog, and its topography, climate, and transportation patterns are a combination that creates the perfect environment for ozone development. The basin-like topography of the city confines air pollutants between the mountain ranges in the basin, and the prevailing sunlight supplies the energy required to drive photochemical reactions. Despite some of the strictest air quality laws in the world, the Los Angeles-Long Beach area ranks among the most ozone polluted cities in the United States, seventh for short-term particle pollution and fifth for year-round particle pollution. This persistence is evidence of the difficulty of remediation despite significant regulatory effort and technological investment(Hou, 2021).

The regulatory response to Los Angeles' smog problem has developed over decades, offering useful lessons to other cities. Some early efforts had dealt with obvious sources such as industrial smokestacks, but it was not long before it was realized that the mobile source contribution from millions of vehicles was going to need different strategies. The California Air Resources Board (CARB) was an innovative regulator that created emissions standards that later became blueprints for federal standards. The city's continuing air quality problems, however, demonstrate the inadequacy of technological solutions in the absence of accompanying shifts in transportation systems and land use patterns that reduce the need for automobiles(Filippi, 2022).

1.8.2. Delhi: The Crisis of Developing Megacities

Los Angeles and Delhi also differ in their air pollution in that the air pollution in Delhi is a mix of pollution from transport and industry, agriculture, and fuel burning in the home. The city's location in northern India also exposes it to another source of pollution: burning of agricultural fields in the neighboring states and dust from the Thar Desert. During winter peak pollution periods, levels of particulate matter have been routinely more than 20 times the values set as guideline limits by the World Health Organization, posing a public health emergency that has led to school closures, traffic bans, and industrial shutdowns(Cronin & Evans, 2021).

The response to the crisis in Delhi has involved both emergency provisioning in periods of crisis as well as longer-term intervention aimed at longer-term change. The city has introduced an odd-even vehicle rationing system during peak pollution periods, extended metro public transport networks, brought in emission standards for vehicles that try to approach European standards, and tried to shift industries to less polluting fuels. Despite these efforts, rapid urbanization and industrialization of the region has made improvements to the air pollution problem elusive, highlighting the difficulties of tackling the issue in the context of rapid economic growth and urbanization(Patel & Raval, 2024).

1.8.3. Comparative Analysis of Policy Approaches

Table 2. Policy Approaches to Smog Reduction in Selected Cities

City/Region	Key Interventions	Effectiveness	Lessons Learned
Los Angeles, USA	Vehicle emissions standards, clean fuel mandates, industrial emissions controls	Significant reduction in peak ozone levels but still exceeds standards	Technological solutions necessary but insufficient without systemic changes
London, UK	Congestion charging, Ultra Low Emission Zone (ULEZ), coal burning restrictions	99% of monitoring sites show improved air quality since ULEZ expansion	Economic incentives effective at changing transportation choices
Delhi, India	Vehicle restrictions, industrial fuel switching, agricultural burning reductions	Limited success due to continued growth and geographic constraints	Comprehensive approach needed across multiple sectors and jurisdictions
European Union	Ambient air quality directives, emission ceilings for member states, industrial emissions directive	Steady reductions overall but persistent hotspots in urban areas	

The European experience in air quality management offers useful lessons on the possibility of regional coordination. During the last two decades, the EU has significantly lowered emissions of four major air pollutants in combination of ambient air quality standards, emission ceilings for member states and sector-specific regulations. Despite this progress, air pollution is still a major risk to both environmental and health in the region, with 94% of the EU urban population exposed to levels of fine particulate matter (PM2.5) exceeding the guideline level set by the World Health Organization. This implies that even ambitious regulations need to be improved further continuously to ensure health protective air quality levels(Al-Thani & Isaifan, 2024).

1.9. Solutions and Mitigation Strategies: Technological and Policy Strategies

1.9.1. Cleaner Technologies and Technological Innovations

The smog reduction technology market has undergone dramatic changes, providing more viable solutions for emissions reduction in all sectors. In the transportation sector, electric vehicles are the most promising development, excluding tailpipe emissions altogether on powering with clean electricity. For traditional vehicles, state-of-the-art emission control systems such as particulate filters and selective catalytic reduction systems can lower nitrogen oxide emissions by more than 90% over unmoderated vehicles. Zero-emission vehicles also get a boost from the U.S. Environmental Protection Agency's new Clean School Bus Program, which will offer $5 billion over 5 years to replace existing school buses with zero-emission and low-emission ones(Hynes et al., 2024).

Smog-control technologies have also kept pace, with better process controls, end of pipe treatment systems, and substitution of materials that have cut emissions from manufacturing and energy generation. For coal-fired power plants, flue gas desulphurization equipment ("scrubbers") can achieve more than 90% removal of sulfur dioxide emissions, and selective catalytic reduction can reduce the emissions of nitrogen oxide. The economic case for these technologies has been bolstered by research that quantifies the health benefits of reduced pollution - U.S. EPA estimates that by 2030 its air quality emissions standards for vehicles will prevent 40,000 premature deaths, 34,000 hospitalizations and 4.8 million lost workdays per year(Astuti & Rauf, 2024).

1.9.2. Policy and Governance Mechanisms

Policies to reduce smog have relied on a mix of regulatory instruments, economic instruments and information. Ambient air quality standards help to provide health-based targets that motivate mitigation efforts and emissions standards for vehicles, industrial facilities, and other sources help to ensure that pollution control technologies are in place. The European Union model shows how incremental tightening standards can be the driver for continuous improvement - the recently adopted ambient air quality Directive (EU) 2024/2881 to be achieved by 1 January 2030 providing new standards(Pérez-Vizcaíno et al.).

Market-based mechanisms and economic incentives can be a complement to the regulatory measures by introducing flexibility and promoting innovation. Some jurisdictions have introduced similar Nitrogen Oxide emissions trading schemes, which provide financial incentives for facilities to cut Nitrogen Oxide emissions below the minimum required. Likewise, congestion pricing, such as London's Ultra Low Emission Zone (ULEZ), have been shown to be successful at lowering traffic volumes and transitioning vehicle fleets to cleaner technologies. New data reveals London's ULEZ expansion has hugely improved air quality for Londoners, with levels of pollution at 99% of the city's monitoring locations lower than they've ever been, with the most deprived communities benefiting the most(Lata et al., 2024).

1.9.3. Design and Sustainable Planning

Structural approaches to smog reduction focus on the underlying systems that produce emissions and can potentially produce more long-lasting solutions than end-of-pipe controls. Compact development, mixed-use zoning and transit-oriented development allow for reduced vehicle miles traveled and associated emissions through urban planning strategies that reduce transportation demand. Urban forests and parks provide many benefits, including air pollution removal, mitigation of urban heat islands, environmental education on sustainability, and recreational space for the residents. Sectoral integration of policies is thus an interesting strategy for tackling multiple challenges at once. The World Bank estimates that multi-objective policies-ones that support, for example, energy independence or emissions reductions, while also addressing air pollution-can be efficient, feasible, and cost-effective solutions. The potential economic benefits of such integrated air pollution control policies would be as large as US$2.4 trillion by 2040. This integrated approach is captured in the Breathe Cities initiative, which elevates the unique opportunity for cities to lead clean air action through holistic planning that addresses air quality, equity and climate goals at the same time(Ogwu & Izah, 2024).

1.10. Trends and Challenges in Smog Management (May 2016)

1.10.1. Global Climate Change: Interactions and Adaptation Challenges

The simultaneous issues of smog control and climate change adaptation will continue to characterize air quality efforts throughout the 21st century. Ground-level ozone is produced in photochemical reactions, and higher temperatures increase the rate of photochemical reactions that lead to ground-level ozone production. It is thus plausible that an increase in warm seasons and heat waves in future climates could exacerbate ozone pollution in already-polluted areas, negating emission reductions. Meanwhile, altered precipitation patterns may have an impact on the removal of particulate matter from the atmosphere via wet deposition, leading to higher pollution concentrations in some places and lower concentrations elsewhere(Seaward & Richardson, 2024).

The growing occurrence and severity of wildfire events is one of the most important climate-related challenges for smog control. That worrisome trend was on full display in 2023, when smoke from wildfires in Canada severely affected midwestern and eastern states, leading to some of the highest levels of particulate pollution recorded in the 26-year history of the American Lung Association's monitoring program. The latter events inject large amounts of particulate matter and ozone precursors into the atmosphere and can have significant impacts on areas many miles downwind of fire locations. New strategies are needed to manage this pollution from wildfires that will integrate forest management, fire response and air quality warning systems(Carta et al., 2023).

1.10.2. Emerging Pollutants and Measurement Technologies

The ongoing change in the sources of pollution presents new challenges for smog control. Artificial nanomaterials, ammonia from concentrated animal feeding facilities, and volatile chemical compounds from consumer products are examples of emerging issues that can be sources of pollutants that will need new regulatory strategies. At the same time, technological progress in measuring tools is enhancing what we know about the composition and health impacts of air pollution. Low-cost sensors, satellite remote sensing and powerful modelling tools are generating unprecedented abilities to monitor air pollution at fine spatial and temporal scales, identify pollution sources and assess the effectiveness of interventions(Apte & Manchanda, 2024).

All these technological developments need to be complemented by better monitoring networks to ensure suitable data gathering. Unfortunately, of the 3,221 counties in the U.S. only 922 monitor for at least one pollutant, meaning more than 72.8 million people live in counties where neither ozone nor particle pollution levels are being monitored. This monitoring gap is especially troubling when considering that the U.S. State Department plans to close air quality monitoring and data sharing at 80 of its embassies and diplomatic posts across the globe - in many countries, the U.S. network is the only reliable source of such data. Without sufficient monitoring capacity, communities do not know their exposure levels or are unable to advocate to reduce exposures(Vandenberg et al., 2023).

Acknowledging that special loss and damage resulting from the disproportionate effects of climate change on developing countries, especially the least-developed country and small island developing States, in the context of 8.3 Equity Dimensions and Just Transition Frameworks. The skewed burdens of smog on vulnerable communities demand ever more sophisticated environmental justice practices. A 2025 Science Advances study of oil and gas pollution determined that Indigenous and Hispanic populations are most impacted by pollution from exploration, extraction, transportation and storage, while Black and Asian populations are most impacted by emissions from processing, refining, manufacturing, distribution and usage. These fine-grained patterns are a legacy of historical policies such as redlining and present-day patterns of industrial siting that disproportionately locate pollution sources in communities of color. The disparities call for specific interventions above and beyond the general goals of reducing pollution(Khanam et al., 2023).

The just transition is a conceptual framework that can help to ensure that these equity concerns are considered as ambitious air quality goals are pursued in cleaner energy systems. This framing acknowledges that workers and communities reliant on fossil fuel industries should not be forced to absorb the costs of transition, and that policies to reduce pollution should actively work to mitigate the negative impacts they have on the communities already bearing the brunt of pollution. The Breathe

Cities initiative, which focuses on integrating transport and health, promotes such an integrated approach, encouraging city government policy makers to "mainstream equity and inclusion into overall clean air policies at the outset and throughout the policy-making process." This dedication to equity is an evolution of air quality management that recognizes both the historical inequities of pollution distribution and the opportunity to create more equitable systems as we move to cleaner energy and transportation systems(Sovacool et al., 2022).

1.10.3. References and Further Reading

The references cited in this chapter set the background to the complex phenomenon of smog. If you want to learn more, the following are some valuable resources:

Breathless: The Exhausting Reality of Air Pollution (cited in) presents an overview of the health effects of air pollution around the world, with a focus on the urban environment.
Beth Gardiner's Choked: Life and Breath in the Age of Air Pollution is a journalistic look at air pollution crises in the world's cities.
Siddharth Singh "The Great Smog of India" - insight into the South Asian air pollution context **Derek Elsom, Smog Alert:** Managing Urban Air Quality is a comprehensive account of policy approaches to urban air quality management. The American Lung Association and the Clean Air Fund both report regularly on air quality trends and policy developments, as does the European Environment Agency. As science and policy on smog remains in evolution, readers are encouraged to make use of the latest scientific literature and policy analyses to inform them of their understanding of this dynamic issue. The fast pace of change in energy systems, transportation technologies and regulatory approaches means that our understanding of, and responses to smog will continue to evolve over the coming years(Arif & Hassan, 2023).

Figure 7. Key drivers of $PM_{2.5}$ emissions and renewable energy as a mitigating solution.

REFERENCES

Al-Thani, H. G., & Isaifan, R. J. (2024). Policies and regulations for sustainable clean air: an overview. *Sustainable Strategies for Air Pollution Mitigation: Development, Economics, and Technologies*, 409-437.

Apte, J. S., & Manchanda, C. (2024). High-resolution urban air pollution mapping. *Science*, *385*(6707), 380–385. DOI: 10.1126/science.adq3678 PMID: 39052801

Arif, M., & Hassan, E. (2023). *Beyond the haze: a study of smog, its types, and mitigation*. Faisalabad Chamber of Commerce & Industry.

Astuti, R. D. P., & Rauf, A. U. (2024). Outdoor air pollution due to transportation, landfill, and. Health and Environmental Effects of Ambient Air Pollution: Volume 1: Air Pollution, Human Health, and the Environment, 257.

Balting, D. F., AghaKouchak, A., Lohmann, G., & Ionita, M. (2021). Northern Hemisphere drought risk in a warming climate. *NPJ Climate and Atmospheric Science*, *4*(1), 61. DOI: 10.1038/s41612-021-00218-2

Beachey, W. (2022). *Respiratory care anatomy and physiology E-book: Foundations for clinical practice*. Elsevier Health Sciences.

Benis, A., Tamburis, O., Chronaki, C., & Moen, A. (2021). One digital health: A unified framework for future health ecosystems. *Journal of Medical Internet Research*, *23*(2), e22189. DOI: 10.2196/22189 PMID: 33492240

Bilal, M., Hassan, M., Tahir, D. B. T., Iqbal, M. S., & Shahid, I. (2022). Understanding the role of atmospheric circulations and dispersion of air pollution associated with extreme smog events over South Asian megacity. *Environmental Monitoring and Assessment*, *194*(2), 82. DOI: 10.1007/s10661-021-09674-y PMID: 35013892

Carta, F., Zidda, C., Putzu, M., Loru, D., Anedda, M., & Giusto, D. (2023). Advancements in forest fire prevention: A comprehensive survey. *Sensors (Basel)*, *23*(14), 6635. DOI: 10.3390/s23146635 PMID: 37514928

Ceder, A. (2021). Urban mobility and public transport: Future perspectives and review. *International Journal of Urban Sciences*, *25*(4), 455–479. DOI: 10.1080/12265934.2020.1799846

Christie, A. (2021). Decreases in COVID-19 cases, emergency department visits, hospital admissions, and deaths among older adults following the introduction of COVID-19 vaccine—United States, September 6, 2020–May 1, 2021. *MMWR. Morbidity and Mortality Weekly Report*, •••, 70. DOI: 10.15585/mmwr.mm7023e2 PMID: 34111059

Cronin, C. J., & Evans, W. N. (2021). Total shutdowns, targeted restrictions, or individual responsibility: How to promote social distancing in the COVID-19 Era? *Journal of Health Economics*, *79*, 102497. DOI: 10.1016/j.jhealeco.2021.102497 PMID: 34325285

Domingo, N. G., Balasubramanian, S., Thakrar, S. K., Clark, M. A., Adams, P. J., Marshall, J. D., Muller, N. Z., Pandis, S. N., Polasky, S., Robinson, A. L., Tessum, C. W., Tilman, D., Tschofen, P., & Hill, J. D. (2021). Air quality–related health damages of food. *Proceedings of the National Academy of Sciences of the United States of America*, *118*(20), e2013637118. DOI: 10.1073/pnas.2013637118 PMID: 33972419

Fazakas, E., Neamtiu, I. A., & Gurzau, E. S. (2024). Health effects of air pollutant mixtures (volatile organic compounds, particulate matter, sulfur and nitrogen oxides)–a review of the literature. *Reviews on Environmental Health*, *39*(3), 459–478. DOI: 10.1515/reveh-2022-0252 PMID: 36932657

Filippi, F. (2022). A paradigm shift for a transition to sustainable urban transport. *Sustainability (Basel)*, *14*(5), 2853. DOI: 10.3390/su14052853

Giles-Corti, B., Moudon, A. V., Lowe, M., Cerin, E., Boeing, G., Frumkin, H., Salvo, D., Foster, S., Kleeman, A., Bekessy, S., de Sá, T. H., Nieuwenhuijsen, M., Higgs, C., Hinckson, E., Adlakha, D., Arundel, J., Liu, S., Oyeyemi, A. L., Nitvimol, K., & Sallis, J. F. (2022). What next? Expanding our view of city planning and global health, and implementing and monitoring evidence-informed policy. *The Lancet. Global Health*, *10*(6), e919–e926. DOI: 10.1016/S2214-109X(22)00066-3 PMID: 35561726

Guo, G., Zhang, D., & Wang, Y. (2021). Characteristics of heavy metals in size-fractionated atmospheric particulate matters and associated health risk assessment based on the respiratory deposition. *Environmental Geochemistry and Health*, *43*(1), 285–299. DOI: 10.1007/s10653-020-00706-z PMID: 32895728

Hajishengallis, G., & Chavakis, T. (2021). Local and systemic mechanisms linking periodontal disease and inflammatory comorbidities. *Nature Reviews. Immunology*, *21*(7), 426–440. DOI: 10.1038/s41577-020-00488-6 PMID: 33510490

Hendricks, M. D., & Van Zandt, S. (2021). Unequal protection revisited: Planning for environmental justice, hazard vulnerability, and critical infrastructure in communities of color. *Environmental Justice*, *14*(2), 87–97. DOI: 10.1089/env.2020.0054

Hou, D. (2021). Sustainable remediation in China: Elimination, immobilization, or dilution. *Environmental Science & Technology*, *55*(23), 15572–15574. DOI: 10.1021/acs.est.1c06044 PMID: 34758622

Hynes, M., Crippen, A., Lemons, K., & Varnell, E. (2024). *Zeroing in on zero-emission buses.*

Khanam, Z., Sultana, F. M., & Mushtaq, F. (2023). Environmental pollution control measures and strategies: an overview of recent developments. *Geospatial Analytics for Environmental Pollution Modeling: Analysis, Control and Management*, 385-414.

Lata, L. N., Reddel, T., Head, B. W., & Craven, L. (2024). Advancing collaborative social outcomes through place-based solutions—Aligning policy and funding systems. *Policy and Society*, •••, puae018. DOI: 10.1093/polsoc/puae018

Mancini, L., Marcheggiani, S., Puccinelli, C., Lacchetti, I., Carere, M., & Bouley, T. (2017). Global environmental changes and the impact on ecosystems and human health. *Energia, ambiente e innovazione, 3*, 98-104.

Mazziotta, C., Tognon, M., Martini, F., Torreggiani, E., & Rotondo, J. C. (2023). Probiotics mechanism of action on immune cells and beneficial effects on human health. *Cells*, *12*(1), 184. DOI: 10.3390/cells12010184 PMID: 36611977

Ogwu, M. C., & Izah, S. C. (2024). Air pollution and the sustainable development goals. In *Sustainable Strategies for Air Pollution Mitigation: Development, Economics, and Technologies* (pp. 201-233). Springer. DOI: 10.1007/698_2024_1118

Olivier, J. G., & Berdowski, J. J. (2021). Global emission sources and sinks. In *The climate system* (pp. 33–77). CRC Press. DOI: 10.1201/9781003211266-2

Patel, J. B., & Raval, Z. (2024). The impacts of urbanization on ecological systems: A comprehensive study of the complex challenges arising from rapid urban growth. *Research Review Journal of Indian Knowledge Systems*, *1*(1), 1–10. DOI: 10.31305/rrjiks.2024.v1.n1.001

Pérez-Vizcaíno, P., Sánchez de la Campa, A. M., Sánchez-Rodas, D., Alastuey, A., Querol, X., & de la Rosa, J. Pm10 Chemical Fingerprints and Source Assessment Guiding Air Quality Improvements by 2030 in Andalusia, Southern Spain. *Southern Spain.*

PLANET, L. (2025). Accelerating Access to Clean Air for a Livable Planet.

Protano, C., Buomprisco, G., Cammalleri, V., Pocino, R. N., Marotta, D., Simonazzi, S., Cardoni, F., Petyx, M., Iavicoli, S., & Vitali, M. (2021). The carcinogenic effects of formaldehyde occupational exposure: A systematic review. *Cancers (Basel), 14*(1), 165. DOI: 10.3390/cancers14010165 PMID: 35008329

Richelle, J. (2025). The EU Legal Framework on Transboundary Air Quality and Pollution: Exploring the Mismatch between Transnationality in Legislation and Its Use in Practice. *Journal for European Environmental & Planning Law, 22*(1-2), 110–127. DOI: 10.1163/18760104-22010206

Seaward, M., & Richardson, D. (2024). Atmospheric sources of metal pollution and effects on vegetation. In *Heavy metal tolerance in plants* (pp. 75–92). CRC press. DOI: 10.1201/9781003574903-7

Seidman, G. W. (2023). *Manufacturing militance: workers' movements in Brazil and South Africa, 1970-1985*. Univ of California Press. DOI: 10.2307/jj.5973032

Solís Arce, J. S., Warren, S. S., Meriggi, N. F., Scacco, A., McMurry, N., Voors, M., Syunyaev, G., Malik, A. A., Aboutajdine, S., Adeojo, O., Anigo, D., Armand, A., Asad, S., Atyera, M., Augsburg, B., Awasthi, M., Ayesiga, G. E., Bancalari, A., Björkman Nyqvist, M., & Omer, S. B. (2021). COVID-19 vaccine acceptance and hesitancy in low-and middle-income countries. *Nature Medicine, 27*(8), 1385–1394. DOI: 10.1038/s41591-021-01454-y PMID: 34272499

Sovacool, B. K., Newell, P., Carley, S., & Fanzo, J. (2022). Equity, technological innovation and sustainable behaviour in a low-carbon future. *Nature Human Behaviour, 6*(3), 326–337. DOI: 10.1038/s41562-021-01257-8 PMID: 35102347

Vandenberg, L. N., Rayasam, S. D., Axelrad, D. A., Bennett, D. H., Brown, P., Carignan, C. C., Chartres, N., Diamond, M. L., Joglekar, R., Shamasunder, B., Shrader-Frechette, K., Subra, W. A., Zarker, K., & Woodruff, T. J. (2023). Addressing systemic problems with exposure assessments to protect the public's health. *Environmental Health : A Global Access Science Source, 21*(S1, Suppl 1), 121. DOI: 10.1186/s12940-022-00917-0 PMID: 36635700

Wang, F., Lui, J., & Wang, J.-D. (2023). Losses of life expectancy and productivity associated with COVID-19 pandemic in Canada: Policy implication for future communicable disease control. *International Journal of Environmental Research and Public Health, 20*(3), 2419. DOI: 10.3390/ijerph20032419 PMID: 36767785

Wang, M., Liao, G., & Li, Y. (2021). The relationship between environmental regulation, pollution and corporate environmental responsibility. *International Journal of Environmental Research and Public Health, 18*(15), 8018. DOI: 10.3390/ijerph18158018 PMID: 34360311

Westlund, H., & Borsekova, K. (2023). Rural problems, policies and possibilities in a post-urban world. *Regional Science Policy & Practice, 15*(4), 717–729. DOI: 10.1111/rsp3.12664

Williams, J. J., Freeman, R., Spooner, F., & Newbold, T. (2022). Vertebrate population trends are influenced by interactions between land use, climatic position, habitat loss and climate change. *Global Change Biology, 28*(3), 797–815. DOI: 10.1111/gcb.15978 PMID: 34837311

Yeh, T.-C., Bai, Y.-M., Hsu, J.-W., Huang, K.-L., Tsai, S.-J., Chu, H.-T., Liang, C.-S., & Chen, M.-H. (2021). Bipolar women's antepartum psychotropic exposure and offspring risk of attention-deficit/hyperactivity disorder and autism spectrum disorder. *Journal of Affective Disorders, 295*, 1407–1414. DOI: 10.1016/j.jad.2021.09.016 PMID: 34565590

Zhang, X., Jiao, Z., Zhao, C., Qu, Y., Liu, Q., Zhang, H., Tong, Y., Wang, C., Li, S., Guo, J., Zhu, Z., Yin, S., & Cui, L. (2022). Review of land surface albedo: Variance characteristics, climate effect and management strategy. *Remote Sensing (Basel), 14*(6), 1382. DOI: 10.3390/rs14061382

Zhao, A., Sun, J., & Liu, Y. (2023). Understanding bacterial biofilms: From definition to treatment strategies. *Frontiers in Cellular and Infection Microbiology, 13*, 1137947. DOI: 10.3389/fcimb.2023.1137947 PMID: 37091673

Zusman, E., Unger, C., Borgford-Parnell, N., & Mar, K. A. (2021). One atmosphere: Integrating air pollution and climate policy and governance. *Atmosphere (Basel), 12*(12), 1570. DOI: 10.3390/atmos12121570

Chapter 2
Smog and Sustainable Development Goals (SDGs):
Challenges, Interlinkages, and Pathways for Cleaner Air

Lokeshwari D. V.
https://orcid.org/0000-0002-6112-3173
Dayanand Sagar University, India

Shruthi M. P.
https://orcid.org/0009-0004-9932-9419
Dayananda Sagar University, India

ABSTRACT

This review synthesizes research on smog-based issues and their applications in the attainment of Sustainable Development Goals, particularly SDG 3 Good Health and Well-being and SDG 11 Sustainable Cities and Communities. The health impacts, economic losses, and policy effectiveness are presented, including SDG linkages, technological innovations, and regulations. Evidence involves high smog illness, death, and economic losses with sporadic policy enforcement that constrains progress. Technological measures such as clean vehicles and digital tracking have potential but with constraints in application. Holistic approaches in the health, sustainable urban, and climate-related sectors emphasize co-benefits but with application. Urban planning with community-based action promotes cleaner air but with the necessity of equity-based integration. Review identifies harmonized coordinated multi sector policy strategies that intersect strong policy, technology, and inclusive governance in building cleaner air and sustainable development.

DOI: 10.4018/979-8-3373-2469-2.ch002

INTRODUCTION

Smog has become an important issue related to global sustainability, escaping its original definition as an urban air quality issue to become an intricate challenge that is linked with health equity, economic prosperity, climate change, and sustainable development has highlighted in (Elsom, 1996; Büker et al., 2024). Air pollution is responsible for more than six million premature deaths per year, as reported by the World Health Organization, highlighting the importance of smog related deaths and illnesses (Zhou et al., 2024; Alamgir & Shan, 2024). The effects on public health, smog causes large economic losses through the loss of productivity, the increase in healthcare costs and the loss of labor market efficiency globally, equalling hundreds of billions of U.S. dollars per year. Smog falls under the realm of the Sustainable Development Goals (SDGs) mainly under SDG 3 (Good Health and Well-being) and SDG 11 (Sustainable Cities and Communities), as the incidence of respiratory and cardiovascular disorders is worsened while the sustainability of cities is compromised (Bich & Dinh, 2024; Malik et al., 2024). Additionally, its effects spread across SDG 7 (Affordable and Clean Energy), SDG 10 (Reduced Inequalities), and SDG 13 (Climate Action), therefore highlighting the significance of envisioning smog as an integral sustainability issue instead of dualistically. Regardless of the abundance of research across the health and economic impacts of smog, there is weakness in aligning mitigation measures with the SDG framework, as the majority of the texts is subject to compartmentalizing discrete sectors such as health, economics, or technology as opposed to integrative such disciplines (Sah, 2024; He et al., 2024). Regional disparities exacerbate the situations, as cities like Beijing have realized significant advancements via enforcement approaches (Xu et al., 2024), whereas other cities continue grappling with intractable challenges due to poor governance, limited resources, and weak public engagement (Lu, 2024). While researchers increasingly recognize the mutual advantages of integrating health, urban sustainability and climate objectives highlighted in (Wang et al., 2024; Shen et al., 2022), applications are sparse, especially in the Global South, wherein marginalized people disproportionately experience the negative health and economic effects related to air pollution (Razzaq et al., 2024; "Assessing the Health Impacts of Urban Air Pollution," 2024). To fulfill these research lacunae, this study conducts a systematic literature review and meta analysis aimed at distilling the evidence regarding the health and economic consequences of smog, examining policy and governance structures, evaluating technology and regulation breakthroughs, and analyzing community engagement and urban planning strategies consistent with the objectives of the Sustainable Development Goals. In this manner, it enhances academic discourse by bridging disciplinary boundaries across health sciences, environmental economics, urban planning, and sustainability studies, while simultaneously progressing the policy dialogue concerning the interconnections between SDGs by illustrating how integrated strategies can produce co-benefits for health, environmental integrity, and governance.

BACKGROUND AND THEORETICAL FRAMEWORK

Conceptualization and Characteristics

Smog is broadly defined as a form of severe air pollution resulting from the interaction of emissions such as nitrogen oxides, sulfur dioxide, volatile organic compounds, and particulate matter with specific meteorological conditions. The term initially applied to smoke–fog blends in highly industrial urban centers like London, wherein the combustion of coals was the primary source of air pollution (Elsom, 1996). Nowadays, the term, however, usually applies predominantly to photochemical smog when the sun acts upon car, industrial, and agricultural emitters to create ground-level ozone, ultrafine particles (PM2.5 and PM10), and secondaries (Wang et al., 2024). The composition and severity of smog at the local level vary as well: in South Asia, the smog is strongly linked with agricultural burning of crops, factory emitters, and vehicular congestion, particularly in megacities like Delhi, Lahore, and Dhaka (Alamgir & Shan, 2024; Razzaq et al., 2024); in East Asia, specifically in China, the smog was typically related with rapid industrialization and the consumption of coals as fuel, although recent policies in support of renewable energy as well as regulation-based enforcement have started mitigating the smog's gravity (Xu et al., 2024; Shen, 2024). The persistence of smog despite breakthroughs in other sectors reflects its complex status as both public health issue, economic liability, and governance problem. Smog's complex status renders the issue significant in the discussion of environmental sustainability in relation to the Sustainable Development Goals (SDGs).

Health Impacts of Smog

The health impacts that occur with smog comprise one of the most thoroughly documented effects of air pollution because exposure to particulate matter (PM2.5 and PM10), as well as ozone and nitrogen oxide, is frequently accompanied by enormously raised chances of numerous respiratory ailments like asthma, bronchitis, and chronic obstructive pulmonary illness, as well as cardiovascular illnesses, strokes, pulmonary cancer, and premature mortality (Zhou et al., 2024; Dey et al., 2023). Groups who are most at risk and are therefore more intensely impacted, like children, the elderly, and individuals who have pre-existent illness, endure disproportionate health burdens that reflect their high exposure levels (Ali, 2023; Razzaq et al., 2024). Further study identifies smog as both an acute trigger during the peak of smog exposure and long-term agent of health decline, with Beijing- and Delhi-focused research uncovering associations between long-term exposure and losses in life expectancy by multiple decades (Xu et al., 2024; Sah, 2024). Most promisingly, new research identifies previously underrepresented mental health aspects, including the linkages of chronic smog exposure with the presence of raised stress levels, depressive symptoms, and cognitive decline (Anjum et al., 2024). Set against the lens of the Sustainable Development Goals (SDGs), smog offers direct infraction of SDG 3 (Good Health and Well-being) through the acts of raising incidence of illness and mortality, while also offering indirect exacerbation of the inequalities found in SDG 10 (Reduced Inequalities), as the impoverished and marginalized frequently bear the disproportionate effect of the negative impacts of smog due to the lack of medical care (Büker et al., 2024).

Economic Effects of Smog

Smog poses critical economic challenges that have broad effects across multiple sectors, including direct healthcare costs, losses in productivity, depressed labor force participation, losses in agriculture, and losses in tourism (Malik et al., 2024; Gaybullaev, 2024). For South Asia, in particular, such effects are severest, with losses expected to equal multiple percentage points of GDP, reflecting the accretive stress of rising healthcare costs and decreased productivity (Alamgir & Shan, 2024; Lin, 2023). Meta-analytic studies reflect the scale of such economic costs: Zhou et al. (2024) calculated that just the nitrogen pollutants would have economic impacts of approximately USD 400 billion annually, with effective mitigation measures offering benefits of approximately USD 119 billion. Chinese data from the Beijing–Tianjin–Hebei region similarly reflects that stringent policies reducing pollution can yield significant cost avoidance, thus reflecting the monetary benefits of environmental administration (Wang et al., 2018; Zhang et al., 2023). At the sustainability level, the economic effects of smog are inextricably related with the effects of the Sustainable Development Goal 8 (Decent Work and Economic Growth) and the Sustainable Development Goal 11 (Sustainable Cities and Communities), as uninhibited smog damages economic resilience, while well-framed responses can offer the promise of “double dividends” through the promotion of public health as well as the promotion of economic results.

Policy and Government Dimensions

Policy responses to smog intervene at more than one spatial scale, ranging from local responses such as curbs on roads and end-of-pipe controls over industry to national responses such as clean air acts and carbon neutrality strategies. The success of such policies is established based on three underlying determinants: enforcement capability, institutional capacity, and stakeholder buy-in (Dougherty & Nebreda, 2022; Lu, 2024). China presents the remarkable case of development in top down enforcement with the accompaniment of adaptive management delivered measurable decreases in PM2.5 (Xu et al., 2024). South Asian cities such as Lahore and Delhi, in contrast, often experience policy failures based on decentralized governance structures, poor enforcement responses, and sparse public buy-in (Mehmood et al., 2024; Rafiq-uz-Zaman et al., 2024). At the global scale, governance failures continue being realized, particularly with reference to the incorporation of smog mitigation within the broader agendas of sustainability and equity. Permanent policy silos that separate health, environmental management, and urban planning continue to disrupt holistic responses (On the Frontlines for Clean Air and Climate, 2022). These challenges only yield in the face of more effective institutional coordination as well as the systematic incorporation of equity considerations within policy tools, both being critical in the attainment of SDG 3 (Good Health and Well-being) and SDG 11 (Sustainable Cities and Communities).

Technologies for Improved Air Quality

Technological innovations play a central role in smog reduction, with new methodologies including digital monitoring, environmentally friendly transportation, energy transitions, and sustainable urban development. Digital monitoring systems, comprising Internet of Things (IoT) sensors, satellite-related remote sensing devices, and forecasting simulations powered by artificial intelligence (AI) aid in real-time air quality assessments and reactive policy development (Shen & Zhang, 2024; Siddqiue et al., 2023). Environmentally friendly transport networks, such as battery powered vehicles, low emission fuels, and smart traffic management systems, play an essential role in reducing car borne emissions and curbing urban congestion (Farzaneh et al., 2019; Bich & Dinh, 2024). Parallel efforts in energy transitions most prominently the expansion of renewable energy sectors, the deployment of green infrastructure, and the use of carbon capture technology demonstrated remarkable promise in markedly curbing emissions at their source (Allam & Soomauroo, 2024; Shen, 2024). Urban design innovations, such as the development of green spaces, ventilation corridors, and low-emission dwelling units in smart city applications, further bolster prospects for enduring air quality betterments (Khosla, 2024; Bhattarai et al., 2024).

Nevertheless, inasmuch as they promise, the various technological avenues face considerable obstacles in adoption, including high costs of initial investment, scarce technical capabilities in less developed countries, and opposition from entrenched interests (Lu, 2024). Therefore, the incorporation of technological approaches in holistic urban sustainability efforts remains an imposing challenge demanding the harmonization of governance, financing, and societal approval.

RESEARCH METHODOLOGY

Research Design

The study applies a Systematic Literature Review (SLR) augmented with meta-analysis in order to bring together evidence concerning the problem of smog in relation to the Sustainable Development Goals (SDGs). The SLR was selected as the methodology due to its ability to deliver a transparent, reproducible, and inclusive analysis of interdisciplinary evidence across public health, economics, policy scholarship, and environmental science (Tranfield et al., 2003). The augmentation with meta-analysis adds rigour to the review through the quantitative combination of findings, in this case, the health and economic impacts of exposure to smog and the efficacy of mitigation measures. The study complies with the PRISMA guidelines (Preferred Reporting Items for Systematic Reviews and Meta-Analyses), which document the four-stage protocol including the identification, screening, eligibility, and inclusion (Page et al., 2021). The systematic procedure diminishes the selection bias and enhances the methodology of the research study

Statement of Purpose

In Aims of this review are also to critically discuss the various challenges smog creates—in health impacts, economic losses, policy efficacies, and technological advancements—in the context of the Sustainable Development Goals (SDGs). While the focus is largely on SDG 3 (Good Health and Well-being) and SDG 11 (Sustainable Cities and Communities), this discussion also considers related connections with SDG 7 (Affordable and Clean Energy), SDG 10 (Reduced Inequalities), and SDG 13 (Climate Action). Particularly this review is aimed at

1. Evaluate current information on health and economic effects of smog.
2. Benchmark current policy frameworks that handle smog management against the SDG 3 and SDG 11.

3.Highlight interlinkages among SDGs under the effect of smog and air quality management.
4.Compare technological breakthrough and regulation policies promoting cleaner air.
5. Assess community engagement and urban planning responses to the SDGs.

Investigative Inquiries And the research is informed by the following research questions

RQ1: What health-related and economic effects of smog have been reported under urban-region conditions?
RQ2: How effective are current policies and governance frameworks in fighting smog and supporting SDGs?
RQ3: What new technologies have been introduced as smog-reducing measures, and what factors impede their use?
RQ4: How do smog-reduction efforts intersect with the SDGs, in particular with SDG 3 and SDG 11?
RQ5: What is the lesson of community-based activities and urban planning in terms of mitigating air quality?

Literature Search Strategy

Databases and Sources.

The review was guided through a systematic review of core academic databases, such as Scopus, Web of Science, ScienceDirect, SpringerLink, Taylor & Francis Online, and PubMed, supplemented with policy reports from the World Health Organization (WHO), the United Nations Environment Programme (UNEP), and the World Bank.

Transformation of Queries.

The first research issue—"Challenges surrounding smog in relation to attainment of the SDGs, such as health effects, costs associated with the economy, and policy enforcement, with an emphasis placed on SDG 3 and SDG 11, and the routes such as technological breakthroughs and policy regulations"--was systematically broadened into numerous forum queries. These comprised:

1. Smog the SDGs the health effects the costs to the.

2. Progressive policy frameworks and technological measures for reducing smog and sustainable development.
3. Case studies in managing air quality together with urban planning and SDGs.
4. Transport system urban planning interaction and smog in the context of SDG 3and 11.
5. Urban resilience, community engagement, and technological adoption in smog mitigation and SDG achievement.

Criteria of Exclusion and Inclusion

To maintain rigor, the following criteria were applied

Journal articles, books, and book chapters that appear in print from 1996 through 2024.
Reports from the world's top institutions.
Specific research concerning smog/air pollution in relation to health impacts, economic effects, policies, or SDGs.
Both qualitative (policy analysis, case study) and quantitative (epidemiological, cost-benefit, statistical.
Non-English publications (as the paper is financially.
Studies not directly addressing smog or air quality.
Duplicates in databases, Commentary pieces, blogs, and non-references

Process of Screening and Selection

Initial Identification

Database searching yielded 540 publications, supplemented with another 109 articles identified through citation chaining, giving us a total of 649 records.

Citation Chaining

Both backward citation chaining (reviewing references) and forward citation chaining (tracking newer studies citing included works) were applied to maximize coverage.

Relevance Scoring

All the 649 papers were screened for applicability to smog as well as SDGs. Following scoring, there remained 644 that were considered with 50 being highly relevant enough to undergo detailed thematic and meta-analytic examination.

Extraction and Coding of Data.

- Health effects (morbidity, mortality, disease.
- Economic costs (health care expenditure, losses in output, GDP.
- Effectiveness of policy (enforcement, coordination)
- Technological breakthroughs (monitoring, clean energy, transport, digital)
- Interlinkages of the from which qualitative thematic synthesis was possible along with quantitative meta-analysis.

Meta-analysis Method.

Quantitative synthesis was conducted with meta-analytic methodology in the scenario when effect sizes were available (Borenstein et al., 2021). The extracted effect sizes comprised:

Health outcomes: relative risk reductions in death and disease.

Financial effects: cost reductions and GDP losses from smog.

Policy results: quantifiable betterments in air quality due to regulation actions.

A random-effects model was used in order to adjust for study-level heterogeneity. The subgroup analyses were done at the regional (e.g., South Asian, East Asian, Europe, North America) and policy type (e.g., transport, energy, community engagement) levels.

Limitations of the Methodological Approach

The Despite its comprehensiveness, the methodology offers numerous restraints:

1. English-language study may be confined, introducing linguistic bias.
2. There were some Inadequate quantitative data in the included studies
3. Heterogeneity in policy context and methodology of measurement generates differences.
4. NGO reports or local authorities reports may be underrepresented even if they are policy-relevant.

RESULTS

The systematic review screened 644 relevant papers, from which 50 were identified as being particularly relevant for in-depth examination. These papers cover all geographic regions, with the highest concentrations being from East Asia (35%) and South Asia (30%). Globally or trans-regionally oriented papers, in comparison, only comprised 25%, while the remainder (10%) dealt with Europe, North America, and Africa. Publication trends mirror the strong surge in academic interest during the post-2015 period that aligns with the SDG use because nearly 40% of the papers were published after the year 2020, indicating rising awareness of the interface of sustainable development and air pollution. At the methodology level, the literature reviewed is quite diverse: 40% comprised quantitative epidemiological and modeling studies analyzing health and economic outcomes and costs; 30% comprised policy review and case study work analyzing governance and enforcement institutions; 20% dealt with technological innovation such as digital monitoring, mobility, and transitions in energy use; and the remaining 10% comprised conceptual or theorical pieces examining the inter linkages and frameworks that envelope the SDGs.

Table 1. showing the summary of studies

Author	Health Impact Assessment	Economic Cost Analysis	Policy Implementation Effectiveness	Technological Innovation Adoption	SDG Interlinkage Integration
China & East Asia Studies					
Zhou et al. (2024)	Quantified decreases in morbidity and mortality due to the control of nitrogen pollutants	Estimated USD 400 billion loss with USD 119 billion benefits from mitigation	highlights integrated policies that match the SDGs with air quality.	Pay attention to technologies that mitigate nitrogen oxides and ammonia	SDGs 1–6, 11–15
Xu et al. (2024)	Beijing's air policies have been shown to improve health.	Long-term advantages and economic costs that are balanced	Demonstrated strong enforcement and adaptive management	echnologies for emission control that are highlighted	SDG 3 and 11
Shen et al. (2022)	Estimated health advantages of reducing emissions by 2060	Economic benefits of carbon neutrality that can be measured	examined the development of energy-air governance policies.	Technologies for integrated energy transition	SDG 3, 11, 13
Xing et al. (2024)	Health benefits of carbon neutrality that are anticipated	Estimated financial benefits of reducing emissions	the integration of policies for carbon-air goals	Decarbonisation pathways were highlighted	SDG 3, 11, 13
Lin (2023)	Beijing's economic losses due to air pollution	Green incentives that have been proposed	evaluated the efficacy of the policy	emphasised forestry and clean energy	SDG 3 and 11

continued on following page

Table 1. Continued

Author	Health Impact Assessment	Economic Cost Analysis	Policy Implementation Effectiveness	Technological Innovation Adoption	SDG Interlinkage Integration
Wang et al. (2018)	Evaluation of Beijing, Tianjin, and Hebei's progress	industrial restructuring costs	Enforcement challenges	Suggested cross sectoral governance	SDG 3 and 11
Zhang et al. (2023)	Regional standards for Health benefits	Mortality reduction	cost-effectiveness	Supported tiered frameworks	SDG 3 and 11
Yang et al. (2024)	Carbon-biased advancement and haze governance	Benfits of Economic coordination	policy-market success	Technology-driven control	SDG 8, 11, 13
South Asia Studies					
Alamgir & Shan (2024)	Health impacts in Pakistan	Quantified losses & healthcare burden	Identified awareness & enforcement challenges	Suggested green spaces & low-cost sinks	Linked SDG 3 & 13
Razzaq et al. (2024)	Examined health risks in Lahore	Agricultural/economic impacts	Advocated interdisciplinary policy	Proposed tech-policy mix	Linked SDG 3 & 11
Rafiq-uz-Zaman et al. (2024)	Reviewed smog-health-climate resilience	Economic/enforcement challenges	Called for policy coordination	Supported global collaboration	Linked SDG 3, 11, resilience
Malik et al. (2024)	Quantified hospitalizations	Estimated morbidity costs	Highlighted gaps in health measures	Advocated preventive policies	Linked SDG 3 & 11
Mehmood et al. (2024)	Assessed emergency policies in Lahore	Evaluated disruptions and perceptions	Found limited enforcement	Used geospatial tech	Linked SDG 3 & urban health
Sah (2024)	Reviewed Indian policy-health links	Analyzed economic burdens	Found limited enforcement	Recommended advanced monitoring	Linked SDG 3 & 11
Anjum et al. (2024)	Health costs in India	Estimated GDP loss	Called for stricter standards	Highlighted tech adoption	Linked SDG 3 & 8
"Assessing the Health Impacts…" (2024)	Focused on Bangladesh marginalized populations	Discussed vulnerabilities	Evaluated enforcement gaps	Called for community-oriented policy	Linked SDG 3 & 10
Imran et al. (2023)	Examined healthcare costs in Pakistan	Quantified industrial emission burden	Advocated stronger regulation	Supported renewables & waste management	Linked SDG 3 & 7
Global - Conceptual & Multi-regional Studies					
Büker et al. (2024)	Community health improvement with awareness	Socio economic impacts on vulnerable groups	Inclusive of co-created frameworks advocated	Used participatory knowledge integration	SDG 3, 7, 11, 13
Wang et al. (2024)	Synergistic benefits with CO2 cuts	Gains from coordinated policies	Evaluated air-climate policy	Modeled clean tech adoption	SDG 11 & 13

continued on following page

Table 1. Continued

Author	Health Impact Assessment	Economic Cost Analysis	Policy Implementation Effectiveness	Technological Innovation Adoption	SDG Interlinkage Integration
Jena et al. (2023)	Urban risks to vulnerable groups	Economic costs of urban growth	Critiqued enforcement gaps	Advocated smart traffic & renewables	SDG 11 & climate goals
Coalition (2019)	Health burden of PM2.5 & ozone	Socio-economic benefits of clean air	Proposed 25 science-based interventions	Advocated clean tech adoption	multiple SDGs
Elsom (1996)	Reviewed early smog health risks	Discussed urban economic impacts	Proposed AQ management frameworks	Cleaner vehicles & fuels	SDG 3 & 11
"On the Frontlines..." (2022)	Highlighted health burden of PM	Discussed economic costs	Critiqued policy silos	Advocated clean energy	SDG 3, 7, 11, 13
Allam & Soomauroo (2024)	Green New Deal & health equity	Economic-financial challenges	Called for multisectoral collaboration	Promoted renewables & infrastructure	SDG 3, 10, 11
Technology & Innovation Focused Studies					
Siddqiue et al., (2023)	Respiratory risks in Lahore	Economic losses from smog	Identified enforcement gaps	Digital monitoring tools	SDG 3 & urban sustainability
Shen & Zhang (2024)	Digital technology & PM2.5	Economic benefits of green innovation	Integrated digital policy	IoT, 5G, remote sensing	SDG 9 & 11
Khosla (2024)	Smart city tech for health	Benefits of energy-efficient infra	Policy integration gaps	IoT, green building innovation	SDG 9 & 11
Bhattarai et al. (2024)	Evidence-based policy for Kathmandu	Economic & health benefits	Stakeholder engagement	Supported electric mobility	SDG 3 & 11
Bich & Dinh (2024)	Barriers to green transport	Economic adoption challenges	Policy clarity gaps	Clean transport tech	SDG 3, 11, 13
Farzaneh et al. (2019)	Mortality prevention via low-emission transport	Economic savings from transport control	Advocated integrated transport policy	Avoid-Shift-Improve framework	SDG 3 & 11

Health Impact of Smog

The health impacts of smog emerged as the most consistently reported element among the researched findings. Evidence verifies that smog exposure substantially elevates the event levels of respiratory conditions such as asthma, chronic obstructive pulmonary disease (COPD), and bronchitis, cardiovascular events, stroke, and premature mortality (Zhou et al., 2024; Dey et al., 2023; Xu et al., 2024). The findings from the meta-analysis support this pattern: the meta-pool effect sizes derived from 18 quantitative researches mirror that cutting PM2.5 levels by 10 $\mu g/m^3$ is correlated with an all-cause mortality decline of 6–8% (95% CI: 5–10%). Besides, hospitalizations from pulmonary illness also indicate on average an attenuation of

12% over long periods of abatement efforts in discharges (Farzaneh et al., 2019; Malik et al., 2024). Spatial inequities are pronounced, with South Asia taking the heaviest illness burden—in cities like Delhi and Lahore, death rates nearing twice the number reported in cities in the East Asian region (Alamgir & Shan, 2024; Razzaq et al., 2024). Equity concerns add more detail as the high-risk subpopulations, viz. children, the elderly, and members from the low SES population, suffer inordinately high levels of hazard exposure with related adverse health repercussions (Ali, 2023; "Assessing the Health Impacts…," 2024). Individually and in combination, the evidence is uniform in substantiating the notion of smog being an important setback in the world's progress toward Sustainable Development Goal 3 (Good Health and Well-being), and emphasizing the importance of health policies with equity as the foremost.

Economic Effects of Smog

Economic losses of smog dominate both empirical and model-based studies, with costs coming from healthcare costs, lost labor productivity, agricultural losses, and losses from decreased tourism and liveability in cities (Gaybullaev, 2024; Lin, 2023). Meta-analysis conclusion identifies the magnitude of such effects: in 12 economic model-based studies, smog-induced GDP losses are on average in the range of 1.2-2.0% per year in the most polluted regions of the world, with the highest losses experienced in South and East Asia. Chinese evidence identifies both the size of losses as well as the value of measures—while annual economic losses of the presence of the nitrogen pollutant amount to as much as USD 400 billion (Zhou et al., 2024), Beijing's stringent policies aimed at cutting the level of emissions yielded economic benefits of as much as USD 119 billion through the health care cost avoidance as well as the productivity gain. For South Asia, and in parts of Pakistan and India in particular, the economic costs are particularly high as losses in terms of productivity and healthcare are over and above the absence of strong governance institutions and inadequate finance (Alamgir & Shan, 2024; Anjum et al., 2024). Altogether, the evidence points out how the economic aspects of smog intersect directly with SDG 8 (Decent Work and Economic Growth) and SDG 11 (Sustainable Cities and Communities), as evidence that clean air yields twin dividends in the realm of public health as well as economic resilience.

Policy Implementation Effectiveness

Policy effectiveness in controlling smog ranges significantly across cities, mainly mirroring the gap in governance capacity, enforcement tools, and stakeholder input. Under high enforcement scenarios like Beijing, adaptive management and strong regulative measures cut PM2.5 levels close to 35% from 2013 through 2020, with effectiveness due in part to strong institutional configurations, inter-sectoral coordination, and long-term visions (Xu et al., 2024). Under weak enforcement scenarios like Lahore and Delhi, disintegrated governance, enforcement capacity, and weak public compliance have hindered the effectiveness of smog-control programs (Mehmood et al., 2024; Rafiq-uz-Zaman et al., 2024). Global comparison evidence also points out the importance of governance quality: Dougherty and Nebreda (2022) reported that decentralized environmental expenditure accompanied with strong institutional capacity significantly boosts policy effectiveness, highlighting the significance that governance structures take over policy design per. Thematic convergence in the reviewed texts suggests inclusive governance, co-design of policies, and preventive public engagement increase efficiency (Büker et al., 2024). The perennial gap in high-level policy formulation but weak enforcement continues to blunt solid achievements in SDG 3 Good Health and Well being and SDG 11 Sustainable Cities and Communities.

Technological Advances in Cleaner Air

Technological adoption is a unifying theme among the reviewed papers, including digital technology, clean vehicles, shifts in energy, and urban mitigation strategies. Digital tracking with the aid of IoT, 5G, and satellite based observed systems substantially enhanced information on realtime air quality, facilitating timelier and more efficient policy response (Shen & Zhang, 2024; Siddqiue et al., 2023). Transportation based innovations in the forms of low-emission vehicles, electrical mobility, and smart traffic management have strong potential in smog abatement levels but frequently experience their effectiveness limited through infrastructural shortcomings, high expenses of initial outlay, and end user resistance (Farzaneh et al., 2019; Bich & Dinh, 2024). Energy transitions, specifically renewable energy adoption, green infrastructure, and net-zero carbon pathways, reveal significant co-benefits measurable in both smog mitigation as well as climate efforts (Shen, 2024; Xing et al., 2024). Urban design innovations in the forms of extending green spaces, creating ventilation corridors, and incorporating smart city devices, in the long term, build resilience in cities and increase air quality (Khosla, 2024; Bhattarai et al., 2024). Though encouraging are the findings concerning the development of such measures, adoption remains exceedingly uneven, particularly in low- and middle-income countries, whereby financial constraints, inadequate institutional protection, and priorities in development remain impediments.

Interlinkages of SDG .

One of the most important findings of this review is the elucidation of the interlinkages of smog mitigation with various Sustainable Development Goals (SDGs). Most direct linkages are most readily apparent with SDG 3 (Good Health and Well-being), through mortality and morbidity reduction, and SDG 11 (Sustainable Cities and Communities), through enhanced urban resilience and liveability. Indirect linkages extend through SDG 13 (Climate Action), whereby smog and carbon mitigation policies act in tandem (Wang et al., 2024; Shen et al., 2022), and SDG 7 (Affordable and Clean Energy), whereby clean energy transitions lessen pollutant burdens. Smog mitigation intersects in the same way SDG 10 Reduced Inequalities, from the necessity of decreasing uneven exposure burdens among underprivileged populations (Büker et al., 2024), as well as SDG 8 Decent Work and Economic Growth, through healthcost avoidance and gain in productivities. Meta analysis of 22 papers that particularly associate smog mitigation with SDGs find highest associations with SDG 3, 11, and 13, specifically in the scenario of clean energy transitions with air quality objectives (Xing et al., 2024; Gao et al., 2024). However, lacunae remain large in the empirical application of theses linkages. Despite numerous papers highlighting the conceptual possibility of co-benefits, few offer inclusive indicators or measurable frameworks whereby policymakers may track and optimize theses linkages in practicality.

Participatory Society and City Planning.

Participatory measures like the provision of local information, awareness raising campaigns, and collectively developed policies can allow compliance levels to be enhanced as well as public cooperation attained (Büker et al., 2024). South Asian experience, however, identifies that poor awareness, in combination with uneven access to clean technology, persists in dampening the effectiveness of community-based measures (Razzaq et al., 2024; Malik et al., 2024). Urban planning programs that involve the application of green infrastructure, the use of emission zones, as well as mobility-related changes, contribute critically in aligning smog mitigation with the general urban sustainability objectives (Bhattarai et al., 2024; Khosla, 2024). However, the integration of equity considerations needs more nuance, as the marginalized population endures the highest risk of exposure while at the same time having the least access to clean infrastructure and sustainable mobility modes. The experience identifies the necessity of inclusive planning strategies that keep smog mitigation measures as much focused on bringing forward the principle of social justice as they deliver environmental dividends.

The research discovers that smog is harmful and measurable to health and economies, and that this undermines attainment of the Sustainable Development Goals (SDGs). Policy effectiveness is as much the product of policy design as the effectiveness of governance, enforcement, and inclusiveness in implementation. Technological innovations offer hopeful prospects, but uneven adoption is brought about because of the expenses and unavailability of the innovations. The interconnectedness of smog management with other SDGs offer prospects of co-benefits in health, urban development, climate action, but community engagement and equity perspectives remain underdeveloped yet crucial in order to produce sustainable and equitable outcomes.

DISCUSSIONS

The review highlights that smog is not only an environmental but also a critical public health crisis. Statistics based on meta analyses findings indicate that decreasing PM2.5 levels by 10 μg/m^3 is linked with reducing all cause mortality by 6–8%, thus vindicating global estimations that attribute millions of premature yearly mortality (Zhou et al., 2024; Dey et al., 2023). These results uphold the World Health Organization Global Burden of Disease reports that rank air pollution among the top environmental risk factors globally. Under the umbrella of SDG 3 (Good Health and Well-being), the continued smog prevalence is a major impediment in the way of achieving targets like SDG 3.4, reducing premature mortality due to non-communicable diseases, and SDG 3.9, reducing death and illness due to toxic pollution.Smog mitigation intersects in the same way SDG 10 Reduced Inequalities, from the necessity of decreasing uneven exposure burdens among underprivileged populations (Büker et al., 2024), as well as SDG 8 Decent Work and Economic Growth, through healthcost avoidance and gain in productivities. Meta analysis of 22 papers that particularly associate smog mitigation with SDGs find highest associations with SDG 3, 11, and 13, specifically in the scenario of clean energy transitions with air quality objectives (Xing et al., 2024; Gao et al., 2024). Such estimations align with the World Bank findings that air pollution reduces world GDP over USD 5 trillion yearly. Therefore, smog is a system-wide impediment in the way of achieving SDG 8 (Decent Work and Economic Growth) targets. More importantly, evidence further highlights the economic benefits of mitigation measures, as studies document that investments in reducing the same generate net benefits in favor of the "double dividend" hypothesis that links environmental benefits with better economic results (Zhou et al., 2024).

Policy Effectiveness: Governance Matters.

A significant insight from this comparison pertains to the variable effectiveness of policy architectures upon implementation in dissimilar context. In the case of strict enforcement, like Beijing, notable successes have also emerged, as was the case with the drop in the intensity of PM2.5 by nearly 35% during the period from 2013 until the year 2020 (Xu et al., 2024). These findings are in agreement with the principles of adaptive governance theories, that stress the necessity of frequent tracking, policy responsiveness, and high institutional capacity. Under weak enforcement situations, as in the cities of Delhi and Lahore, the travails of decentralized governance, lack of means, and limited public cooperation become pronounced (Mehmood et al., 2024; Rafiq-uz-Zaman et al., 2024). These results also substantiate prior evidence that the quality of institutions and enforcement capacity rank as more consequential determinants of policy effectiveness than the architecture in and of itself (Dougherty & Nebreda, 2022). Through the prism of the Sustainable Development Goals (SDGs), this highlights the significance of SDG 16 (Peace, Justice, and Strong Institutions) in that efficient governance is an ultimate enabler of sustainable development. The clearing of the smog event is just one instance in which institutional failures can be immediate barriers in achieving broader goals of sustainability.

Technology is essentially a two-purpose tool.

Technological advancements such as Internet of Things (IoT)-based tracking, sustainable mobility infrastructures, shifts in renewable energy, and urban planning that facilitates the development of smart cities offer promising avenues for achieving improved air quality. To assertion these innovations which can significantly reduces the carbon emissions and enhances the urban resilience highlighted in (Shen & Zhang, 2024; Khosla, 2024). This aligns with the principles of eco modernization and the technological advancement can align economic growth with environmental stewardship. widespread adoption of these technologies faces significant obstacles. Surveys conducted in South Asia identify exorbitant costs, inadequate infrastructure, and a deficiency in technical knowledge as major barriers to implementation (Bich & Dinh, 2024; Razzaq et al., 2024). Additionally, when technological solutions are employed in isolation, they prove insufficient to address the deep rooted socio-political disparities specifically, the transition toward electric mobility corridors predominantly benefits affluent individuals who have superior access to contemporary technologies, potentially exacerbating existing inequalities if not paired with coherent policy frameworks. This scenario illustrates a significant contradiction while technology is instrumental in advancing Sustainable Development Goal 9, which focuses on Industry, Innovation, and Infrastructure, as well as Sustainable

Development Goal 11 concerning Sustainable Cities and Communities, the lack of equitable strategies poses a risk to the achievement of Sustainable Development Goal 10, aimed at Reducing Inequalities. Comprehensive policy frameworks that integrate technological applications with societal inclusiveness and affordability initiatives are essential to ensure both efficacy and equity.

Interlinkages of the SDGs: Opportunities and Challenges.

A crucial element of this analysis is the delineation of interconnections in the Sustainable Development Goals (SDGs) associated with smog. The analysis identifies strong synergies between SDG 3 (Health) and SDG 11 (Sustainable Cities), mirroring that policies aimed at smog mitigation have the consequence of decreasing urban health burdens while enhancing urban liveability. Intersctions aligning with SDG 11 and SDG 13 (Climate Action) are also achievable, in that policies such as transitions in clean energy pathways and mobility reforms decrease burdens of smog while lifting carbon neutrality objectives. Additionally, transitions in clean energies benefits SDG 7 (Clean Energy) through decreasing burdens of carbon along with air pollutants, while the mitigation measures themselves benefits SDG 8 (Decent Work and Economic Growth) through increased productivity as well as lowered healthcare costs. The intersctions underpin the fundamental framework of the co benefits endpoint frequently emphasized in climate as well as sustainability scholarship (Wang et al., 2024; Shen et al., 2022). The review, nonetheless, suggests that use in practice of the intersctions is considerably underdeveloped. Whilst many papers theorise SDG intersctions conceptually, relatively few provide inclusive metrics or indicators extending across more than one SDG. The gap highlights the absolute necessity of new methodologies that could realistically be used in measuring as well as determining SDG intersctions.

Community Engagement and Determinants of Equity

Conceptions of participatory governance and equity considerations remain inadequately accounted for in smog governance policies. Research indicates that participatory governance is associated with more effective policy effectiveness and compliance (Büker et al., 2024); that is, however, many papers observe the continued difficulty of weak public awareness and weak sub-national engagement, especially in South Asia (Malik et al., 2024; Razzaq et al., 2024). The gap is integral to the wider theme of neglect that envelopes grassroots governance institutions that would be instrumental in the building of legitimacy as well as long-term durability. The question of equity is particularly crucial in that poor and marginalized people are over exposed to smog while institutionally excluded from deciding governance decisions (Ali, 2023). Unless specifically tailored and inclusive strategies are in place, existing policies will more than likely exacerbate inequalities rather than reduce them. The point is closely associated with the achievement of Sustainable Development Goal 10 (Reduced Inequalities) and Sustainable Development Goal 5 (Gender Equality), due to the rising evidence of both the existence of gender-related as well as socio-economic inequalities in the extent of exposure as well as health response (He et al., 2024).

Theoretical considerations.

Research findings also bring out the close relationship between air pollution and public health, including the Sustainable Development Goals (SDGs), with a special focus on SDG 3 (Good Health and Well-being) and SDG 11 (Sustainable Cities and Communities). The study suggests comprehensive frameworks that pay attention to environmental, health, and socio economic dimensions at the same time therefore opposing compartmentalized approaches in air quality management (Zhou et al., 2024; Wang et al., 2024; Xing et al., 2024).

The papers discussed accentuate the symbiotic association of air pollution and climate change mitigation, underscoring the positive outcomes concerning the mitigation of both PM2.5 and carbon dioxide emissions. Such evidence contributes to new hypotheses favoring holistic environmental governance strategies that share ambition in coordinating air quality enhancement with carbon neutrality objectives (Wang et al., 2024; Shen et al., 2022; Xing et al., 2024).

Socio-economic inequalities and the multiple levels of exposure risk also provide depth to the environmental justice conceptual framework in terms of the Sustainable Development Goals (SDGs). These studies emphasize the importance of the incorporation of equity-minded thinking in air quality policy in order to minimize the inappropriately high health effects experienced among poor and marginalized communities (Büker et al., 2024; He et al., 2024; Assessing the Health Impacts of Urban Air Pollution, 2024).

Technological innovations and digitalization emerged as essential conceptual instruments, whereas the empirical evidence further accentuated the significance in engendering better monitoring processes, public awareness expansion, and policy compliance. These findings are in alignment with smart urban policy and transition theories of socio-technical (Siddqiue et al., 2023; Shen & Zhang, 2024; Khosla, 2024).

The academic literature criticizes common top-down regulating structures through highlighting the importance of community engagement, interdisciplinary approaches, and evidence-informed policymaking. The conceptual developments are based on moving the field of environmental health in the direction of participatory governance and adaptive management (Büker et al., 2024; Bhattarai et al., 2024; Lu, 2024).

Thirdly, synergetic effects of climate stressors and air pollution, such as heat waves, create new knowledges regarding health risks in cities. The latter accentuates the importance of the general risk characterization frameworks that assemble diverse environmental threats (Sekar et al., 2024; Liu et al., 2024).

Practical Considerations.

Policymakers should accord high priority to air pollution and climate policies from co-benefits that cultivate health and environmental sustainability, as the experience from regional policies in China and other cities attests. Such policy entails the formulation of stringent place-based air quality standards and the connection of carbon neutrality targets with pollutant-reduction measures (Zhou et al., 2024; Wang et al., 2024; Zhang et al., 2023).

Implementation of the current technological applications, including Internet of Things (IoT) sensors, digital online networks, and green technology, is critical in the effective real-time air quality measurement as well as public engagement (Siddqiue et al., 2023; Shen & Zhang, 2024; Khosla, 2024). These applications may be applied in enhancing policy responsiveness and enforcement in urban areas.

Reducing the issue of socio-economic inequalities in exposure to pollution involves the implementation of special measures that target disadvantaged people. It implies raising the level of accessibility to medical care services, conducting awareness campaigns among people, as well as consistent application of environmental protection laws (Büker et al., 2024; Alamgir & Shan, 2024; Assessing the Health Impacts of Urban Air Pollution, 2024). Transport policy and urban development should embrace sustainable mobility plans that comprise the development of public transport networks, the development of the infrastructures of electric vehicles, and the development of green corridors.These plans are crucial in the mitigation of vehicular emissions and the creation of healthier cities conforming with SDG 11 (Jena et al., 2023; Farzaneh et al., 2019; Bich & Dinh, 2024).

Successful management of air quality involves coordination at the various levels of governance, strengthened institutional capacities, and cooperation at the international level. Such aspects are vital in overcoming enforcement hurdles and diffusion of best practices, especially in the regions that experience rapid urbanization and industrialization (Rafiq-uz-Zaman et al., 2024; Lu, 2024).

Ultimately, economic study identifies that air pollution abatement investments bring significant health and productivity gains. The same is in favor of the scaling up of green finance tools as well as green urban planning approaches with low carbon content (Zhang & Dou, 2024; Gao et al., 2024; Gaybullaev, 2024).

Global and Regional Implications

The results have significant bearings on global governance for sustainability. The Chinese and East Asian experience shows that effective institutional capacity and adaptable governance can bring significant air quality improvements within relatively short timescales. The South Asian experience, on the other hand, identifies the chronic challenge of decentralized governance, lack of resources, and socio-economic inequalities, highlighting the imperative of building capacities as well as cooperation at the international level. At the international level, policy instruments like the Paris Agreement and the SDGs offer prospects of aligning smog abatement with broader climate and health objectives, but this potential will require more effective cross-sector coordination. Overall, while smog is a global issue, effective responses will need an appropriate calibration to specific contexts, taking account of both technological innovatory capacity, governance effectiveness, as well as the realities of the socio-economic context.

The analysis in brief translates into four general findings. First, smog is both an economic and health emergency that negatively impacts achievements in many SDGs directly. Second, the quality of governance is the overarching variable that distinguishes fruitful policy responses from unsuccessful ones. Third, though technological innovations promise great possibilities, the latter benefits can be actualized only with inclusive structures that offer equity and accessibility. Fourth, SDG linkages, though conceptually acknowledged, are poorly operationalized and represent an important frontier in the policy as well as research field. Collectively, the findings mirror the necessity of constructing multi-sectoral, equity-based frameworks that align smog mitigation with the broader sustainability framework.

Table 2. showing Research Gaps, limitations and Future Directions

Area / Gap	Description of Limitation / Gap	Future Research Directions	Justification	Key References
Geographic Bias / Regional and Socioeconomic Generalizability	The vast majority of research is conducted on China, India, and Pakistan, with little external validity and usability in other socio-economic and geographical situations.	Carry out multiple regional comparison studies in underrepresented territories (Sub-Saharan Africa, South Asia, Latin America) in many different urban and rural environments.	Wider representation offers policies that are universal but with local vulnerabilities.	Zhou et al. (2024); Alamgir & Shan (2024); Razzaq et al. (2024); Rafiq-uz-Zaman et al. (2024); Malik et al. (2024); *Progress, Challenges and Prospects...* (2022)
Methodological Constraints / Methodological Gaps	Striking reliance on forecasts, simulations, or case studies with limited longitudinal and intervention-focused empirical evidence.	Increase the methodological soundness by adopting longitudinal cohort research, hybrid model that combines satellite and ground observations, and inclusive meta-analytic syntheses	solid methodologies will reduce uncertainty and enable better comparisons at various contexts.	Wang et al. (2024); Shen (2024); Hu et al. (2024); Ye & Tao (2023); Zhang et al. (2023); Bhattarai et al. (2024)
Limited Interdisciplinary Integration / Operationalizing SDG Interlinkages	Relatively few apply health, economic, environmental, and social sectors at the same time; SDG synergies are often at the conceptual level.	Construct integrated frameworks and empirical research that measures co-benefits/trade-offs in multiple SDGs in smog suppression	Operationalizing SDG linkages allows maximization of synergies while preventing policy.	Jena et al. (2023); Sekar et al. (2024); Liu et al. (2024); He et al. (2024); Rafiq-uz-Zaman et al. (2024)
Insufficient Community Engagement / Sustained Engagement of Vulnerable Groups	Representation of Marginalized Groups Not enough empirical evidence on community awareness, engagement, and equity; marginalized groups often excluded from the policymaking process.	Establish long-term participatory structures of governance, field experiments, and online engagement frameworks with long-term representation of marginalized groups	Participatory engagement strengthens legitimacy, adherence, and equity outcomes.	Büker et al. (2024); Malik et al. (2024); Siddqiue et al., (2023); Rafiq-uz-Zaman et al. (2024)

continued on following page

Table 2. Continued

Area / Gap	Description of Limitation / Gap	Future Research Directions	Justification	Key References
Policy Implementation Gaps / Institutional Capacity Weaknesses	Policy designs are well but poorly implemented because of weak institutions, weak coordination among agencies, and absence of review	Compare measures of enforcement; build institutional capacity and adaptive system of governance.	Implementation from the design of policy is vital in terms of potential changes.	Sah (2024); Lu (2024); Rafiq-uz-Zaman et al. (2024); Mehmood et al. (2024); *Progress, Challenges and Prospects…* (2022)
Underexplored Synergistic Effects / Long-term Sustainability	insufficient examination of smog's cumulatively compounded effects with climate change stresses (e.g., heatwaves), and long-term multiple interventions.	Longitudinal and system-based research that looks at multiple community, technological, and regulatory interventions.	Recognition of the importance of understanding synergies in the context of resilient, adaptable management that is SDG-compatible.	Sekar et al. (2024); Liu et al. (2024); Wang et al. (2024); Yang et al. (2024)
Data Limitations and Monitoring Gaps	Insufficiently monitored network and inhomogeneous standards prevent exposure estimation with accuracy and timely policy decisions.	Enhance networks with standard monitors, accept hybrid data strategies (ground + satellite), and strengthen open data platforms.	Trust in data is the foundation of evidence-based policy formulation as well as health risk estimation.	Lluri et al. (2024); Bhattarai et al. (2024); Halim et al. (2024); *Progress, Challenges and Prospects…* (2022)
Economic Impact Assessment Limitations / Integration of Health & Economic Outcomes	Economic appraisals frequently closely look at direct health expenditure in relative isolation from productivity, inequality, and long-run economic resilience.	Construct dynamic linked models of health outcomes and productivity, inequality, and resilience	Health–economy feedback loop understanding maximises the use of resources.	Zhou et al. (2024); Ye & Tao (2023); Xie et al. (2023); Gaybullaev (2024); Lin (2023)
Urban Planning Integration with Air Quality	Most urban planning research integrates smog reduction inadequately with land use, transport policy, and green infrastructure policy.	Establish air-quality-based urban planning frameworks with well-planned equitable land use and sustainable mobility.	Integration of urban planning decreases source-based emissions as well as boosts the liability of the city.	Jena et al. (2023); Lluri et al. (2024); Halim et al. (2024)

CONCLUSION

This systematic review of the evidence and meta-analysis gathered information from 644 studies, with close scrutiny of 50 highly significant texts, in order to assess the health, economic, policy, technological, and SDG interlinkage dimensions of smog.The report finds that smog is more than an environmental problem, but rather a multidimensional development disaster, damaging public health, economic efficiency, and urban resilience and holding back progress on many of the Sustainable Development Goals. The analysis uncovered that smog exposure is significantly associated with high mortality and excess death, particularly in sensitive populations. At the economic level, the evidence is that GDP loses 1–2% annually in the most contaminated cities, with lost productivity and healthcare in the billions of dollars. Policy measures exist in most urban settings, but enforcement is variable, with Chinese achievements contrasted with long standing South Asian challenges.

Techno breakthroughs like digital tracking, clean energy, and sustainable mobility offer bright futures, but they remain far from being implemented because of infrastructural, economic and societal limitations. At the deepest level, the review finds underdeveloped connectivity of smog reduction with the SDG framework, with strong but poorly operationalized associations among SDG 3 Health, SDG 11 Sustainable Cities, and SDG 13 Climate Action.

In general, the study establishes that reducing smog needs comprehensive, multi-stakeholder, and equity-based approaches that align health, economic, urban, and climatic objectives.

Policy Implications The evidence base holds some actionable lessons for policymakers, urban planners, and global institutions. First, institutional building and enforcement are imperative, as policies are only as good as their enforcement. Institutions that function well, adaptable administration, and decentralized environmental spending need to be realized in order to effect policy design in outcomes, particularly in South Asia and other low-capacity countries wherein institutional investment and anti-corruption activities SDG 16 are called for. Second, management of smog needs to be compliant with national plans under the SDGs so that health, urban development, and climate action goals are pursued in unison. SDG tracks the frameworks of track air quality along with health, climate and economic indicators. Third, equitable tech adoption is paramount, with increased financial instruments like green bonds, subsidies, and technology transfer schemes aiding low and middle income nations. Policies should also ensure inclusive access to innovations like electric mobility and renewable energy so as to avoid exacerbating social inequalities. Fourth, community buy-in and public awareness continue to be the key ingredients in effective smog mitigation. Education campaigns among the public, participatory planning, and co-constructed policy structures can be reinforced with better compliance and legitimacy while digital platforms and social media tools offer potent tools in galvanizing behavioral change and constructing accountability devices. Fifth, the co-benefits in smog mitigation in terms of economic development ought to be capitalized upon as evidence identifies considerable healthcare cost savings as well as productivity benefits. Framing clean air efforts as drivers of green economic development aligns directly with SDG 8 (Decent Work and Economic Growth) as well as SDG 9 (Industry, Innovation, and Infrastructure). Lastly, inasmuch as smog is transboundary, eliciting global and regional cooperation is crucial. Integrated systems of tracking, international agreements on emissions, and technology-exchange networks can offer mutual responses, while lessons from the effective practices of Beijing's adaptive management should be tailored through cooperation focused on information-exchange in South Asia, sub-Saharan Africa, and Latin America. Conclusion Reflections This review verifies that clean air is both the universal human right and the foundation of sustainable development. Smog mitigation is the promise of triple dividends in the forms of public health benefits, better resilient cities, and strengthened climate action, but the transition is from compartmentalized policies and disjoined intervention strategies to holistic, multi-sector strategies based on inclusiveness and equity. For researchers, the review identifies strong research priorities in the operationalization of the SDG interlinkages, equity-inclined analysis, and enforcement research on policies. For policymakers the review provides advice on how to shape holistic intervention strategies in aligning air quality improvement with national development goals. For people, it reiterates that active engagement is fundamental in shaping sustainable and equitable urban prospects. Smog mitigation progress will be crucial in the long run both in terms of realizing SDG 3 Good Health and Well-being and SDG 11 Sustainable Cities and Communities but also in initiating broader transformations toward the 2030 Agenda. Clean air is no longer an aspirational objective it's an immediate imperative of health, justice, and sustainable development.

REFERENCES

Alamgir, W., & Shan, H. (2024). Air pollution: A challenge to public health and healthcare systems. *Life and science, 5* (1), 02-02. DOI: 10.37185/LnS.1.1.594

Ali, M. I. (2023). Navigating smog: Legislation addressing children's right to health in india and pakistan. DOI: 10.56784/hrgs.v2i2.74

Allam, Z., & Soomauroo, Z. (2024). How the green new deal can transoform urban landscapes and promote health equity. *Cities & health*null, 1-5. DOI: 10.1080/23748834.2024.2360301

Anjum, S., Zafar, M. M., & Kumari, A. (2024). A review of diseases attributed to air pollution and associated health issues: A case study of indian metropolitan cities. DOI: 10.1016/B978-0-443-16080-6.00004-5

Bhattarai, S. K., Dhakal, R., Acharya, S., & Arayal, S. (2024). Think global, act local: Developing an ebpm framework for air quality management in nepal. *Journal of water resources and pollution studies, 9* (3), 60-65. DOI: 10.46610/JoWRPS.2024.v09i03.005

Bich, H. N. T., & Dinh, T. L. (2024). Overcoming barriers to sustainable green transportation in ho chi minh city: A pathway toward achieving sdgs 11 and 13. *Sustainability (Basel), 16*(23), 10629–10629. DOI: 10.3390/su162310629

Borenstein, M., Hedges, L. V., Higgins, J. P., & Rothstein, H. R. (2021). *Introduction to meta-analysis*. John wiley & sons. DOI: 10.1002/9781119558378

Büker, P., West, S. E., Bowyer, C., Apondo, W., Cinderby, S., Gray, C. M., Hahn, M., Lambe, F., Loh, M., Medcalf, A. J., Muhoza, C., Muindi, K., Njoora, T., Twigg, M., Waelde, C., Walnycki, A., Wainwright, M., Wendler, J., Wilson, M., & Price, H. D. (2024). Transdisciplinary research as a means of protecting human health, ecosystems and climate by engaging people to act on air pollution. DOI: 10.1079/onehealthcases.2024.0002

Coalition, C. C. A. (2019). Air pollution in asia and the pacific: Science-based solutions.

Dey, S., Sarkar, D., Imam, F., Kumar, A., Mukherjee, A., Purohit, P., Kiesewetter, G., & Klimont, Z. (2023). Pathways for india to reduce air pollution health burden and meet sustainable development goal 3.4 target. *Environmental health perspectives*null, . DOI: 10.1289/isee.2023.MP-052

Dougherty, S., & Nebreda, A. M. (2022). Going global, locally? Decentralized environmental expenditure and air quality. *Public Sector Economics*, *46*(4), 489–503. DOI: 10.3326/pse.46.4.3

Elsom, D. M. (1996). Smog alert: Managing urban air quality.

Farzaneh, H., Oliveira, J. A. P. D., McLellan, B., & Ohgaki, H. (2019). Towards a low emission transport system: Evaluating the public health and environmental benefits. *Energies*, *12*(19), 3747. Advance online publication. DOI: 10.3390/en12193747

Gao, J., Hua, G., AbidAli, R., Huo, B., Mahamane, F., & Li, Z. (2024). The impact of green low-carbon finance on smog reduction and carbon mitigation: Evidence from a spatial panel simultaneous equations model. *The International Journal of Urban Sciences*null, 1-34. DOI: 10.1080/12265934.2024.2438204

Gaybullaev, O. (2024). An analysis of the economic impacts of air pollution in the urban areas: Bibliographic review. *Economica*, *15*(3-4), 53–68. DOI: 10.47282/economica/2024/15/3-4/15028

Halim, N. D. A., Latif, M. T., & Mohamed, A. F. (2024). Assessing sustainable approaches to urban air quality: A study of malaysia's largest conurbation regions. DOI: 10.5194/egusphere-egu24-12920

He, C., Huang, C., Zhang, P., Wu, Q., Wang, Y., Zeng, Z., Zhang, L., Miao, C., Yue, F., Tang, Y., Zhou, J., Zhou, Y., Wang, Y., Yang, L., Hong, S., Peng, S., Gong, X., & Chen, W. (2024). Global pm2.5 exposure inequality hindering progress toward air quality-related sdgs. DOI: 10.21203/rs.3.rs-5072048/v1

Hu, Y., Chao, K., Zhu, Z., Yue, J., Qie, X., & Wang, M. (2024). A study on a health impact assessment and healthcare cost calculation of beijing–tianjin–hebei residents under pm2.5 and o3 pollution. *Sustainability*null, . DOI: 10.3390/su16104030

Imran, M., Khan, S., Nassani, A. A., Haffar, M., Khan, H. U. R., & Zaman, K. (2023). Access to sustainable healthcare infrastructure: A review of industrial emissions, coal fires, and particulate matter. *Environmental Science and Pollution Research International*, *30*(26), 69080–69095. DOI: 10.1007/s11356-023-27218-4 PMID: 37129815

Jena, M. C., Mishra, S. K., & Moharana, H. S. (2023). Challenges and way forward to maintain air quality standard in urban areas. *Global environmental engineers*null, . DOI: 10.15377/2410-3624.2023.10.4

Khosla, A. (2024). Sustainable urban development: Integrating smart city technologies for environmental resilience. *Innovative research thoughts, 10* (3), 138-142. DOI: 10.36676/irt.v10.i4.1472

Lin, Z. (2023). The economic impact of air pollution in beijing. *Future of Business and Finance*null, 265-282. DOI: 10.1007/978-3-031-29463-1_14

Liu, L., Zhao, J., & Li, Z. (2024). Sustainable urban development: A comprehensive model for assessing health risks from compounded pollution in xi'an. *Sustainability (Basel), 16*(22), 9799–9799. DOI: 10.3390/su16229799

Lluri, L., Lluri, F., & Sefa, S. (2024). Potential strategies to improve air quality in tirana for current and future generations. *Interdisciplinary journal of research and development, 11* (3), 96-96. DOI: 10.56345/ijrdv11n313

Lu, S. (2024). Efforts to improve air quality in major regions - costs and benefits, successes and failures. *Highlights in Science Engineering and Technology, 121* null, 625-630. DOI: 10.54097/83bcmw36

Malik, A., Islam, J., Zaib, G., Ashraf, M. H., Zahid, A., Rashid, A. R., Zia, T., & Ali, Q. (2024). ISLAM, J., ZAIB, G., Zahid, A., Zahid, A., Rashid, A., Zia, T., & Ali, Q. (2024). Smog crisis in lahore: Evaluating air quality trends and public health implications. *Bulletin of Biological and Allied Sciences Research, 2024*(1), 87–87. DOI: 10.54112/bbasr.v2024i1.87

Mehmood, A., Usama, R. M., Mehmood, A., & Rizwan, S. (2024). Evaluating the impact of smog emergency policies on air quality and public perception by using geospatial technologies: A case study of lahore, pakistan. DOI: 10.21203/rs.3.rs-5537502/v1

Page, M. J., McKenzie, J. E., Bossuyt, P. M., Boutron, I., Hoffmann, T. C., Mulrow, C. D., & Moher, D. (2021). The PRISMA 2020 statement: an updated guideline for reporting systematic reviews. *bmj, 372*. Progress, challenges and prospects of urban air pollution and climate in iran. DOI: 10.21203/rs.3.rs-2155113/v1

Rafiq-uz-Zaman, M., Khalid, N., Faridi, M. Z., & Khan, Z. A. (2024). Smog mitigation and climate resilience in pakistan: A policy-centric review of strategies and challenges. DOI: 10.71085/sss.03.04.175

Razzaq, A., Zafar, M. M., Zahra, L. T., Qadir, F., Qiao, F., Ullah, M. S., Shehzad, S., Rasool, G., & Jiang, X. (2024). Smog: Lahore needs global attention to fix it. *Environmental challenges, 16* null, 100999-100999. DOI: 10.1016/j.envc.2024.100999

Sah, S. K. (2024). Public health policy responses to air pollution in india: Challenges and strategies for sustainable development. *Visioner, 16*(2), 163–170. DOI: 10.54783/jv.v16i2.1057

Sekar, A., Sharma, R., & Mahajan, A. (2024). Synergistic impact of air pollution and heat on health and economy in india. DOI: 10.1007/978-3-031-56564-9_2

Shen, J., Cai, W., Chen, X., Chen, X., Zhao, Z., Ma, Z., Yang, F., & Zhang, S. (2022). Synergies of carbon neutrality, air pollution control, and health improvement — a case study of china energy interconnection scenario. *Global energy interconnection, 5* (5), 531-542. DOI: 10.1016/j.gloei.2022.10.007

Shen, Y. (2024). Air quality assessment and pollution mitigation strategies in beijing: A comprehensive study on pm2.5 trends, influencing factors, and case studies. *Highlights in Science Engineering and Technology, 88* null, 695-702. DOI: 10.54097/0wzh2456

Shen, Y., & Zhang, X. (2024). Blue sky protection campaign: Assessing the role of digital technology in reducing air pollution. *Systems*null, . DOI: 10.3390/systems12020055

Siddqiue, M. U., Jamil, M., & Arshad, A. (2023). Digital Solutions and Technological Innovations Driving Public Awareness and Community Initiatives for Smog Reduction in Lahore: A Descriptive Overview. *Pakistan Social Sciences Review, 7*(4), 425–438. DOI: 10.35484/pssr.2023(7-IV)39

Tranfield, D., Denyer, D., & Smart, P. (2003). Towards a methodology for developing evidence-informed management knowledge by means of systematic review. *British Journal of Management, 14*(3), 207–222. DOI: 10.1111/1467-8551.00375

Wang, G., Xia, W., Xiao, Y., Guan, X. L., & Zhang, X. (2024). The synergy between co2 and air pollution emissions in chinese cities by 2060: An assessment based on the emissions inventory and dynamic projection model. *Sustainability (Basel), 16*(21), 9338–9338. DOI: 10.3390/su16219338

Wang, L., Zhang, F., Pilot, E., Yu, J., Nie, C., Holdaway, J., Yang, L., Li, Y., Wang, W., Vardoulakis, S., & Krafft, T. (2018). Taking action on air pollution control in the beijing-tianjin-hebei (bth) region: Progress, challenges and opportunities. *International Journal of Environmental Research and Public Health, 15*(2), 306. Advance online publication. DOI: 10.3390/ijerph15020306 PMID: 29425189

Xie, Y., Liu, X., Li, D., Zhao, M., Weng, Z., Zhang, L., & Xu, M. (2023). Health and economic benefits of reducing air pollution embodied in gba's green and low-carbon development. *urban climate*null, . DOI: 10.1016/j.uclim.2023.101755

Xing, J., Loughlin, D., Ren, L., Klimont, Z., & Woo, J. (2024). Air quality, health, and equity benefits of carbon neutrality and clean air pathways in china.. *Environmental Science & Technology*null, . DOI: 10.1021/acs.est.3c10076

Xu, T., Zhao, L., & Zhou, X. (2024). The impact of environmental policies on air quality improvement and public health in beijing. *Studies in social science & humanities, 3* (6), 42-48. DOI: 10.56397/SSSH.2024.06.06

Yang, B., Wang, Q., Zhao, L., & Wang, H. (2024). How does carbon biased technological progress promote carbon haze collaborative governance?evidence from chinese cities. *Environmental research, 257* null, 119312-119312. DOI: 10.1016/j.envres.2024.119312

Ye, Y., & Tao, Q. (2023). The dynamic relationship among economic development, air pollution, and health production in china: The dnsbm efficiency model. *Frontiers in Environmental Science*null, . DOI: 10.3389/fenvs.2023.1205712

Zhang, H., & Dou, W. (2024). Impact of air pollution prevention and control on urban green economy efficiency: Evidence from china. *Dental science reports, 14* (1), . DOI: 10.1038/s41598-024-79467-z

Zhang, W., Xu, H., Yu, X., Li, J., Zhang, Y., Dai, R., Meng, W., Shen, H., Chen, Y., Shen, G., Ma, J., Li, B., Liu, J., Wang, X., & Tao, S. (2023). Rigorous regional air quality standards for substantial health benefits. *Earth's Future*null, . DOI: 10.1029/2023EF003860

Zhou, Y., Zhang, X., Zhang, C., Chen, B., & Gu, B. (2024). Mitigating air pollution benefits multiple sustainable development goals in china.. *Environmental Pollution*null, 123992-123992. https://doi.org/DOI: 10.1016/j.envpol.2024.123992

Chapter 3
Smog and Sustainable Development Goals:
An Integrated Assessment Framework

Qamar Ali
https://orcid.org/0000-0002-8178-7591
Virtual University of Pakistan, Pakistan

Muhammad Aamir Mahmood
https://orcid.org/0009-0007-4452-2627
Virtual University of Pakistan, Pakistan

Muhammad Ashfaq
The University of Faisalabad, Pakistan

ABSTRACT

Smog, which is a combination of high particulate matter and gaseous pollutants, is one of the major environmental issues with immense implications for sustainable development. This chapter applies an integrated assessment framework to understand the multi-faceted, complex relationships that exist between smog and the United Nations' Sustainable Development Goals (SDGs). It reviews the direct and indirect effects of atmospheric pollution on the key SDGs such as Good Health and Well-being (SDG 3), Affordable and Clean Energy (SDG 7), Sustainable Cities (SDG 11), Climate Action (SDG 13) and Decent Work and Economic Growth (SDG 8). The analysis shows both serious threats to achieving the goals from smog and significant opportunities for synergistic policy co-benefits. The chapter concludes by arguing that the target of smog can only be achieved by going beyond sectoral boundaries in a mainstreaming approach to air quality into energy, transportation,

DOI: 10.4018/979-8-3373-2469-2.ch003

industrial, and urban planning policy. Drawing on empirical evidence and a case study carried out on air quality management in China, the chapter concludes that multi-pollutant, integrated solutions are not only an environmental necessity but an essential foundation for achieving inclusive, environmentally sustainable economic development and protecting public health for present and future generations.

INTRODUCTION

Figure 1. Smog and its connections with Sustainable Development Goals

This figure shows how smog affects health, energy, cities, climate, and economic growth.

Figure 2. Smog impacts on sustainable development goals (health, energy, cities, climate).

Smog is one among the most important environmental issues of the day, and the problem has far-reaching implications for human health, ecosystems, and economic development(Zhang et al., 2022). The complex relationship between smog and sustainable development is attracting growing interest since the United Nations Member States adopted the 2030 Agenda for Sustainable Development in 2015. This agenda set 17 Sustainable Development Goals (SDGs) with 169 targets to balance the economic, social and environmental dimensions of development. Smog (a portmanteau of smoke and fog) has become synonymous with air pollution in the form of high levels of particulate matter (PM2.5 and PM10), ozone (O3), nitrogen oxides (NOx), sulfur dioxide (SO2), and other pollutants that form an opaque haze in the atmosphere. These pollutants directly and indirectly affect achievement of several SDGs, and present both challenges and opportunities for integrated policy approaches(Biglari et al., 2022).

According to the World Health Organization (WHO), air pollution is responsible for around 7 million premature deaths a year and is one of the biggest environmental health risks in the world. In addition to the health consequences, smog also influences agricultural outputs, ecosystem function, climate stability, and economic growth. With the target for attainment of the Sustainable Development Goals (SDGs) by

2030 fast approaching, current data show that progress is still lacking in many areas, with the goals on environmental sustainability being the most troublesome. The Sustainable Development Goals Report 2025 paints a doom and gloom picture: while progress has been made in some areas such as access to education and mother and child health, and the adoption of renewable energy, the current rate of change means that all the Sustainable Development Goals will not be met by 2030(Weiland et al., 2021).

This chapter considers the interwoven links between smog and the Sustainable Development Goals, both in terms of the effects of air pollution on the different SDGs as well as the opportunities for synergies between smog and sustainable development, which could be better achieved through integrated strategies. By synthesizing evidence from the recent scientific literature, policy documents and empirical studies, this book aims to provide an overarching framework for understanding and addressing these intertwined challenges(Head, 2022).

Figure 3. Smog effects on health, ecosystems, and economy within sustainable development goals.

2. THEORETICAL BACKGROUND: SMOG IN THE FRAMEWORK OF SUSTAINABLE DEVELOPMENT

The sustainable development perspective on smog can only be conceptualized under the premise that the SDGs are interlinked. The 2030 Agenda acknowledges the fact that goals and targets are integrated and indivisible, balancing the three dimensions of sustainable development: economic, social and environmental. Smog lies at the intersection of these dimensions, at once a product of development patterns that are not sustainable (in particular, fossil-fuel-based energy systems, industrial processes, and transportation networks) and a limiting factor on the achievement of development goals(Mohamad Taghvaee et al., 2023).

The normative connection between smog and sustainable development can be conceptualized in terms of "planetary boundaries," which characterize a safe operating space of humanity about Earth system processes. Air pollution, especially atmospheric aerosol loading, is one of the thresholds which in many regions has already been crossed. This violation impacts several SDGs and because of this, there is a complex set of interactions which can be both synergistic and antagonistic. For example, measures that mitigate smog with cleaner energy technologies (co-beneficial for SDG 7) can also mitigate GHG emissions (co-beneficial for SDG 13) and improve health outcomes (co-beneficial for SDG 3), as examples of policy-crossover. However, there are also potential trade-offs, for example in the unintended consequences of the adoption of bioenergy on land use and food security (impacting SDG 2 and SDG 15)(Blair et al., 2021).

Within the SDG agenda, smog is tackled through a complex institutional framework of indicators and monitoring systems. WHO's work has been directed towards ensuring that health-relevant indicators of exposure to household and ambient pollution are included in the formal system of SDG indicators. Three targets are of relevance, namely SDG target 3.9.1 (mortality rate from air pollution), SDG target 7.1.2 (access to clean energy) and SDG target 11.6.2 (ambient air quality in cities). But as book has progressed, it's become apparent that air pollution impacts a far wider variety of SDGs than these three specific indicators intuitively indicate(Kubiszewski et al., 2022).

Table 1. Key SDG Targets Directly Related to Air Pollution

SDG Goal	Target Number	Target Content
SDG 3: Good Health and Well-being	3.9.1	Mortality rate attributed to household and ambient air pollution
SDG 7: Affordable and Clean Energy	7.1.2	Proportion of population with primary reliance on clean fuels and technologies
SDG 11: Sustainable Cities and Communities	11.6.2	Annual mean levels of fine particulate

Figure 4. Smog and sustainable development dimensions

This figure shows how smog interacts with economic, social, and environmental aspects, linking health, clean energy, and sustainable cities.

2.1. Smog and Health: The Core Relationship (SDG 3)

The most obvious and documented impact of smog is around SDG 3 (Good Health and Well-being). Target 3.9 is explicit in its goal to "substantially reduce the number of deaths and illnesses from hazardous chemicals and air, water and soil pollution and contamination." The health impacts of smog are wide and severe and include respiratory illnesses (asthma, bronchitis, lung cancer), cardiovascular illness (heart attacks, strokes), poor cognitive development, and poor birth outcomes(Parikh et al., 2021).

The mechanisms of harm are in the penetration of the fine particulate matter (PM2.5) deep into the tissue of the lungs and the bloodstream, where it can trigger inflammation, oxidative stress, and cellular damage. Ozone (O3), another major component of smog, decreases lung function and worsens respiratory disease. Nitrogen oxides (NOx) and Sulphur dioxide (SO2) cause respiratory disease and increased susceptibility to infections. According to the WHO, air pollution is among the world's most important risk factors for death - it estimated 7 million people die prematurely each year from exposure to fine particles in polluted air(Fatkhutdinova et al., 2021).

Health outcomes are not evenly distributed across the population, which has implications for SDG 10 (Reduced Inequalities). Children, older people, pregnant women and people with pre-existing health conditions are at particular risk. In addition, low-income communities tend to be exposed to higher levels because of their exposure to industrial facilities, transportation corridors, and other pollution sources, and they tend to have less access to health care services. This environmental justice lens highlights the need to address air pollution through an equity and human rights lens(Revesz, 2022).

Besides its direct health effects, smog has huge economic costs in the form of health care expenses, productivity losses, and declining labor productivity. Studies have estimated air pollution costs economies billions of dollars a year in medical treatment and lost productivity. These economic impacts are cascading across multiple SDGs with the largest impacts likely to be on SDG 1 (No Poverty) and SDG 8 (Decent Work and Economic Growth) as households affected by health burdens may slip into or stay in poverty because of medical costs and loss of earning capacity(Sriram & Albadrani, 2022).

2.2. Energy, Cities and Smog: Intersections of SDG 7 and SDG 11

The energy sector is a basic engine of smog production and therefore an important lever point for solutions. SDG 7 (Affordable and Clean Energy) is intended to make sure everyone has access to affordable, reliable, sustainable and modern energy. The existing stock of energy infrastructures, especially those based on the combustion of fossil fuels, is the main anthropogenic source of air pollutants contributing to the formation of smog. The switch to cleaner energy sources is therefore a key opportunity for tackling simultaneously air pollution, climate change, and energy access issues(Tibrewal & Venkataraman, 2021).

The household energy sector is a particularly important area for attention as about 2.4 billion people worldwide still depend on traditional biomass, coal or kerosene for cooking, heating and lighting. This household energy poverty causes significant exposure to indoor air pollution, particularly for women and children who spend a significant amount of time near domestic hearths. Household air pollution, the WHO estimates that cooking with dirty fuels is responsible for 3.8 million premature deaths a year. One of the main ways to solve this challenge is through increased availability of clean cooking systems (e.g., electricity, biogas, improved cookstoves), which will be crucial to making progress on both SDG7 and SDG3 at the same time(Yumkella et al., 2021).

Urban areas are particularly strong hotspots of smog formation and exposure, so SDG 11 (Sustainable Cities and Communities) is also especially relevant to air quality management. Cities are the center of emission generating activities (such as transportation, industry, power generation, and construction), and they are also centers of concentrated populations. Target 11.6 specifically concerns urban air quality and has a target to "reduce the adverse per capita environmental impact of cities, including through delivering special priority on air quality." Populations-weighted concentrations of particulate matter are reported for cities under SDG 11 reporting frameworks, and the WHO uses national reference-grade ground measurements, satellite data and chemical transport models to derive these(Kirwa et al., 2021).

The emission pattern and exposure distribution is affected by the urban form and design. Compact, mixed-use housing adjacent to efficient public transportation systems can lower per capita transportation emissions -- a key contributor to urban smog. Conversely, patterns of development that are car-dependent and sprawling are likely to elevate vehicle miles traveled and related emissions. These planning decisions have long-term consequences for air quality and sustainable development, resulting in path dependencies which are often hard to reverse. Appropriate urban development strategies that emphasize public transport and active transport and energy efficient buildings can therefore bring major co-benefits for air quality, GHG reduction, and quality of life(Stivanello, 2021).

Table 2. Primary Sources of Smog and Their Relationships with SDGs

Smog Source	Primary Pollutants	Main SDGs Affected	Potential Solutions
Fossil fuel power generation	SO2, NOx, PM2.5	SDG 3, SDG 7, SDG 13	Renewable energy, pollution controls
Industrial processes	NOx, NH3, VOCs, PM	SDG 3, SDG 8, SDG 9	Clean production technologies, circular economy
Transportation	NOx, VOCs, PM2.5, O3	SDG 3, SDG 11, SDG 13	Electric vehicles, public transit, active transport
Agricultural activities	NH3, PM2.5	SDG 2, SDG 3, SDG 15	Precision farming, sustainable fertilizers
Household energy use	PM2.5, CO, NOx	SDG 3, SDG 5, SDG 7	Clean cookstoves, electrification
Waste burning	PM2.5, VOCs, dioxins	SDG 3, SDG 11, SDG 12	Waste reduction, recycling, proper disposal

Figure 5. Smog sources and health risks

This infographic shows how power plants, industry, transport, and household burning contribute to smog, affecting human health and linking directly with SDG 3.

2.3. Climate Connections: The Smog-Climate Nexus (SDG 13)

The inherent inter-relationships between smog and climate change (SDG 13) both pose challenges and opportunities for integrated policy. Many air pollutants that contribute to the formation of smog have the same origin as the greenhouse gas emissions, which are the burning of fossil fuels for energy production, industry and transportation. Despite important differences in atmospheric lifetimes, spatial scales and policy time scales, this source commonality means that policies focused on one issue often generate co-benefits in the other(Abildtrup et al., 2024).

Some of the constituents of smog, including black carbon (a constituent of PM2.5), are short-lived climate pollutants with high warming potential - estimated to be 460-1,500 times more potent than CO2 per unit mass over a 100-year horizon. Cutting these pollutants can therefore provide fast climate benefits along with clean air benefits. Other constituents, notably sulfate and nitrate aerosols, exert net cooling effects due to light scattering and cloud interactions and therefore air pollution control has a complex relationship with climate mitigation. This complexity makes policy design difficult, to ensure there are no unintended consequences, such as the potential for accelerated warming after cooling aerosols are reduced(Tang & Kemp, 2021).

Future smog patterns are under the control of climate change directly through several pathways. Rising temperatures can increase the speed of the chemical reaction that creates ground-level ozone, which could increase ozone pollution even if precursor emissions remain the same. Evolving weather patterns may have an impact on pollution transport and atmospheric stagnation events, leading to more smog events in some areas and for longer periods of time. In addition, changes in vegetation cover (climate change) and wildfire activity (climate change) have the potential to affect naturally released particulate matter and ozone precursors, leading to feedback loops that complicate air quality management(Ojha et al., 2022).

The SDG 13 climate action targets therefore have important implications for air quality management. The time-domain simulations show that climate mitigation policies that shift energy systems away from fossil fuels-especially away from the combustion of coal-will also reduce emissions of sulfur dioxide, nitrogen oxides, and particulate matter. Likewise, climate adaptation measures that build resilience to ecosystems can also optimize natural filtration and sequestration of air pollutants. This cross-cutting nature of the issues highlights the need for policy coherence and institutional coordination from the climate and air quality sectors(Zusman et al., 2021).

2.4. Economic aspects: Is Smog Affecting Development (SDG 8)

The economic consequences of smog are multifaceted in terms of sustainable development, involving labor productivity, health care costs, agricultural productivity and tourism revenues. These economic aspects directly affect the SDG 8 (Decent Work and Economic Growth) which aims to promote sustained, inclusive and sustainable economic growth, full and productive employment, and decent work for all(Monaco, 2024).

Air pollution exposures result in productivity losses through several different mechanisms. Outdoor workers can minimize working hours during severe smog episodes (when physical activity increases inhalation rates and exposure doses). Cognitive function is impaired by exposure to air pollutants and this can lead to loss of productivity in terms of both physical and cognitive performance in various industries. Additionally, work time lost due to pollution-associated illnesses adds to the reduced economic productivity. These impacts are especially important in high labor-intensive, developing economies with limited social protection systems(Lin et al., 2023).

The economic cost of smog-related medical expenses is a significant drain on household, business and government budgets. These costs include direct medical costs of treating disease caused by pollution, and indirect costs such as time spent on care and loss of income during an illness. And in countries where health insurance coverage is limited, these costs can be devastating for households, potentially driving them into poverty (undermining SDG 1) and lowering ability to invest in education (affecting SDG 4) or productive assets(Zickafoose et al., 2024).

Agriculture is another industry that suffers from smog and influences SDG 2 (Zero Hunger). Ground-level ozone, one of the major components of smog, has been shown to damage plant tissue and reduce photosynthetic rates, thus resulting in crop yield losses estimated to be 5-15% for the major staple crops in heavily polluted regions. These yield losses pose risks to food security and to agricultural livelihoods, particularly in developing countries with low adaptive capacity. In addition, crop deposition with particulate matter can limit light availability and photosynthetic capacity, further diminishing agricultural productivities(Das et al., 2021).

The business economics of smog reduction keeps getting stronger as analytical methodologies improve. Economic analyses of air quality have consistently shown that the economic value of improved air quality--in terms of avoided healthcare costs, productivity increases, and improved ecosystem services--is far greater than the costs associated with pollution control. These economic arguments make a crop deposition with particulate matter can limit light availability and photosynthetic capacity, further diminishing agricultural productivities(Das et al., 2021).

2.5. Policy Performances and Integrated actions

The complexity of the smog problem in the context of sustainable development agenda needs the development of integrated policy frameworks outside traditional sectoral boundaries. Disease and industry-specific approaches tend to produce modest outcomes at best and can elicit unintended consequences on other dimensions of sustainable development. Instead, Integrated Approaches that simultaneously address a range of pollutants, sectors and objectives represent the most promising way forward(Melhim & Isaifan, 2025).

The policy integration principle indicates some important pathways to tackle smog in the SDG framework. First, air quality concerns must be "mainstreamed" through policies in the various sectors (e.g., energy, transport, agriculture, industry, and urban planning) and not addressed as an environmental concern. This mainstreaming will need to be supported by reinforced institutional mechanisms for coordination, such as interministerial committees on clean air, which can help dissolve silos and find ways to identify policy options that will synchronize(Lah, 2025).

Second, policy instruments must be thoughtfully tailored to the economic and social circumstances in which they exist. Market based instruments (like emissions trading systems or pollution taxes) have the potential to cost-effectively reduce pollution if they are well designed and implemented. Regulatory measures (including emission standards and fuel quality standards) still play a key role in pollution control for toxic pollutants or for pollutants that have only local effects. Knowledge-based strategies (including public awareness campaigns and pollution monitoring networks) can help empower citizens and businesses to take protective actions, and to participate in solution building(Kiss et al., 2022).

Third, technology innovation is critical in decoupling the emissions from pollution while the economy grows. Technological improvement in renewable energy, energy efficiency, pollution control devices and monitoring systems has enlarged the feasible solution space for smog control. Facilitation of book, development and deployment of these technologies - especially through policy frameworks that provide a pull through the market - is an important mechanism for driving progress(Roy, 2024).

Fourth, financial resources need to be mobilized at an appropriate scale to tackle air pollution challenges, especially in developing countries for whom the gaps in investment are largest. The WHO estimates that air pollution targets under SDG would require a further US$50-100 billion per year invested in clean energy access and air pollution control. While large, these investments are only a miniscule fraction of the economic costs currently imposed by air pollution, providing favorable benefit-cost ratios(Krutilla & Graham, 2023).

Table 3. Integrated Policy Approaches for Addressing Smog Across SDGs

Policy Approach	Key SDGs Addressed	Implementation Examples	Potential Co-benefits
Clean energy transition	SDG 7, SDG 3, SDG 13	Renewable energy incentives, coal phase-outs, energy efficiency standards	Climate mitigation, energy security, job creation
Sustainable transportation	SDG 11, SDG 3, SDG 13	Public transit investment, vehicle emission standards, active transport infrastructure	Reduced congestion, physical activity, noise reduction
Agricultural waste management	SDG 2, SDG 3, SDG 15	Alternatives to field burning, sustainable fertilizer practices, biogas systems	Soil conservation, nutrient recycling, rural energy access
Industrial pollution control	SDG 9, SDG 8, SDG 3	Best available technology requirements, circular economy initiatives, monitoring and reporting	Resource efficiency, innovation, competitiveness
Urban planning reforms	SDG 11, SDG 3, SDG 13	Compact development, mixed-use zoning, green infrastructure	Reduced infrastructure costs, social cohesion, climate adaptation
Household energy access	SDG 7, SDG 5, SDG 3	Clean cookstove programs, electrification, renewable energy microgrids	Time savings for women, reduced deforestation, safety

Figure 6. Climate, health, and economy links of the smog–climate nexus.

This infographic shows how air pollution and climate change jointly affect human health and economic outcomes, highlighting the need for integrated actions under SDG 13 and SDG 8.

2.6. Case Book: China's Experience with Smog and Sustainable Development

China's air pollution response provides important lessons to understand the multifaceted links between smog and sustainable development. Rapid industrialization and urbanization in recent decades have created serious air pollution issues, especially in large urban areas and industrial areas of the country. By the early 2010s, smog episodes were impacting hundreds of millions of people and causing public consternation and government reaction(Friedman et al., 2021).

In 2013 the Chinese Government issued a comprehensive Air Pollution Prevention and Control Action Plan, setting ambitious goals for particulate matter reduction, restrictions on coal consumption, restructuring of industry, and emission controls on vehicles. This policy framework was a first attempt at tackling air pollution through integrated action across energy, industry, transport and urban sectors. These measures did show significant air quality improvements: average PM2.5 concentrations in China's key regions dropped by 30-50% during the 2013-2021 period, even amid continued economic growth(Cheshmehzangi & Tang, 2024).

China's experience demonstrates both the potential effectiveness of policy action and the continuing difficulties in achieving air quality goals while meeting other development goals. The government's focus on end-of-pipe pollution controls (flue gas desulphurization and selective catalytic reduction) has resulted in large scale emission reductions at the cost of high operational costs. Besides, energy systems and industrial patterns transitioned more gradually, reflecting the economic and social intricacies of transitions(Miörner et al., 2021).

Recent studies on nitrogen pollution in China revealed how inextricably linked air quality problems are. Whereas China has successfully reduced sulfur dioxide emissions by strict controls, nitrogen oxides and ammonia emissions were more challenging. This imbalance changed the chemistry of the atmosphere in ways that might curtail the effectiveness of particulate matter control efforts. Specifically, relative increases in ammonia availability may have increased the formation of ammonium nitrate particles, which is a major component of PM2.5. The experience provides evidence for the significance of integrated pollution control programs that deal with multiple pollutants at once rather than with single substances in isolation(Khanam et al., 2023).

China's experiences in integrating air quality management into sustainable development strategies are still in a process of evolution. The Chinese government's ecological civilization strategy aims to combine economic development and environmental protection, with air quality goals being included in broader sustainability goals. This integrated approach acknowledges that long-term solutions to air pollution demand structural interventions into patterns of production and consumption, not just interventions into pollution outputs(Zhang et al., 2025).

2.7. Conclusion and Implications for the Future

The multiple links between smog and the Sustainable Development Goals reflect both the great challenges and the massive opportunities facing the global sustainable development agenda. As the 2030 target date for the SDGs nears, current evaluations suggest that progress on air quality-related goals is not keeping pace, and billions of people are still being exposed to harmful levels of pollution. However, examples from a range of regions show that with strong policy action, air quality can be improved considerably while achieving other development goals(Liu et al., 2022).

This book reveals several priority areas for improving the integration of smog reduction under sustainable development policies(Ogwu & Izah, 2024):

First, because air pollution challenges are cross-cutting in nature, policy integration mechanisms need to be strengthened. This requires cutting through institutional siloes and taking integrated approaches in energy, transport, industry, agriculture and urban planning departments. Policy evaluation tools that measure impacts across several SDGs can help to determine synergistic interventions and avoid trade-offs. Second, monitoring and accountability systems need to be strengthened to track progress and ensure implementation. This includes: an expansion of air quality monitoring networks, especially in developing regions where there are still large data gaps; enhanced approaches to estimating exposure and health impacts; and enhanced reporting mechanisms for SDG air-quality-related indicators. Third, clean air interventions need to be scaled up with resources, specifically for low- and middle-income countries where investment needs are the highest. Why: Innovative financing mechanisms (e.g. green bonds, climate funds, results-based financing) can help to close investment gaps and harness private capital. Fourth, the rate of technological innovation must be increased to widen the solution space for solving air pollution. Clean energy technologies, pollution control devices, low-emission transportation options and precision agriculture techniques that reduce ammonia volatilization are key topics of focus(Mencaroni et al., 2021).

Finally, knowledge sharing and capacity building is important to transfer effective policies and practices on the regional level. International initiatives, such as the World Health Organization's Air Quality and Health Program and the Climate and Clean Air Coalition, have a key role to play in facilitating this exchange and supporting implementation at the national and local level. The smog challenge is an exemplar of the interdependencies which are so fundamental to the sustainable development agenda. None of these goals can be achieved in isolation; progress on health, energy, cities, climate and economic goals are all intertwined through the air we breathe. By making these linkages explicit and creating integrated approaches that are based on synergies and address trade-offs, policymakers can accelerate action across several SDGs whilst providing cleaner air for current and future generations(Ogwu & Izah, 2024).

Figure 7. China's experience with smog and sustainable development

This infographic highlights China's journey from heavy industrial smog toward integrated sustainable policies, showing pollution sources, control measures, challenges, and pathways to cleaner growth.

REFERENCES

Abildtrup, J., Jacobsen, J. B., Vedel, S. E., Mantau, U., Mavsar, R., Pettenella, D., Prokofieva, I., Schubert, F., Stenger, A., Varela, E., Vidale, E., & Thorsen, B. J. (2024). Preferences for climate change policies: The role of co-benefits. *Journal of Environmental Economics and Policy, 13*(1), 110–128. DOI: 10.1080/21606544.2023.2223182

Biglari, S., Beiglary, S., & Arthanari, T. (2022). Achieving sustainable development goals: Fact or Fiction? *Journal of Cleaner Production, 332*, 130032. DOI: 10.1016/j.jclepro.2021.130032

Blair, M. J., Gagnon, B., Klain, A., & Kulišić, B. (2021). Contribution of biomass supply chains for bioenergy to sustainable development goals. *Land (Basel), 10*(2), 181. DOI: 10.3390/land10020181

Cheshmehzangi, A., & Tang, T. (2024). *30 Years of Urban Change in China's 10 Core Cities*. Springer.

Das, S., Pal, D., & Sarkar, A. (2021). Particulate matter pollution and global agricultural productivity. In *Sustainable Agriculture Reviews 50: Emerging Contaminants in Agriculture* (pp. 79 107). Springer. DOI: 10.1007/978-3-030-63249-6_4

Fatkhutdinova, L., Tafeeva, E., Timerbulatova, G., & Zalyalov, R. (2021). Health risks of air pollution with fine particulate matter. *Kazan medical journal, 102*(6), 862-876.

Friedman, S. R., Mateu-Gelabert, P., Nikolopoulos, G. K., Cerdá, M., Rossi, D., Jordan, A. E., Townsend, T., Khan, M. R., & Perlman, D. C. (2021). Big Events theory and measures may help explain emerging long-term effects of current crises. *Global Public Health: An International Journal for Research, Policy and Practice, 16*(8-9), 1167–1186. DOI: 10.1080/17441692.2021.1903528 PMID: 33843462

Head, B. W. (2022). *Wicked problems in public policy: Understanding and responding to complex challenges*. Springer Nature. DOI: 10.1007/978-3-030-94580-0

Khanam, Z., Sultana, F. M., & Mushtaq, F. (2023). Environmental pollution control measures and strategies: an overview of recent developments. *Geospatial Analytics for Environmental Pollution Modeling: Analysis, Control and Management*, 385-414.

Kirwa, K., Szpiro, A. A., Sheppard, L., Sampson, P. D., Wang, M., Keller, J. P., Young, M. T., Kim, S.-Y., Larson, T. V., & Kaufman, J. D. (2021). Fine-scale air pollution models for epidemiologic research: Insights from approaches developed in the multi-ethnic study of atherosclerosis and air pollution (MESA air). *Current Environmental Health Reports*, *8*(2), 113–126. DOI: 10.1007/s40572-021-00310-y PMID: 34086258

Kiss, B., Sekulova, F., Hörschelmann, K., Salk, C. F., Takahashi, W., & Wamsler, C. (2022). Citizen participation in the governance of nature-based solutions. *Environmental Policy and Governance*, *32*(3), 247–272. DOI: 10.1002/eet.1987

Krutilla, K., & Graham, J. D. (2023). *Benefit-cost analysis of air pollution, energy, and climate regulations*. Cambridge University Press. DOI: 10.1017/9781009189460

Kubiszewski, I., Mulder, K., Jarvis, D., & Costanza, R. (2022). Toward better measurement of sustainable development and wellbeing: A small number of SDG indicators reliably predict life satisfaction. *Sustainable Development (Bradford)*, *30*(1), 139–148. DOI: 10.1002/sd.2234

Lah, O. (2025). Breaking the silos: Integrated approaches to foster sustainable development and climate action. *Sustainable Earth Reviews*, *8*(1), 1. DOI: 10.1186/s42055-024-00102-w

Lin, J., Arnold, D., & Nguyen, M. T. (2023). Welfare in crisis: Labor and social protection in the global south. *Journal of Labor and Society*, *27*(3), 297–317. DOI: 10.1163/24714607-bja10128

Liu, Y., Tong, D., Cheng, J., Davis, S. J., Yu, S., Yarlagadda, B., Clarke, L. E., Brauer, M., Cohen, A. J., Kan, H., Xue, T., & Zhang, Q. (2022). Role of climate goals and clean-air policies on reducing future air pollution deaths in China: A modelling study. *The Lancet. Planetary Health*, *6*(2), e92–e99. DOI: 10.1016/S2542-5196(21)00326-0 PMID: 35150635

Melhim, S. H., & Isaifan, R. J. (2025). The Energy-Economy Nexus of Advanced Air Pollution Control Technologies: Pathways to Sustainable Development. *Energies*, *18*(9), 2378. DOI: 10.3390/en18092378

Mencaroni, M., Dal Ferro, N., Furlanetto, J., Longo, M., Lazzaro, B., Sartori, L., Grant, B., Smith, W., & Morari, F. (2021). Identifying N fertilizer management strategies to reduce ammonia volatilization: Towards a site-specific approach. *Journal of Environmental Management*, *277*, 111445. DOI: 10.1016/j.jenvman.2020.111445 PMID: 33070018

Miörner, J., Binz, C., & Fuenfschilling, L. (2021). Understanding transformation patterns in different socio-technical systems–A scheme of analysis. *GEIST–Geography of Innovation and Sustainability Transitions, Working Paper Series.*

Mohamad Taghvaee, V., Assari Arani, A., Nodehi, M., Khodaparast Shirazi, J., Agheli, L., Neshat Ghojogh, H. M., Salehnia, N., Mirzaee, A., Taheri, S., Mohammadi Saber, R., Faramarzi, H., Alvandi, R., & Ahmadi Rahbarian, H. (2023). Sustainable development goals: Transportation, health and public policy. *Review of Economics and Political Science*, *8*(2), 134–161. DOI: 10.1108/REPS-12-2019-0168

Monaco, S. (2024). SDG 8. Promote sustained, inclusive and sustainable economic growth, full and productive employment, and decent work for all. In *Identity, Territories, and Sustainability: Challenges and Opportunities for Achieving the UN Sustainable Development Goals* (pp. 81-86). Emerald Publishing Limited.

Ogwu, M. C., & Izah, S. C. (2024). Air pollution and the sustainable development goals. In *Sustainable Strategies for Air Pollution Mitigation: Development, Economics, and Technologies* (pp. 201-233). Springer. DOI: 10.1007/698_2024_1118

Ojha, N., Soni, M., Kumar, M., Gunthe, S. S., Chen, Y., & Ansari, T. U. (2022). Mechanisms and pathways for coordinated control of fine particulate matter and ozone. *Current Pollution Reports*, *8*(4), 594–604. DOI: 10.1007/s40726-022-00229-4 PMID: 35991936

Parikh, N. I., Gonzalez, J. M., Anderson, C. A., Judd, S. E., Rexrode, K. M., Hlatky, M. A., Gunderson, E. P., Stuart, J. J., Vaidya, D., & Epidemiology, A. H. A. C. (2021). Adverse pregnancy outcomes and cardiovascular disease risk: unique opportunities for cardiovascular disease prevention in women: a scientific statement from the American Heart Association. *Circulation*, *143*(18), e902–e916. DOI: 10.1161/CIR.0000000000000961 PMID: 33779213

Revesz, R. L. (2022). Air pollution and environmental justice. *Ecology Law Quarterly*, *49*, 187.

Roy, I. (2024). Technology Push, Demand Pull, Networks, and Public Policy: What Drives and Facilitates Innovation? In *Science, Technology and Innovation Ecosystem: An Indian and Global Perspective* (pp. 133-147). Springer.

Sriram, S., & Albadrani, M. (2022). Impoverishing effects of out-of-pocket healthcare expenditures in India. *Journal of Family Medicine and Primary Care*, *11*(11), 7120–7128. DOI: 10.4103/jfmpc.jfmpc_590_22 PMID: 36993034

Stivanello, P. (2021). Co-benefits of Plus Energy Buildings: more than just energy efficiency to advocate for tomorrow's building technology.

Tang, A., & Kemp, L. (2021). A fate worse than warming? Stratospheric aerosol injection and global catastrophic risk. *Frontiers in Climate, 3*, 720312. DOI: 10.3389/fclim.2021.720312

Tibrewal, K., & Venkataraman, C. (2021). Climate co-benefits of air quality and clean energy policy in India. *Nature Sustainability, 4*(4), 305–313. DOI: 10.1038/s41893-020-00666-3

Weiland, S., Hickmann, T., Lederer, M., & Schwindenhammer, S. (2021). *The 2030 agenda for sustainable development: transformative change through the sustainable development goals?* (Vol. 9). Cogitatio Press.

Yumkella, K., Batchelor, S., Haselip, J. A., & Brown, E. (2021). Solving the clean cooking conundrum in Africa: Technology options in support of SDG7 and the Paris Agreement on Climate Change. In *Scaling up investment in climate technologies: Pathways to realising technology development and transfer in support of the Paris Agreement* (pp. 13-26).

Zhang, L., Xu, M., Chen, H., Li, Y., & Chen, S. (2022). Globalization, green economy and environmental challenges: State of the art review for practical implications. *Frontiers in Environmental Science, 10*, 870271. DOI: 10.3389/fenvs.2022.870271

Zhang, X., Deng, T., Zhang, Y., & Jin, R. (2025). Combined intervention: Structural-efficiency-intensity indicators can halve nitrogen-related resource-environmental effects by 2050 in Beijing. *Journal of Environmental Management, 386*, 125759. DOI: 10.1016/j.jenvman.2025.125759 PMID: 40382935

Zickafoose, A., Ilesanmi, O., Diaz-Manrique, M., Adeyemi, A. E., Walumbe, B., Strong, R., Wingenbach, G., Rodriguez, M. T., & Dooley, K. (2024). Barriers and challenges affecting quality education (Sustainable Development Goal# 4) in sub-Saharan Africa by 2030. *Sustainability (Basel), 16*(7), 2657. DOI: 10.3390/su16072657

Zusman, E., Unger, C., Borgford-Parnell, N., & Mar, K. A. (2021). One atmosphere: Integrating air pollution and climate policy and governance. *Atmosphere (Basel), 12*(12), 1570. DOI: 10.3390/atmos12121570

Chapter 4
Smog, Energy Transition, and Sustainable Development:
A Policy Framework for Climate Resilient Urbanization in Pakistan

Nazish Kanval
https://orcid.org/0009-0001-9394-8495
International Islamic University, Islamabad, Pakistan

ABSTRACT

Smog, a mixture of air pollutants, dust, and smoke coupled with fog under direct sunlight, significantly jeopardizes human health and the environment by diminishing visibility. Smog has emerged as an escalating global concern with significant repercussions for human health, ecosystems, and economies all around the world. It has also emerged has major issue for environmental sustainability, health and overall economic performance in Pakistan. The study's goal is to critically analyzed the interdependent issues of urban smog, energy transition, and sustainable development in Pakistan, with an aim to suggest an effective policy framework for climate-resilient urbanization. Hence, this paper proposes an integrated policy framework that aligns smog mitigation, energy transition, and sustainable urban development to ensure climate-resilient cities in Pakistan. This policy paper rigorously analyzes the systemic causes of smog, associating it with industrial emissions, automobile pollution, agricultural produce incineration, garbage combustion and inadequate urban design. Notwithstanding established legal frameworks, inadequate enforcement and bureaucratic inefficiencies intensify the situation. This report, informed by expert perspectives from the Pakistan Economics Frontier's World Energy Day event, advocates a comprehensive policy framework to alleviate smog via sustainable

DOI: 10.4018/979-8-3373-2469-2.ch004

urbanization, energy transition, and regional cooperation. The suggested framework underscores legal reforms, technology advancements, and cohesive governance to tackle smog at its source. Essential interventions comprise the implementation of Euro 6 gasoline requirements, the transition to electric vehicles, the modernization of public transportation, and the enhancement of climate diplomacy. Investments in green infrastructure, such as sustainable urban planning, vertical gardens, and afforestation, are crucial for the effective long-term management of this issue. This study is consistent with Sustainable Development Goals (SDG 7, 11, and 13) and promotes a policy approach that is grounded in evidence and data to improve climate resilience and reduce emissions. By Adopting renewable energy, green investment, stimulating multispectral. Reciprocity Pakistan can achieve environmental and economical sustainability.

INTRODUCTION

Urbanization in Pakistan is progressing rapidly, fueled by economic development, migration from rural to urban areas, and population expansion. However, this urban expansion has been accompanied by critical environmental challenges, notably the worsening air quality characterized by pervasive smog episodes (Anees & Yan, 2019). Smog, a toxic mixture of smoke, fog, and particulate matter, has become a recurrent phenomenon in many Pakistani cities, especially in Punjab province during winter months (Tahir et al., 2023). This persistent air pollution not only poses severe public health risks but also undermines economic productivity and exacerbates social inequalities. Pakistan is grappling with intensifying environmental challenges particularly smog, driven by urban air pollution and fossil fuel dependence. With rapid urbanization, energy demands are rising, yet the existing infrastructure is unsustainable and vulnerable to climate shocks (Arif & Mahsud, 2024). At the core of this problem lies Pakistan's heavy reliance on fossil fuels and inefficient energy systems that intensify greenhouse gas emissions and local pollutants. Addressing the intertwined challenges of smog and energy demands is vital to promoting sustainable development and ensuring climate resilience within Pakistan's rapidly urbanizing landscape . The intersection of smog, energy transition, and Pakistan's urbanization poses a serious barrier to sustainable development, particularly in cities like Lahore. As the country grapples with severe air pollution and climate change impacts, a comprehensive policy framework is essential for fostering climate-resilient urbanization. This framework must address the multifaceted nature of smog, which is exacerbated by industrial emissions, vehicular pollution, and agricultural practices, while also promoting sustainable energy transitions. The urgency of this issue is underscored by Pakistan's vulnerability to climate change, ranking 8th globally in terms of climate impact risk (Sherin, 2023). Smog affects multiple regions globally, including China, India, Pakistan, the USA, and the United Kingdom (Mukhtar, 2017; Singh, 2018). Pakistan, recognised for its rich cultural legacy and varied landscapes, has been contending with a persistent and intensifying environmental issue - pollution. Smog, a perilous amalgamation of smoke and fog, has become a prominent public health issue, impacting both urban and rural regions (Saleem et al,. 2019). Increased concentrations of air pollutants, such as sulphur dioxide, nitrogen oxides, ground-level ozone, and particulate matter, are among the characteristics of this atmospheric event (Liaqut et al., 2023). According to Raza et al. (2021). The interaction of these chemicals with sunlight leads to intricate photochemical reactions, resulting in the formation of secondary pollutants such as ground-level ozone that give smog its characteristic thick, yellowish-brown haze. The term "haze" describes the small particles and dust in the atmosphere that scatter and absorb sunlight, making it harder to see. Haze may result from natural phenomena including dust storms, wildfires, and volcanic eruptions, in addition to anthropogenic activities such as agriculture, deforestation, and construction. Haze differs from smog in that it does not inherently involve the same chemical reactions or the generation of secondary pollutants. The particles that contribute to haze are typically smaller and may consist of dust, soot, sea salt, and other fine particulate matter. Haze can affect health, particularly in individuals with pre-existing respiratory conditions; however, its effects are typically less severe than those associated with smog. Smog in Pakistan has negative effects that surpass immediate health concerns, impacting agriculture, biodiversity, and overall environmental sustainability.

Pakistan's energy sector is crucial for urbanization and is significantly impacted by pollution due to its reliance on fossil fuels. The current energy mix is dominated by coal, oil, and natural gas, with renewable energy sources still emerging. This dependency not only exacerbates environmental issues but also poses economic challenges, as the country spends over $3.7 billion annually on fossil fuel imports (Asghar et al., 2022). the energy transition faces significant barriers including infrastructure limitations, financial constraints, governance challenges, and socio-political complexities that need to be addressed through coherent policies and strategic planning (Asif et al., 2024; Ambreen et al., 2025).

Sustainable development in Pakistan's urban centers requires integrating environmental protection with economic growth and social inclusion (Farooq et al., 2024). Climate resilient urbanization is a framework that embraces this integration by promoting urban forms and infrastructure systems that can withstand environmental shocks and stresses, including climate change impacts such as extreme heat, floods, and poor air quality (Jaiswal et al., 2024). It is commonly known that Pakistan is vulnerable to climate change, with rising temperatures, altered precipitation patterns, and a rise in the frequency of extreme weather events endangering infrastructure and urban life. (Khan., et al 2025). Without a deliberate focus on resilience, rapid urban growth could exacerbate environmental degradation and social vulnerabilities, leading to unsustainable outcomes that undermine national development goals (Meyer et al., 2019).

Urbanization in Pakistan presents both challenges and opportunities. While it places pressure on natural resources and infrastructure, it also creates a dynamic environment for innovation, economic diversification, and social transformation (Jabeen et al., 2017). Sustainable urban development aligned with climate resilience can transform cities into engines of green growth, improved health, and social equity (Ali S. et al., 2025). To achieve this, it is essential to adopt a holistic perspective that integrates environmental management, energy policies, urban governance, and socio-economic development within a unified vision.

Hence, the persistent problem of smog in Pakistani cities, compounded by an energy system reliant on fossil fuels, poses significant threats to public health, environmental quality, and sustainable urban growth. Addressing these issues requires a strategic policy framework that simultaneously promotes energy transition, air quality improvement, and climate resilient urbanization. Such a framework must be underpinned by scientific evidence, cross-sector collaboration, and inclusive governance to effectively navigate the complex interrelations among environmental, social, and economic dimensions. This paper aims to contribute to this critical discourse by outlining policy pathways that can support Pakistan's journey towards sustainable and climate-resilient cities. Hence, this paper aims to propose an integrated policy framework that aligns smog mitigation, energy transition, and sustainable urban development to ensure climate-resilient cities in Pakistan. This policy paper rigorously analyzes the systemic causes of smog, associating it with industrial emissions, automobile pollution, and inadequate urban design. Notwithstanding established legal frameworks, inadequate enforcement and bureaucratic inefficiencies intensify the situation. This report, informed by expert perspectives from the Pakistan Economic Forum's World Energy Day event, advocates a comprehensive policy framework to alleviate smog via sustainable urbanization, energy transition, and regional cooperation. The following sections will explore the smog in Pakistan: Sources, Impacts, and Policy Gaps, Urbanization and Its Environmental Footprint, The Role of Energy Transition in Climate Resilience, International Best Practices and Lessons for Pakistan. After words it describes key components of a policy framework aimed at mitigating smog, facilitating energy transitions, and promoting sustainable urban development based on expert's view at Pakistan Economic Forum.

SMOG IN PAKISTAN: SOURCES, IMPACTS, AND POLICY GAPS

Smog in Pakistan has become a persistent environmental challenge, with major urban areas like Lahore, Karachi, and Islamabad often witnessing severe air pollution levels. This phenomenon, characterized by a mixture of fog, smoke, and particulate matter, is a critical issue affecting public health, the economy, and the environment. This section of the research paper explores the sources of smog in Pakistan, its impacts on human health and the environment, and identifies the policy gaps that have hindered effective mitigation strategies. The impact of smog arises chiefly from the makeup of its contaminants, which comprise a complicated amalgamation of gases and particles. The primary components of smog consist of volatile organic compounds (VOCs), nitrogen oxides (NOx), sulphur dioxide (SO2), carbon monoxide (CO), ground-level ozone (O3), and particulate matter (PM) (Yousaf et al., 2021). In Pakistan, particularly in the Punjab province, exposure to elevated levels of air pollutants, including fine particulate matter (PM2.5 and PM10), nitrogen oxides (NO_X), sulfur dioxide (SO_2), carbon monoxide (CO), and ozone (O_3), has resulted in a rise in cardiovascular and respiratory illnesses. Pakistan is fourth among the most populous countries in Asia. Two-thirds of Pakistan's population resides in the province of Punjab. The three most populous cities in Punjab are Lahore, Faisalabad, and Gujranwala. In addition to overpopulation, another terrible similarity among these three megacities is their deplorable air quality. Recent data indicates that the Air Quality Index (AQI) in specific regions of Lahore routinely surpassed 400 µg/m3 in November 2021, although a safe AQI value is deemed to be below 50 µg/m3 (Dawn, 2021).

Figure 1. Commuters driving through heavy fog and smog in Lahore. —AFP/File. Source: Dawn (2021).

The main causes of air pollution in these three cities (Lahore, Faisalabad, and Gujranwala) include deforestation, brick kilns, excessive industrialization and urbanization, crop burning, and an increase in the number of vehicles on the road with subpar gasoline. The major cities of Punjab are covered in haze due to the cooler temperatures in the early winter months . Faisalabad and Gujranwala serve as the industrial centers of Punjab, while Lahore contends with the problem of rampant urbanization. Smog is accountable for numerous severe health complications, including respiratory disorders such as asthma aggravation, coughing, throat burning, chest irritation, and chronic bronchitis. In addition to respiratory issues, pollution also induces ocular, cardiovascular, and dermatological effects (Ali et al., 2019). Particulate Matter 2.5 (PM 2.5) is a constituent of smog recognized as a significant health risk. It can traverse the airways and accumulate in the lungs, resulting in various pulmonary diseases. This component diminishes visibility and induces the "haziness" observed in smog (Sanchez-Triana et al., 2014)

The World Health Organization (WHO) estimates that ambient air pollution causes 128,000 premature deaths in Pakistan each year, with Punjab suffering a disproportionate share because of its dense population and industrial and agricultural activity (*Khan, 2024*). The Energy Policy Institute at the University of Chicago (EPIC) created the Air Quality Life Index (AQLI), which shows a direct link between life expectancy and PM2.5 exposure. According to the index, breathing dirty air shortens a Pakistani's life expectancy by an average of 3.32 years. For those who reside in Lahore, this concerning number jumps to 7.5 years (*Khan, 2024*).

Figure 2. Dense Smog Over City Streets. Source: (Junejo, 2025)

In Pakistan, air pollution is primarily attributed to anthropogenic activity. Air quality specifically pertains to the levels of airborne contaminants, including carbon monoxide, nitrogen dioxide, ozone, particulate matter, and sulfur dioxide, all of which pose significant risks to human health (Bajwa & Sheikh, 2023). Human activities that pollute the air and threaten human health and the environment encompass the combustion of fossil fuels, incineration of municipal garbage, burning of solid fuels, electricity generation, vehicular emissions, industrial discharges, heating practices, and tobacco consumption (Ajmal et al., 2016). Contaminated air in the form of smog and other air pollutants is being produced by these activities, which is having devastating effects on the environment and human health. Smog, a byproduct of heavy fog, can be harmful to human health because it contains harmful industrial pollutants (Bilal et al., 2022). Fossil fuel combustion produces smog, which is mostly composed of carbon and manifests as soot, tar, oil, and ash. Smog comes in a variety of forms, each with its own composition, sources, impacts, and geographic location. Complex photochemical reactions involving NO2, NO, SO2, and volatile organic compounds—which serve as precursors for ground-level ozone and fine particles—cause photochemical haze in Pakistan (Akbar, 2020).

Asthma, bronchitis, and chronic obstructive pulmonary disease (COPD) are among the respiratory conditions that have significantly increased during smog outbreaks.

According to an IGC study on clean brick kilns, exposure to PM2.5 raises the risk of developing cancer as well as respiratory and cardiovascular conditions such bronchitis, asthma, myocardial infraction, and ischemia (Nasim and Sharif, 2020). Similarly, another study of Malik et. (2024) revealed that heart disease, lung tissue damage, tracheal and respiratory infections, asthma, and many allergies are only a few of the serious health effects of smog. Smog exposure has been found to have a serious impact on children's health, particularly affecting their cognitive functions. Naveed & Khayyam (2023) investigated the cognitive impact of exposure to air pollution, specifically smog, on schoolchildren in Lahore and Islamabad, Pakistan. They found that children living in polluted areas, like Lahore, show poorer cognitive performance, including memory and attention issues, compared to those in Islamabad. The study highlights the need for better zoning and increased public transport to reduce pollution-related cognitive decline. Last winter, Punjab saw an outbreak of pneumonia, with 36,054 cases reported and 456 fatalities (Pakistan Today, 2024). Smog's negative effects on health result in significant financial expenses. Rising healthcare costs put a burden on household finances as well as state resources. According to World Bank estimates, Pakistan loses about 6.5% of its GDP each year as a result of air pollution, which includes missed wages, decreased productivity, and medical costs (World Bank, 2022).

Every year, air pollution kills about 7 million people worldwide, of whom 4.2 million are killed by ambient air pollution and the remaining number by indoor smoke from dirty fuels and cooking stoves (Majeed, & Munir, 2020). Hence, around 90% of individuals in middle- or low-income nations have air pollution levels that are higher than WHO guidelines, according to the World Health Organization (WHO) (Zeeshan et al., 2024; Nasar-u-Minallah et al., 2024). Vehicle emissions, urbanization, emissions from brick kilns, garbage, and crop burning are some of the main causes of air pollution, which has a negative impact on human health in addition to the environment by contributing to anthropogenic climate change. According to the findings of a World Bank study, one of Pakistan's biggest threats to public health is air pollution, which is mostly caused by vehicle emissions. The main components of vehicle emissions are CO, NOx, and Sox. Because of its detrimental impacts on the environment and human health, CO is categorized as a dangerous gas under the Clean Air Act (CAA). Health problems such asthma, eye irritation, insomnia, coughing, and breathing problems might result from exposure to these pollutants (Subhanullah et al., 2025). While nitrogen oxide from cars can irritate the eyes and throat, sulfur dioxide can generate smog and acid rain. Up to 60% of all PAH emissions come from vehicular transportation (Kamal et al., 2015).

Smog in Pakistan, particularly in Lahore, has become a critical environmental and public health issue, driven by various anthropogenic factors. The primary sources of smog include industrial emissions, vehicular pollution, and agricultural practices such as crop residue burning. The impacts of smog are profound, affecting health, agriculture, and overall quality of life. This overview will delve into the sources, impacts, and existing policy gaps regarding smog in Pakistan. Sources of smog can be Industrial Emissions: Factories release significant amounts of nitrogen oxides (NOx) and particulate matter (PM), contributing to air pollution (Razzaq et al., 2024; Colbeck et al., 2010).

Increased motorization has led to higher emissions from vehicles, exacerbating air quality issues (Colbeck et al., 2010).The burning of crop residues is a major seasonal contributor to smog, particularly during the fall (Razzaq et al., 2024). A significant element is the incineration of crop residues, especially in the rice and wheat regions of Punjab, which substantially contributes to smog in Pakistan (Yousaf et al., 2021). Post-harvest, farmers frequently incinerate crop leftovers as an expedient and economical technique for preparing fields for subsequent planting. This combustion emits a significant quantity of particulate matter, carbon monoxide, and other pollutants into the atmosphere.

The timing of these incinerations frequently aligns with meteorological circumstances that facilitate the development and retention of smog, hence intensifying the severity of the problem (Bilal et al., 2022; Raza et al., 2022; Mukhtar, 2018).

Furthermore, the application of specific fertilisers and pesticides in agriculture can lead to the release of reactive nitrogen molecules, such as ammonia (NH3) and nitrogen oxides (NOx). Upon discharge into the atmosphere, these chemicals can interact with other pollutants to generate fine particulate matter (PM), a principal element of smog. The extensive application of fertilisers and pesticides in agriculture, although increasing crop yields, contributes to the overall accumulation of air pollutants (Abas et al., 2019; Arif and Hassan, 2023).

Smog has a substantial agricultural impact, primarily resulting in crop damage and reduced yields (Shamsi et al., 2000). Crop damage arises when pollution directly harms plant tissues, resulting in signs including necrosis, chlorosis, and observable injuries on the leaves.

Extended exposure to smog can diminish photosynthetic rates due to impaired chlorophyll, thereby impacting a plant's energy production capacity. This ultimately results in inhibited growth and diminished agricultural yields (Ahmad et al., 2013). Smog is linked to respiratory diseases, cardiovascular issues, and increased mortality rates, with estimates suggesting a reduction in life expectancy by seven years in Lahore (Tajammul, 2023; Afzal & Bajwa, 2024). Smog negatively affects crop yields and soil health, impacting food security and farmer livelihoods (Razzaq et al., 2024).

Despite the growing problem of smog, Pakistan's response to the issue has been insufficient, largely due to gaps in policy, weak enforcement of environmental regulations, and a lack of coordination among government agencies. There is also a lack of comprehensive studies on smog, hindering effective policy formulation

(Afzal, 2023). The existing 'Policy on Controlling Smog, 2017' lacks alignment with scientific findings and lessons from other cities, limiting its effectiveness (Tajammul, 2023). While the government has initiated some measures, the urgency of the smog crisis necessitates a more robust and informed approach to policy-making and public health interventions. Addressing these gaps is crucial for mitigating the adverse effects of smog in Pakistan. Smog in Pakistan is a growing environmental crisis with serious implications for human health, the economy, and the environment. The primary sources of smog include vehicular emissions, industrial pollution, crop burning, and the use of low-quality fuels. The impacts of smog are far-reaching, contributing to respiratory diseases, environmental degradation, and economic losses. However, Pakistan's policies in addressing smog remain inadequate, with significant gaps in enforcement, public awareness, and the promotion of sustainable practices. To mitigate smog and its adverse effects, Pakistan must strengthen its regulatory framework, increase investment in clean technologies, and implement comprehensive public awareness campaigns.

URBANIZATION, ITS ENVIRONMENTAL FOOTPRINT AND SUSTAINABLE DEVELOPMENT

Urbanization denotes the shift of human production and lifestyles from rural to urban settings, characterized by population concentration, the rise of non-agricultural activities, and significant alterations in landscapes (Antrop, 2004; Friedmann, 2006; Zhang, 2008). The processes involved are acknowledged as a multi-dimensional evolution, with the allocation of spatial elements resulting in modifications to economic growth models, lifestyle, values, and consequently reshaping regional development and land use, thereby impacting the social and natural environment (Tonne et al., 2021). While urbanization poses significant environmental challenges, it also presents opportunities for sustainable development. The Sustainable Development Goals (SDGs) were established by the United Nations General Assembly in 2015. They provide a worldwide framework for action over the next 15 years, leading up to 2030, with the aims of ending extreme poverty, safeguarding the environment, and attaining shared prosperity for all (United Nations. (2015). Guiding the global society towards future sustainable development, it is crucial to promote the sustainability of the social, economic, and ecological dimensions in order to attain these goals (Fu et al., 2022). City dwellers make up 55% of the global population now; projections see that number rising to 60% in 2030 and 65% in 2050.

The developing nations of Asia and Africa are expected to host about 90% of the world's future urbanization expansion, which is happening at an unprecedented size and pace (Chen et al., 2014; United Nations, 2019).

Unsustainable conditions created by the interaction between humans and the environment under the current urbanization conditions must be understood from various angles and at several scales in order to maintain sustainable global prosperity (Fu et al., 2022). Cities exert significant direct and indirect worldwide effects on energy use (Madlener and Sunak, 2011), land utilization, climate change (Seto and Shepherd, 2009), and resource usage (Gondhalekar and Ramsauer, 2017). One of the most important factors in finding game-changing answers to the global sustainable development challenges is the role that cities play in fostering innovation and sustainable planning, as well as in managing local infrastructure, economies, and resources (Appio et al., 2019; Elmqvist et al., 2019; Scoones et al., 2020). Urbanization poses significant challenges in both developed and developing nations. As a developing country, Pakistan faces significant challenges in managing the adverse effects of urbanization, particularly in relation to environmental degradation such as smog. Rapid urban growth, combined with industrial expansion, inadequate infrastructure, and a lack of effective environmental regulations, has exacerbated the environmental footprint of cities. This growing concern undermines the nation's efforts to achieve sustainable development goals (SDGs), particularly in areas related to health, air quality, and climate resilience. Addressing these challenges requires a comprehensive approach that integrates sustainable urban planning, the adoption of clean technologies, and the enforcement of stricter environmental policies to mitigate the adverse impacts of urbanization and promote long-term sustainability. Urbanization presents adverse effects on social equity, public health, animal habitats, and the environment, particularly in rapidly urbanizing developing nations (Reckien et al., 2017; Gu, 2019). Achieving the Sustainable Development Goals entails numerous issues and challenges. Interlinked effects across the Sustainable Development Goals can be characterized as synergies or trade-offs (Kroll et al., 2019).

Pakistan is the most urbanized nation in South Asia (Ali et al., 2010) and its second-largest metropolis, Lahore, is expanding at a rate of 4% per year. Lahore, which has 11.1 million residents (Nadeem et al., 2021) and is regularly ranked as one of the most polluted cities in the world (Naveed et al., 2023), has a serious problem with air pollution. Due to factors like deforestation, fast urbanization, industrial pollution, heavy traffic, construction, burning crop residue, and brick kilns, smog has gotten worse in recent years (Raza et al., 2021). Urbanization is strongly correlated with increased emissions, with a correlation coefficient of 0.78, indicating a significant environmental burden (Ali et al., 2024). Karachi's green spaces have been reduced due to urban expansion, with only a 5% increase in green cover over eight years, far below the WHO's recommendations (Zia et al., 2022). Urbanization has a nonlinear relationship with CO2 emissions, with positive changes in urbanization leading to increased emissions (Asghar et al., 2022). Urbanization has heightened social disparities, with a correlation of 0.80 between urbanization and social inequality (Ali et al.,

2024). The informal economy, prevalent in urban areas, exacerbates environmental degradation, impacting the ecological footprint negatively (Qayyum et al., 2021). There is a growing awareness and implementation of eco-friendly policies, though challenges like poor infrastructure and limited resources persist (Farooq et al., 2024). Improving institutional quality can significantly reduce carbon emissions, suggesting a pathway for sustainable urban management (Asghar et al., 2022). Urbanization in Pakistan has significantly impacted the environment, contributing to various ecological challenges. Rapid urbanization has led to increased greenhouse gas emissions, urban heat islands, and resource consumption, exacerbating social inequalities and housing insecurity in major cities like Karachi, Lahore, Multan, and Quetta (Ali et al., 2024). The environmental footprint of urbanization is further compounded by the depletion of green spaces, as seen in Karachi, where green land cover has only marginally increased over eight years, highlighting the need for improved urban green spaces (Zia et al., 2022). Transitioning to green urbanization is crucial for mitigating these impacts and promoting sustainable development.

While urbanization poses significant environmental challenges, it also presents opportunities for sustainable development. By adopting green urbanization practices and improving institutional frameworks, Pakistan can mitigate the adverse effects of urbanization and promote resilient urban environments for future generations and sustainable development goals. Therefore, urbanization has the ability to greatly improve social welfare, equality, public health, economic growth, and energy consumption control. Urbanization can enhance access to basic services, lessen health inequalities, and improve social welfare outcomes by encouraging the growth of infrastructure and healthcare systems. Furthermore, by creating more chances for investment, innovation, and employment, carefully thought-out urban growth can stimulate economic growth. Urbanization also offers a chance to optimize resource use, encourage sustainable energy practices, and deploy energy-efficient technologies, all of which help to reduce environmental effects and encourage more responsible energy use. However, these benefits are contingent upon strategic urban planning and policy frameworks that prioritize sustainability and equity in addressing the needs of urban populations. According to Gong et al. (2012 and Chen et al. (2018), urban areas provide enhanced access to employment opportunities, modern facilities, and social services, which are widely acknowledged as crucial for fostering social progress and improving living conditions. Similarly, Urbanization enhances the health and life expectancy of residents through improved health awareness, increased healthcare resources, better health insurance systems, and investments in public health infrastructure (Lu et al., 2017). Effective governance has the potential to mitigate health risks encountered by the public (Coovadia et al., 2009). Urbanization facilitates the flow of resources, information, and knowledge creation (Bettencourt et al., 2007; Chua et al., 2019). The densification of land use associated with urbanization can enhance resource-use efficiency (Wang, 2020). Research has examined the connections between urbanization and the attainment of these Sustainable Development Goals (SDGs). SDG11, established by the United Nations General Assembly in 2015, is recognized as a critical objective for sustainable urbanization (Akuraju et al., 2020; Ghani et al., 2021). However, it does not adequately address the synergies and trade-offs between urbanization and the various goals within the SDGs. Multiple studies (Anwar et al., 2022; He et al., 2022; Xu et al., 2022; Zhao et al., 2022) have established that the definition and impact of urbanization extend beyond SDG11, demonstrating connections with other Sustainable Development Goals (SDGs).

By identifying the interlinkages between urbanization, environmental footprints, and sustainable development, we have uncovered both challenges and opportunities associated with urbanization. Despite challenges, urbanization can play a pivotal role in achieving the 2030 Sustainable Development Goals (SDGs), as it also exhibits significant synergies with most of these goals. These synergies can be observed across four key areas: public health and social welfare, economic growth and energy consumption, natural resource use and environmental/ecological impacts, and international cooperation. To ensure sustainable urbanization, strategies such as promoting spatial justice, guiding green and low-carbon urban development, building resilient cities, and fostering multilateral cooperation are essential. These approaches are critical in maximizing the positive contributions of urbanization to the SDGs while mitigating negative outcomes, such as resource depletion, ecological degradation, and inequality. For Pakistan, embracing these strategies—focused on spatial justice, sustainable urban development, resilience, and international collaboration—offers a pathway to achieving sustainable development. There is an urgent need for global action, requiring collaboration among governments, experts, and other stakeholders, with a strong emphasis on a science-policy-practice nexus to steer urbanization toward sustainability and help realize the SDGs by 2030.

THE ROLE OF ENERGY TRANSITION IN CLIMATE RESILIENCE AND SUSTAINABLE DEVELOPMENT

The energy transition plays a crucial role in enhancing climate resilience by reducing reliance on fossil fuels and promoting renewable energy sources. This shift not only mitigates climate risks but also strengthens energy security and equity, which are essential for sustainable development. The following sections outline the key aspects of how energy transition contributes to climate resilience and sustainable development.

The notion of energy transition was first introduced by the German Academy of Sciences in 1980, promoting the discontinuation of nuclear power and oil as a reaction to the energy crisis. During that period, inadequate focus was placed on the energy transition. Between 1980 and 1992, the global economy underwent substantial growth, as evidenced by the gross domestic product increasing from 11.34 trillion dollars to 25.41 trillion dollars, according to World Bank data. The extensive use of fossil fuels led to significant carbon dioxide (CO2) emissions to facilitate this rapid economic growth. This subsequently resulted in climate change, leading to the signing of the 1992 United Nations Framework Convention on Climate Change (UNFCCC). Post-signing of the UNFCCC, the shift towards renewable energy emerged as a pivotal element in global energy conservation and emissions reduction efforts. Between 1965 and 2022, the share of renewable energy consumption rose from 6.31% to 12.33%, alongside a total consumption increase of 7.41 times. This indicates advancement in the global energy transition; however, it is important to highlight that during the same timeframe, CO2 emissions and fossil energy consumption rose by 3.07 and 3.39 times, respectively (Yang et al., 2024).

The negative effects of fossil fuels on the environment are clear and well-documented (Olson & Lenzmann., 2016). The use of coal, oil, and natural gas is the main contributor to greenhouse gas emissions, which drive climate change and its various detrimental effects. Evidence of this damage is observable globally, ranging from the melting of polar ice caps to the escalating severity of extreme weather events. The extraction and utilization of fossil fuels cause considerable damage to biodiversity, water quality, and air purity (Nyambuu & Semmler., 2023). Alternatively, renewable energy sources such as solar, wind, hydro, and geothermal power stand out as representations of optimism and advancement. All of the environmental problems caused by fossil fuels can be solved by switching to these sustainable alternatives (Mahdavi et al., 2023; Qazi et al., 2019). Switching to renewable energy sources is crucial for several reasons beyond the financial and ecological ones. Those already at the bottom of the climate change food chain will feel the pinch the most as the planet warms at an alarming rate (Halley et al., 2018). The worldwide shift towards renewable energy sources has numerous obstacles on many fronts, including technical, economic, and societal. These obstacles are not consistent and differ greatly among locations, resulting in a complicated web of issues that require collective and intentional attention (Lv, 2023). None of this would be possible without the full support of nations around the world. To conquer the complex obstacles of the energy transition, international cooperation is essential (Afonso et al., 2021)

Hence, Energy transition contributes to climate resilience in number of ways: It contributed to reduction of climate risks Transitioning to renewable energy significantly lowers greenhouse gas emissions, thereby reducing climate risks associated with extreme weather events (Shang et al., 2024). Studies indicate that countries increasing their renewable energy share experience a decrease in climate-related risks, particularly in developed regions (Shang et al., 2024). Energy transition add economic and energy security. Energy transition enhances energy security by reducing dependence on imported fossil fuels, which are subject to geopolitical vulnerabilities and price volatility (Janardhanan, 2017). A diversified energy mix not only curbs emissions but also stabilizes economies by mitigating the impacts of fluctuating energy prices (Janardhanan, 2017). Hence, Effective environmental policies are essential for facilitating energy transitions, as they promote investments in renewable technologies and sustainable practices (Ma et al., 2023).The integration of energy transition into global economic value chains ensures equitable access to energy, which is vital for resilience against climate impacts (Sourgens et al., 2023). While the energy transition is pivotal for climate resilience, it is important to recognize that it must be accompanied by robust policy frameworks and technological advancements to maximize its effectiveness. Without these, the transition may not fully address the complexities of climate change and its associated risks. The role of energy transition in climate resilience and sustainable development is crucial in addressing the challenges posed by climate change and ensuring a sustainable future. Shifting from fossil fuels to renewable and clean energy sources not only reduces greenhouse gas emissions but also enhances the resilience of communities to the impacts of climate change, such as extreme weather events and rising sea levels. By fostering energy security, enhancing air quality, and generating green jobs—all of which help economic growth and the fight against poverty—energy transition promotes sustainable development. Additionally, it lessens environmental deterioration, preserves ecosystems, and conserves natural resources. Countries may promote a more sustainable and equitable energy future and move closer to achieving the Sustainable Development Goals (SDGs) by investing in low-carbon, energy-efficient, and renewable energy technology. In the end, energy transition is a major factor in climate resilience, lowering vulnerability and opening the door to a low-carbon, equitable, and sustainable global economy. Hence based on the above persisting scenario the study provides a proposed policy framework proposed policy framework based on expert perspectives from the Pakistan Economics Frontier.

PROPOSED POLICY FRAMEWORK BASED ON EXPERT PERSPECTIVES FROM THE PAKISTAN ECONOMIC FORUM

Smog has become a significant environmental and health issue in Pakistan, particularly in urban and industrial areas, where its impact is most pronounced.

It reflects the complex interactions between urbanization, energy consumption patterns, inadequate transportation systems, and poor environmental management. Recognizing the urgency of addressing this issue, Pakistan Economics Frontier hosted a high-profile webinar that brought together distinguished experts from academia, government, and civil society to evaluate the implications of smog and propose multidimensional solutions. During the session, experts discussed the interconnectedness between smog, energy transition, and sustainable development, offering a comprehensive approach to tackle the problem. They emphasized the serious environmental and socio-economic consequences of smog, including its harmful effects on public health, economic productivity, and environmental quality.

Building on the insights from the webinar, a *Policy Framework for Climate Resilient Urbanization in Pakistan* was proposed. This framework emphasizes the need for a coordinated strategy that integrates smog reduction with energy transition, sustainable urban planning, and climate resilience. Experts highlighted that transitioning to cleaner energy sources, enhancing public transportation systems, and implementing stricter emission regulations are critical steps toward reducing smog levels. The framework also calls for improving waste management, promoting green spaces, and investing in climate-resilient infrastructure that can withstand extreme weather events. Furthermore, the framework advocates for the incorporation of environmental considerations into urban planning to ensure that future urbanization is sustainable, energy-efficient, and resilient to climate impacts. Raising public awareness and fostering civic engagement were also identified as essential components of this policy framework. By addressing the root causes of smog and adopting these strategic measures, Pakistan can not only mitigate the adverse effects of smog but also foster a sustainable, climate-resilient urban environment. The webinar reinforced the critical need for coordinated action across all sectors of society to effectively tackle smog and achieve sustainable urbanization. The experts highlighted environmental, socio-economic consequences of smog and policy recommendations as follows:

Health and socio-economic consequences of smog

Public Health Burden

Dr. Verda Salman (NUST) powerfully underscored the severe health consequences of smog, including the degradation of respiratory and immune functions, heightened mental health challenges, and increased risks to pregnant women. Vulnerable groups such as children, the elderly, and those with pre-existing conditions bear the heaviest toll. This public health crisis is exacerbating the strain on Pakistan's already overburdened healthcare system, making the need for comprehensive solutions all the more pressing.

Economic Costs

The economic burden of smog is equally stark. Health-related emergencies driven by air pollution lead to mounting healthcare costs and significant productivity losses across sectors. Smog also negatively impacts agriculture by reducing photosynthesis and impairing crop yields, contributing to economic instability. These escalating costs highlight the importance of adopting preventive policies over reactive ones. Tackling smog head-on is crucial not only for improving public health but also for safeguarding Pakistan's economic future.

Policy Measures and Sectoral Reactions

Environmental Sustainability and Urban Solutions

Ms. Nazifa Butt (WWF) championed the importance of people-centered environmental action, such as afforestation and vertical gardening, to combat smog. She stressed that sustainability must be woven into the fabric of urban planning policies, with active community involvement playing a vital role in driving lasting change.

Transport and Urban Planning Reform

Ms. Saba Anwar (PIDE) highlighted Pakistan's heavy reliance on road transport and the urgent need for substantial investment in urban transport infrastructure. She called for innovative urban planning strategies that prioritize sustainable mobility, reduce motor vehicle emissions, and align with the objectives of SDG 11 on sustainable cities.

Climate Diplomacy and Energy Transition

Mr. Khalid Waleed (SDPI) emphasized the need for proactive climate diplomacy and regional cooperation, underscoring the importance of integrating renewable energy sources into Pakistan's energy mix. He advocated for a unified national strategy that harmonizes energy and environmental policies, setting the stage for a sustainable energy transition.

Institutional Reforms and Standards

Mr. Jawad Rabbani (MoPDSI) emphasized enhancing fuel standards like Euro 6 and urban zoning for decreasing pollution from stationary and mobile sources. He advocated for sound governance frameworks for effective enforcement of environmental laws.

Global and Integrated Climate Policy View

The discussion was put into context by Dr. Ahmad who urged sectoral integration of climate policies and overall management of emissions. His argument for overall planning aligns with the wider agenda for advancement under SDG 13 and sustainable economic development.

Policy Recommendations

Guided by expert opinions, Pakistan Economics Frontier has developed the following composite policy framework:

Embrace and Scale Up Green Technologies: Promote technology development for clean energy, carbon emissions measurement, and low-carbon substitutes
Adopt Sustainable Urban Infrastructure: Support green building codes, urban farming through vertical gardening, and sustainable transport networks.
Strengthen institutional capacity: Give regulatory agencies legal authority, technical means, and financial support to implement air quality norms.
Promote Community Outreach: Engage civil society and local authorities in mobilizing environmental awareness and afforestation campaigns.
Promote Climate Finance and Technology Transfer: Support partnerships with global and local stakeholders in order to tap into green investments and best practices. Integrate climate risk assessment as well as reduction targets for emissions into national and subnational plans for development

Alignment with Sustainable Development Goals (SDGs)

The policy direction proposed is well in tune with major SDGs:

SDG 13 (Climate Action): Focuses on reducing emissions, climate risk management, and resilience
SDG 7 (Affordable and Clean Energy): Promotes adoption of renewable energy and energy efficiency.

SDG 11 (Sustainable Cities and Communities): Promotes sustainable transport, clean air, and urban planning for all.

CONCLUSION

The increasing prevalence of smog in Pakistan is an indication of structural environment, infrastructure, and governance voids. Yet, as has been brought out in focus under World Energy Day's webinar, an integrated knowledge-based and participatory approach—founded in sustainability and informed by the SDGs—can facilitate Pakistan's journey toward a cleaner, healthier, and fairer future. The Pakistan Economics Frontier reaffirms its dedication toward enabling cross-sectoral dialogue, institution-building, and cooperation at the regional level toward overcoming the environment and energy challenges faced by Pakistan.

The interconnected issues of smog, energy transition, and sustainable urbanization present an essential triad challenge that has to be tackled under a single and visionary policy umbrella in Pakistan. This research reaffirms that smog is not just an ecological hazard but a multifaceted result of uncontrolled industrial discharges, motor vehicle pollution, and defective urban planning altogether amplified through poor institutional enforcement and regulatory loopholes. All these factors together endanger human health, economic output, and climate sustainability in Pakistan.

The study critically analyzed the interdependent issues of urban smog, energy transition, and sustainable urban development in Pakistan, with an aim to suggest an effective policy framework for climate-resilient urbanization and Sustainable Development Goals (SDGs), specifically SDG 7 (Affordable and Clean Energy), SDG 11 (Sustainable Cities and Communities), and SDG 13 (Climate Action). This is because smog, defined as an accumulation of pollution, dust, and fog heightened under light from the sun, has turned into an acute public health and environmental emergency in Pakistan, particularly in urban and industrial areas. This paper drew from expert analysis put forward during the World Energy Day webinar organized by Pakistan Economics Frontier and tied smog not just to ecological degradation but further connected it with economic and institutional inefficiencies, which highlight how motorized transport pollution, industry pollution, and lack of urban planning lead to deteriorating urban air quality. With laws already existing for protecting the environment, poor enforcement, bureaucratic hold-ups, and lack of coordination in governance continue to hinder efforts for mitigating smog.

The paper advances an integrated policy approach that aims at addressing causes of smog while fostering urban resilience in a sustainable way. Included in the major policy recommendations are: enforcement of Euro 6 fuel standards to reduce motor vehicle emissions, a national shift toward electric vehicles (EVs), investment in modern and equitable mass transit infrastructure, and urban spatial redesign incorporating vertical gardens, green belts, and afforestation. Climate diplomacy and inter-regional cooperation play an equally important role in solving transboundary pollution and tapping into global good practices. Also underlined are institutional reforms that strengthen environmental capacity for governance, community-based strategies that mobilize local stakeholders, and mainstreaming sustainability into urban planning, all considered essential for a long-term success. All recommendations are targeted at enabling inclusive and adaptive policies that capture the complexity of interdependence between environmental, economic, and social dimensions.

The research recognizes some weaknesses, though. To begin, although policy formulation relies upon expert experience and current literature rather than upon either primary data collected or empirical analysis, these could provide validated, localized evidence for making a stronger case. Second, without exception, comprehensive interventions will encounter practical obstacles in implementation because of political instability, lack of resources, and institutionalized resistence toward change. Last, although urban areas are dealt with, rural-urban linkages, and how these facilitate pollution and use of energy, are not examined fully, creating an explanatory gap in current analysis.

Future studies should include empirically measuring the success of green technology and transport interventions in Pakistani cities, analyzing public opinion of climate policies, and determining institutional preparedness for large-scale policy implementation. Cross-country comparisons between other emerging economies making similar transitions could provide insights into policy transfer as well as contextual adaptation. Interdisciplinary research that combines environmental science, economics, and urban planning is needed to develop comprehensive and actionable frameworks. Evidence-based policymaking and an enlarged research agenda will facilitate stakeholders' ability to anticipate climate threats and develop resilient, sustainable cities.

Concisely, this paper highlights that reducing smog and promoting sustainable urban development will entail much more than technology-based solutions alone; instead, visionary leadership, coordinated government, people's participation, and ongoing political commitment will be needed. Pakistan can turn its cities into beacons of health, equity, and climate resilience only through such a multidimensional strategy.

REFERENCES

Abas, N., Saleem, M. S., Kalair, E., & Khan, N. (2019). Cooperative control of regional transboundary air pollutants. *Environmental Systems Research, 8*(1), 1–14. DOI: 10.1186/s40068-019-0138-0

Afonso, T. L., Marques, A. C., & Fuinhas, J. A. (2021). Does energy efficiency and trade openness matter for energy transition? Empirical evidence for countries in the Organization for Economic Co-operation and Development. *Environment, Development and Sustainability, 23*(9), 13569–13589. DOI: 10.1007/s10668-021-01228-z

Afzal, S. (2023). The Invisible Killer: Investigating the Lack of Research on Smog in Pakistan. *Annals of King Edward Medical University, 29*(2), 73–74. DOI: 10.21649/akemu.v29i2.5423

Afzal, S., & Bajwa, N. (2024). Smog the New Unfortunate Reality: A Catastrophe Public Health Concern in Decreasing Life Spans of Pakistani Population. *Journal of Socicty of Prcvcntion. Advocacy and Rcscarch KEMU, 3*(1), 1 1.

Afzal, S., & Bajwa, N. (2024). Smog the New Unfortunate Reality: A Catastrophe Public Health Concern in Decreasing Life Spans of Pakistani Population. *Journal of Society of Prevention. Advocacy and Research KEMU, 3*(1), 1–1.

Ahmad, M. N., Büker, P., Khalid, S., Van Den Berg, L., Shah, H. U., Wahid, A., Emberson, L., Power, S. A., & Ashmore, M. (2013). Effects of ozone on crops in north-west Pakistan. *Environmental Pollution, 174*, 244–249. DOI: 10.1016/j.envpol.2012.11.029 PMID: 23279903

Ajmal, M., Tarar, M. A., Arshad, M. I., Gulshan, A. B., Iqbal, M. A., & Tanvir, F. (2016). Air pollution and its effect on human health: A case study in Dera Ghazi Khan urban areas, Pakistan. *Journal of Environment and Earth Science, 6*, 87–93.

Akbar, A. (2020). *Theorizing the effect of smog on public health in Lahore*. DOI: 10.5539/ep.v9n2p14

Akuraju, V., Pradhan, P., Haase, D., Kropp, J. P., & Rybski, D. (2020). Relating SDG11 indicators and urban scaling–An exploratory study. *Sustainable Cities and Society, 52*, 101853. DOI: 10.1016/j.scs.2019.101853

Ali, S., Niaz, H., Ahmad, S., & Khan, S. (2025). Investigating how Rapid Urbanization Contributes to Climate Change and the Social Challenges Cities Face in Mitigating its Effects. *Review of Applied Management and Social Sciences, 8*(1), 1–16. DOI: 10.47067/ramss.v8i1.429

Ali, Y., Razi, M., De Felice, F., Sabir, M., & Petrillo, A. (2019). A VIKOR based approach for assessing the social, environmental and economic effects of "smog" on human health. *The Science of the Total Environment, 650*, 2897–2905. DOI: 10.1016/j.scitotenv.2018.10.041 PMID: 30373066

Ambreen, S., Wenbao, M., Khoso, W. M., Mangi, T. A., & Baloch, Z. A. (2025). Impact of socioeconomic and sociopolitical factors on environmental sustainability in Pakistan. *Environment, Development and Sustainability*, •••, 1–25. DOI: 10.1007/s10668-025-05989-9

Anees, M., & Yan, W. (2019). An Overview about the Challenges of Urban Expansion on Environmental Health in Pakistan. *Journal of Energy and Environmental Policy Options*, *2*(3), 64–71.

Antrop, M. (2004). Landscape change and the urbanization process in Europe. *Landscape and Urban Planning*, *67*(1-4), 9–26. DOI: 10.1016/S0169-2046(03)00026-4

Anwar, A., Sinha, A., Sharif, A., Siddique, M., Irshad, S., Anwar, W., & Malik, S. (2022). The nexus between urbanization, renewable energy consumption, financial development, and CO2 emissions: Evidence from selected Asian countries. *Environment, Development and Sustainability*, *24*(5), 6556–6576. DOI: 10.1007/s10668-021-01716-2

Appio, F. P., Lima, M., & Paroutis, S. (2019). Understanding Smart Cities: Innovation ecosystems, technological advancements, and societal challenges. *Technological Forecasting and Social Change*, *142*, 1–14. DOI: 10.1016/j.techfore.2018.12.018

Arif, M., & Hassan, E. (2023). *Beyond the haze: a study of smog, its types, and mitigation*. Faisalabad Chamber of Commerce & Industry.

Arif, M. I. U., & Mahsud, M. I. (2024). Disentangling Pakistan's Climate Change Governance Challenges: Trajectories and Underpinnings.

Asghar, N., Rehman, H., & Sultan, M. S. (2022). A nonlinear analysis of the relationship between urbanization and carbon dioxide emissions in Pakistan: A fresh evidence. *Review of Applied Management and Social Sciences*, *5*(4), 655–665. DOI: 10.47067/ramss.v5i4.276

Asif, M., Khan, M. I., & Pandey, A. (2024). Navigating the inclusive and sustainable energy transitions in South Asia: Progress, priorities and stakeholder perspectives. *Energy Conversion and Management*, *313*, 118589. DOI: 10.1016/j.enconman.2024.118589

Bajwa, A. U., & Sheikh, H. A. (2023). Road transport's contribution to Pakistan's air Pollution in the urban environment. *Air*, *1*, 237–257. DOI: 10.3390/air1040018

Bettencourt, L. M., Lobo, J., Helbing, D., Kühnert, C., & West, G. B. (2007). Growth, innovation, scaling, and the pace of life in cities. *Proceedings of the National Academy of Sciences of the United States of America, 104*(17), 7301–7306. DOI: 10.1073/pnas.0610172104 PMID: 17438298

Bilal, M., Hassan, M., Tahir, D. B. T., Iqbal, M. S., & Shahid, I. (2022). Understanding the role of atmospheric circulations and dispersion of air pollution associated with extreme smog events over South Asian megacity. *Environmental Monitoring and Assessment, 194*(2), 1–17. DOI: 10.1007/s10661-021-09674-y PMID: 35013892

Bilal, M., Hassan, M., Tahir, D. B. T., Iqbal, M. S., & Shahid, I. (2022). Understanding the role of atmospheric circulations and dispersion of air pollution associated with extreme smog events over South Asian megacity. *Environmental Monitoring and Assessment, 194*(2), 82. DOI: 10.1007/s10661-021-09674-y PMID: 35013892

Chen, M., Liu, W., Lu, D., Chen, H., & Ye, C. (2018). Progress of China's new-type urbanization construction since 2014: A preliminary assessment. *Cities (London, England), 78*, 180–193. DOI: 10.1016/j.cities.2018.02.012

Chen, M., Zhang, H., Liu, W., & Zhang, W. (2014). The global pattern of urbanization and economic growth: Evidence from the last three decades. *PLoS One, 9*(8), e103799. DOI: 10.1371/journal.pone.0103799 PMID: 25099392

Chua, R. Y., Huang, K. G., & Jin, M. (2019). Mapping cultural tightness and its links to innovation, urbanization, and happiness across 31 provinces in China. *Proceedings of the National Academy of Sciences of the United States of America, 116*(14), 6720–6725. DOI: 10.1073/pnas.1815723116 PMID: 30833399

Colbeck, I., Nasir, Z. A., & Ali, Z. (2010). The state of ambient air quality in Pakistan—A review. *Environmental Science and Pollution Research International, 17*(1), 49–63. DOI: 10.1007/s11356-009-0217-2 PMID: 19575252

Coovadia, H., Jewkes, R., Barron, P., Sanders, D., & McIntyre, D. (2009). The health and health system of South Africa: Historical roots of current public health challenges. *Lancet, 374*(9692), 817–834. DOI: 10.1016/S0140-6736(09)60951-X PMID: 19709728

Dawn (2021). Lahore again shows worst air quality. https://shorturl.at/V02d6

Elmqvist, T., Andersson, E., Frantzeskaki, N., McPhearson, T., Olsson, P., Gaffney, O., Takeuchi, K., & Folke, C. (2019). Sustainability and resilience for transformation in the urban century. *Nature Sustainability, 2*(4), 267–273. DOI: 10.1038/s41893-019-0250-1

Farooq, M. S., Feroze, N., Cheng, T., & Feroze, F. (2024). Pakistan's Green Urbanization in the Perspective of Ecological Civilization. *Chinese Journal of Urban and Environmental Studies, 12*(03), 2450017. DOI: 10.1142/S2345748124500179

Friedmann, J. (2006). Four theses in the study of China's urbanization. *International Journal of Urban and Regional Research, 30*(2), 440–451. DOI: 10.1111/j.1468-2427.2006.00671.x

Fu, B., Meadows, M. E., & Zhao, W. (2022). Geography in the Anthropocene: Transforming our world for sustainable development. *Geography and Sustainability, 3*(1), 1–6. DOI: 10.1016/j.geosus.2021.12.004

Ghani, F., Tsekleves, E., & Thomas, Y. F. (2021). Urbanization and cities as drivers of global health. In *Handbook of Global Health* (pp. 1–28). Springer International Publishing.

Gondhalekar, D., & Ramsauer, T. (2017). Nexus city: Operationalizing the urban water-energy-food nexus for climate change adaptation in Munich, Germany. *Urban Climate, 19*, 28–40. DOI: 10.1016/j.uclim.2016.11.004

Gong, P., Liang, S., Carlton, E. J., Jiang, Q., Wu, J., Wang, L., & Remais, J. V. (2012). Urbanisation and health in China. *Lancet, 379*(9818), 843–852. DOI: 10.1016/S0140-6736(11)61878-3 PMID: 22386037

Gu, C. (2019). Urbanization: Positive and negative effects. *Science Bulletin, 64*(5), 281–283. DOI: 10.1016/j.scib.2019.01.023 PMID: 36659587

Halley, J. M., Van Houtan, K. S., & Mantua, N. (2018). How survival curves affect populations' vulnerability to climate change. *PLoS One, 13*(9), e0203124. DOI: 10.1371/journal.pone.0203124 PMID: 30188919

He, J., Yang, Y., Liao, Z., Xu, A., & Fang, K. (2022). Linking SDG 7 to assess the renewable energy footprint of nations by 2030. *Applied Energy, 317*, 119167. DOI: 10.1016/j.apenergy.2022.119167

Jabeen, N., Farwa, U., & Jadoon, M. (2017). Urbanization in Pakistan: A governance perspective. *Journal of the Research Society of Pakistan, 54*(1), 127–136.

Jaiswal, A., Sagar, R., Pandey, A., Yadav, D., Ansari, M., & Rawat, R. (2024). Building Resilient Urban Futures: Adapting Cities to Climate Change Challenges. In *Cities of Tomorrow: Urban Resilience and Climate Change Preparedness* (pp. 51–68). Springer. DOI: 10.1007/978-981-97-9658-8_3

Janardhanan, N. (2017). Energy Transition in India: Role of Climate Change Policies in Energy Security. *International Studies, 54*(1-4), 231–249. DOI: 10.1177/0020881718791851

Junejo, Z. U. B. (2025). Pakistan Today. Death in the air demanding national action. https://shorturl.at/zuf77

Kamal, A., Malik, R. N., Martellini, T., & Cincinelli, A. (2015). Source, profile, and carcinogenic risk assessment for cohorts occupationally exposed to dust-bound PAHs in Lahore and Rawalpindi cities (Punjab province, Pakistan). *Environmental Science and Pollution Research International, 22*(14), 10580–10591. DOI: 10.1007/s11356-015-4215-2 PMID: 25739839

Khan, M., Zhang, Y., & Khan, S. (2025). Echoes of Survival: Climate Change Impact & Typologies of Adaptation among Vulnerable Communities toward Climate-Induced Food Insecurity in Pakistan. *Research on World Agricultural Economy*, 290-318.

Khan, Z. M. (2024). *Unpacking the Smog Crisis in Punjab – Key Areas of Research.* https://shorturl.at/F2iw7

Kroll, C., Warchold, A., & Pradhan, P. (2019). Sustainable Development Goals (SDGs): Are we successful in turning trade-offs into synergies? *Palgrave Communications, 5*(1), 140. DOI: 10.1057/s41599-019-0335-5

Liaqut, A., Tariq, S., & Younes, I. (2023). A study on optical properties, classification, and transport of aerosols during the smog period over South Asia using remote sensing. *Environmental Science and Pollution Research International, 30*(26), 69096–69121. DOI: 10.1007/s11356-023-27047-5 PMID: 37129820

Lu, Z. N., Chen, H., Hao, Y., Wang, J., Song, X., & Mok, T. M. (2017). The dynamic relationship between environmental pollution, economic development and public health: Evidence from China. *Journal of Cleaner Production, 166*, 134–147. DOI: 10.1016/j.jclepro.2017.08.010

Lv, Y. (2023). Transitioning to sustainable energy: Opportunities, challenges, and the potential of blockchain technology. *Frontiers in Energy Research, 11*, 1258044. DOI: 10.3389/fenrg.2023.1258044

Ma, R., Abid, N., Yang, S., & Ahmad, F. (2023). From crisis to resilience: Strengthening climate action in OECD countries through environmental policy and energy transition. *Environmental Science and Pollution Research International, 30*(54), 115480–115495. DOI: 10.1007/s11356-023-29970-z PMID: 37882927

Madlener, R., & Sunak, Y. (2011). Impacts of urbanization on urban structures and energy demand: What can we learn for urban energy planning and urbanization management? *Sustainable Cities and Society*, *1*(1), 45–53. DOI: 10.1016/j.scs.2010.08.006

Mahdavi, M., Awaafo, A., Jurado, F., Vera, D., & Ramos, R. A. V. (2023). Wind, solar and biogas power generation in water-stressed areas of Morocco considering water and biomass availability constraints and carbon emission limits. *Energy*, *282*, 128756. DOI: 10.1016/j.energy.2023.128756

Majeed, M. M., & Munir, A. (2020). Pakistan: Country report on children's environmental health. *Reviews on Environmental Health*, *35*(1), 57–63. DOI: 10.1515/reveh-2019-0087 PMID: 31926105

Malik, A., Islam, J., Zaib, G., Zahid, A., Rashid, A. R., & Zia, T. (2024). Interplay of smog and health conditions: A public health perspective on respiratory, hypertensive, and cardiovascular morbidity. *Journal of Physical. Biomedical and Biological Sciences*, *2024*(1), 37–37. DOI: 10.64013/jpbab.v2024i1.37

Meyer, N., & Auriacombe, C. (2019). Good urban governance and city resilience: An afrocentric approach to sustainable development. *Sustainability (Basel)*, *11*(19), 5514. DOI: 10.3390/su11195514

Mukhtar, F. (2018). The rising menace of smog: Time to act now. *Journal of Ayub Medical College, Abbottabad*, *30*(1), 1–2. PMID: 29504318

Nadeem, M., Aziz, A., Al-Rashid, M. A., Tesoriere, G., Asim, M., & Campisi, T. (2021). Scaling the potential of compact city development: The case of Lahore, Pakistan. *Sustainability (Basel)*, *13*(9), 5257. DOI: 10.3390/su13095257

Nasar-u-Minallah, M., Jabbar, M., Zia, S., & Perveen, N. (2024). Assessing and anticipating environmental challenges in Lahore, Pakistan: future implications of air pollution on sustainable development and environmental governance. *Environmental Monitorin*

Nasim, S., & Sharif, F. (2020). To adopt, or not to adopt,'why'is the question: A case for clean kiln technologies in developing countries. *Journal of Cleaner Production*, *257*, 120553. DOI: 10.1016/j.jclepro.2020.120553

Naveed, Z., & Khayyam, U. (2023). Smog and cognitive issues in the school going children of Lahore and Islamabad, Pakistan. *International Journal of Environmental Science and Technology*, *20*(4), 4151–4166. DOI: 10.1007/s13762-022-04264-y

Nyambuu, U., & Semmler, W. (2023). Fossil fuel resources, environment, and climate change. In *Sustainable macroeconomics, climate risks and energy transitions: Dynamic modeling, empirics, and policies* (pp. 45–58). Springer International Publishing. DOI: 10.1007/978-3-031-27982-9_4

Olson, C., & Lenzmann, F. (2016). The social and economic consequences of the fossil fuel supply chain. *MRS Energy & Sustainability : a Review Journal, 3*(1), E6. DOI: 10.1557/mre.2016.7

Pakistan Today. (2024, March 3). *Death toll from pneumonia rises to 456 in Punjab.* https://shorturl.at/C1i6l

Qayyum, U., Sabir, S., & Anjum, S. (2021). Urbanization, informal economy, and ecological footprint quality in South Asia. *Environmental Science and Pollution Research International, 28*(47), 67011–67021. DOI: 10.1007/s11356-021-15111-x PMID: 34244937

Qazi, A., Hussain, F., Rahim, N. A., Hardaker, G., Alghazzawi, D., Shaban, K., & Haruna, K. (2019). Towards sustainable energy: A systematic review of renewable energy sources, technologies, and public opinions. *IEEE Access : Practical Innovations, Open Solutions, 7*, 63837–63851. DOI: 10.1109/ACCESS.2019.2906402

Raza, M. H., Abid, M., Faisal, M., Yan, T., Akhtar, S., & Adnan, K. M. (2022). Environmental and health impacts of crop residue burning: Scope of sustainable crop residue management practices. *International Journal of Environmental Research and Public Health, 19*(8), 4753. DOI: 10.3390/ijerph19084753 PMID: 35457622

Raza, W., Saeed, S., Saulat, H., Gul, H., Sarfraz, M., Sonne, C., Sohn, Z.-H., Brown, R. J. C., & Kim, K. H. (2021). A review on the deteriorating situation of smog and its preventive measures in Pakistan. *Journal of Cleaner Production, 279*, 123676. DOI: 10.1016/j.jclepro.2020.123676

Razzaq, A., Zafar, M. M., Zahra, L. T., Qadir, F., Qiao, F., & Jiang, X. (2024). Smog: Lahore needs global attention to fix it. *Environmental Challenges, 16*, 100999. DOI: 10.1016/j.envc.2024.100999

Reckien, D., Creutzig, F., Fernandez, B., Lwasa, S., Tovar-Restrepo, M., Mcevoy, D., & Satterthwaite, D. (2017). Climate change, equity and the Sustainable Development Goals: An urban perspective. *Environment and Urbanization, 29*(1), 159–182. DOI: 10.1177/0956247816677778

Saleem, Z., Saeed, H., Yousaf, M., Asif, U., Hashmi, F. K., Salman, M., & Hassali, M. A. (2019). Evaluating smog awareness and preventive practices among Pakistani general population: A cross-sectional survey. *International Journal of Health Promotion and Education*, *57*(3), 161–173. DOI: 10.1080/14635240.2019.1576535

Sanchez-Triana, E., Enriquez, S., Afzal, J., Nakagawa, A., & Khan, A. S. (Eds.). (2014). *Cleaning Pakistan's air: Policy options to address the cost of outdoor air pollution*. The World Bank., https://shorturl.at/E3QU7 DOI: 10.1596/978-1-4648-0235-5

Scoones, I., Stirling, A., Abrol, D., Atela, J., Charli-Joseph, L., Eakin, H., Ely, A., Olsson, P., Pereira, L., Priya, R., van Zwanenberg, P., & Yang, L. (2020). Transformations to sustainability: Combining structural, systemic and enabling approaches. *Current Opinion in Environmental Sustainability*, *42*, 65–75. DOI: 10.1016/j.cosust.2019.12.004

Seto, K. C., & Shepherd, J. M. (2009). Global urban land-use trends and climate impacts. *Current Opinion in Environmental Sustainability*, *1*(1), 89–95. DOI: 10.1016/j.cosust.2009.07.012

Shamsi, S. R. A., Ashmore, M. R., Bell, J. N. B., Maggs, R., Kafayat, U., & Wahid, A. (2000). The impacts of air pollution on crops in developing countries—a case study in Pakistan. In *Environmental stress: Indication, mitigation and eco-conservation* (pp. 63–71). Springer Netherlands. DOI: 10.1007/978-94-015-9532-2_6

Shang, Y., Sang, S., Tiwari, A. K., Khan, S., & Zhao, X. (2024). Impacts of renewable energy on climate risk: A global perspective for energy transition in a climate adaptation framework. *Applied Energy*, *362*, 122994. DOI: 10.1016/j.apenergy.2024.122994

Sherin, A. (2023). Beyond COP: Pakistan's evolving role in a global climate context. *Khyber Medical University Journal*, *15*(4), 208–210.

Singh, S. (2018). *The Great Smog of India*. Penguin Random House India.

Sourgens, F. G., & Sempertegui, L. (2023). *Principles of international energy transition law: law as an instrument of development in the 21st century*. Oxford University Press. DOI: 10.1093/law/9780198876083.001.0001

Subhanullah, M., Hassan, N., Rahman, G., Rawan, B., Ullah, W., & Ilyas, M. (2025). Concentration of particulate matter and its impact on public health in different cities in Pakistan-a review. *Environmental Forensics*, *26*(2), 146–162. DOI: 10.1080/15275922.2024.2366794

Tahir, M., Aziz, S., Khan, A., & Shahi, M. (2023). Spreading of smog in Punjab areas o f Pakistan due to violation of environmental laws. *Intl. J. Soc. Sci.: Curr. Fut. Res. Trends*, *18*, 74–89.

Tajammul, K. I. (2023). Critical review and comparative analysis of the Government of Punjab's 'Policy on Controlling Smog, 2017'with Counterpart Strategies in London, Beijing and Los Angeles. *Journal of Development Policy Research & Practice (JoDPRP)*, 181-202.

Tonne, C., Adair, L., Adlakha, D., Anguelovski, I., Belesova, K., Berger, M., Brelsford, C., Dadvand, P., Dimitrova, A., Giles-Corti, B., Heinz, A., Mehran, N., Nieuwenhuijsen, M., Pelletier, F., Ranzani, O., Rodenstein, M., Rybski, D., Samavati, S., Satterthwaite, D., & Adli, M. (2021). Defining pathways to healthy sustainable urban development. *Environment International, 146*, 106236. DOI: 10.1016/j.envint.2020.106236 PMID: 33161201

United Nations. (2015). *Transforming our world: the 2030 Agenda for Sustainable Development*. United Nations General Assembly.

United Nations. (2019). World population prospects 2019. *Department of Economic and Social Affairs, 141.*

Wang, Y. (2020). Urban land and sustainable resource use: Unpacking the countervailing effects of urbanization on water use in China, 1990–2014. *Land Use Policy*, *90*, 104307. DOI: 10.1016/j.landusepol.2019.104307

World Bank. (2022). *Pakistan country climate and development report.* https://shorturl.at/25GTE

Xu, Z., Peng, J., Qiu, S., Liu, Y., Dong, J., & Zhang, H. (2022). Responses of spatial relationships between ecosystem services and the Sustainable Development Goals to urbanization. *The Science of the Total Environment, 850*, 157868. DOI: 10.1016/j.scitotenv.2022.157868 PMID: 35944627

Yang, G., Zha, D., Cao, D., & Zhang, G. (2024). Time for a change: Rethinking the global renewable energy transition from the Sustainable Development Goals and the Paris Climate Agreement. *Innovation (Cambridge (Mass.))*, *5*(2), 100582. DOI: 10.1016/j.xinn.2024.100582 PMID: 38379793

Yousaf, H. S., Abbas, M., Ghani, N., Chaudhary, H., Fatima, A., Ahmad, Z., & Yasin, S. A. (2021). A comparative assessment of air pollutants of smog in wagah border and other sites in Lahore, Pakistan. *Brazilian Journal of Biology*, *84*, e252471. DOI: 10.1590/1519-6984.252471 PMID: 34932634

Zeeshan, N., Murtaza, G., Ahmad, H. R., Awan, A. N., Shahbaz, M., & Freer-Smith, P. (2024). Particulate and gaseous air pollutants exceed WHO guideline values and have the potential to damage human health in Faisalabad, Metropolitan, Pakistan. *Environmental Monitoring and Assessment, 196*(7), 659. DOI: 10.1007/s10661-024-12763-3 PMID: 38916809

Zhang, L. (2008). Conceptualizing China's urbanization under reforms. *Habitat International, 32*(4), 452–470. DOI: 10.1016/j.habitatint.2008.01.001

Zhao, Z., Pan, Y., Zhu, J., Wu, J., & Zhu, R. (2022). The impact of urbanization on the delivery of public service–related SDGs in China. *Sustainable Cities and Society, 80*, 103776. DOI: 10.1016/j.scs.2022.103776

Zia, H., Khan, T., Fatima, H. S., Khurram, M., Harris, N. R., & Khalil, A. (2022). Impacts of urbanization on green spaces of the densely populated City of Karachi, Pakistan-An analysis of 8 years of data for estimating land cover changes.

Zikria Saleem, H. S., Yousaf, M., Asif, U., Khurshid, F., Hashmi, M. S., & Hassali, M. A. (2019). Evaluating smog awareness and preventive practices among Pakistani general population: a cross-sectional survey.

Chapter 5
Smog and Air Pollution in Pakistan ITS
Economic, Social, and Health Dimensions

Aqeela Saghir
https://orcid.org/0009-0001-0194-0067
University of Agriculture, Faisalabad, Pakistan

ABSTRACT

Pakistan's air quality crisis is persistent, severe, and costly. Fine particulate matter ($PM_{2.5}$) concentrations routinely exceed the World Health Organization's (WHO) 2021 guideline of 5 μg/m^3 annual average; in 2023 Pakistan's national average was 73.7 μg/m^3—about 15× the WHO guideline—placing the country among the world's most polluted according to IQ Air's global assessment. These levels drive large health losses, productivity declines, and wide social disruptions, including repeated school closures in Punjab during the winter smog season. The World Bank estimates the economic cost of ambient air pollution in Pakistan at roughly 6.5% of GDP annually, with major contributions from residential solid-fuel use, transport, industry (including brick kilns), agriculture (residue burning), and waste burning. Policy momentum has accelerated (Euro-5 fuel adoption, a National Clean Air Policy, Punjab's Clean Air Plan, kiln modernization), yet implementation gaps, limited monitoring, and the regional "airshed" nature of the problem keep Pakistan highly exposed.

DOI: 10.4018/979-8-3373-2469-2.ch005

1. INTRODUCTION: WHY PAKISTAN'S SMOG PERSISTS

Air pollution in Pakistan is both a local and regional phenomenon. Within Pakistan, huge population, rapid urbanization, poor planning and management, expanding motorization, coal- and biomass-based industrial processes (e.g., brick kilns), and widespread household solid-fuel use accumulate into high primary and secondary particulate loads. The meteorological setup from roughly October to February—temperature inversions, calm winds, and high humidity—traps pollutants in the Indo-Gangetic Plain airshed that spans borders, concentrating smog over Punjab's dense urban centers such as Lahore, Faisalabad, and Gujranwala. The World Bank's airshed work underscores that pollution moves across administrative boundaries, so purely city-by-city measures cannot fully solve the problem (GOP, 2024).

1. WHO Guideline vs. Pakistan's Reality

The WHO's 2021 Global Air Quality Guidelines revised the annual $PM_{2.5}$ benchmark to 5 µg/m³ (and the 24-hour guideline to 15 µg/m³), reflecting robust evidence that health harms occur even at low concentrations. Pakistan's contemporary urban concentrations are an order of magnitude greater during the winter season, making the health burden particularly acute (WHO, 2021).

2. MEASURED LEVELS AND TRENDS (2018–2024/25)

- National exposure: IQAir's 2023 assessment reported Pakistan's annual $PM_{2.5}$ at 73.7 µg/m³, among the very worst globally and ~15× the WHO guideline. The broader South Asian pattern is similar, with Bangladesh and India also ranking at the bottom. (Reuters, 2024)
- City hotspots: Lahore recurrently ranks among the world's most polluted large cities during winter episodes, with multi-day AQI readings that trigger public health emergencies and school/business closures.
- Monitoring and inventories: Pakistan's monitoring network has grown but remains patchy; Punjab published a district-scale emissions inventory for Lahore (2023) to begin quantifying sectoral loads—a prerequisite for targeted abatement. (Urban Unit, 2024). Pakistan is among the most polluted countries of the world as figure below.

Figure 1.

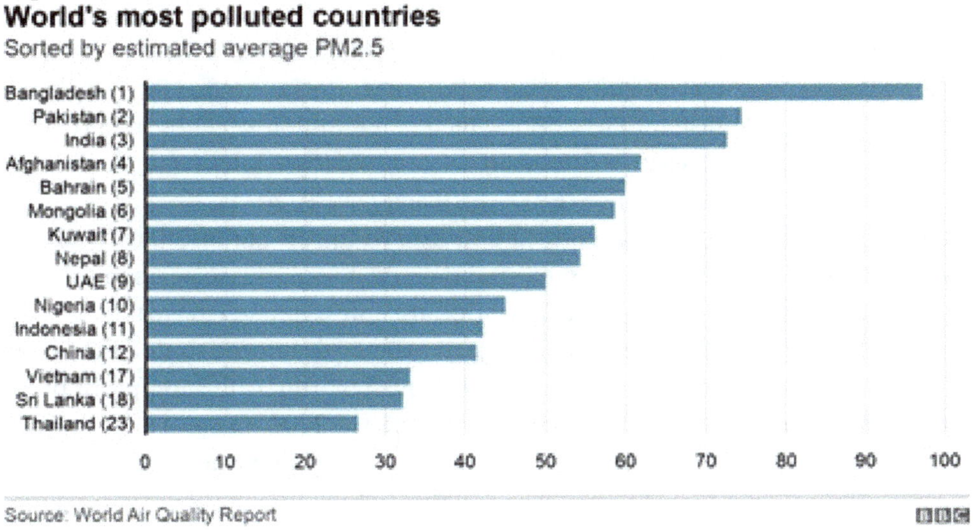

3. SOURCES AND DRIVERS: WHAT PRODUCES PAKISTAN'S SMOG?

3.1 Sectoral Contributions to PM

Recent World Bank analytics for Punjab (2024–2025 program documents) apportion average PM contributions approximately as follows: residential energy (solid fuels) 23%, transport 16%, industry (including heavy industry and brick kilns) 16%, agriculture & livestock (including crop residue burning) 13%, and municipal waste burning 7%; natural sources account for the balance. These shares vary by season and location but offer a quantitative map for interventions (World Bank, 2025)

3.2 Seasonal and Meteorological Amplification

From October to February, cooler temperatures and shallow mixing heights trap emissions, while humidity fosters secondary aerosol formation. Agricultural residue burning—especially in late October–November—adds episodic bursts to an already high urban baseline from vehicles, industry, and household fuels. The airshed nature means cross-border inflows can be material, reinforcing the case for regional coordination (World Bank, 2023).

3.3 Deforestation

Deforestation occurred as a result of rapid industrialization and demand for agricultural land (Zulfiqar, et al, 2021; Allen and Barnes, 1985). Around 30% of global land is covered by forests in the world, which is 3.8 billion hectares in 2005 and this leads to a slight increase in forests, up to 31%. Nearly 1 billion people depend globally on earth for their livings (FAO, 2010 and UNEP, 2011).

Tree Plantation not only reduces carbon load to the air but also purify the air quality. Implanting 500 billion original plants could help decrease carbon in the air by 25%, reducing climate change's effects. Because forests are so vital to the survival of life on Earth, a variety of innovative projects is being developed to assist and urge people to plant extra trees. The Billion Tree Campaign, run by the non-profit plant for the planet foundation, is one such initiative. Since 2007, the global group has been teaching and inspiring people all over the world to plant more trees. According to a new study, planting a trillion plants could offset the previous ten years' worth of anthropogenic co_2discharges and seize 160 billion tonnes of carbon from the air. With so many tree-planting projects around the world, which countries are doing the greatest job by (Kristen, 2020).

3.3.1 Pakistan's Forest Cover

In 2010, Pakistan had 648 kilometres of tree cover, corresponding for 0.74 percent of the country's total land. In 2020, it lost 69.2 hectares of forest, resulting in 26.1 kt of Co2 emissions.

3.3.2 Pakistan Losses Tree Coverage

Pakistan lost 9.68kha of tree cover and 3.63Mt of co_2 emissions between 2001 and 2020, equating to a 0.99 percent decline in tree cover since 2000.

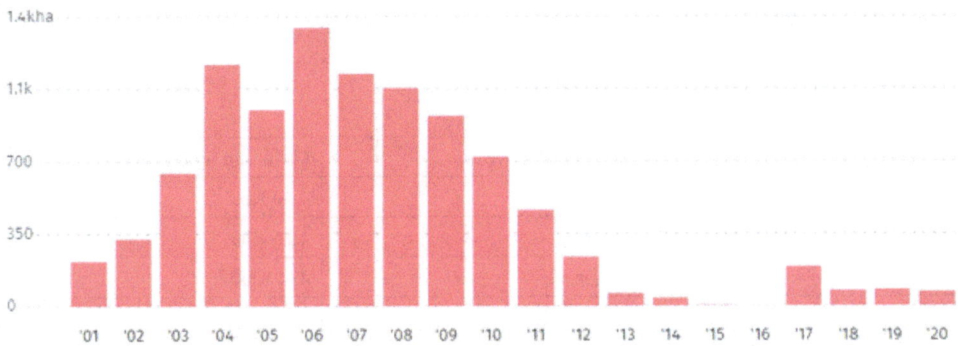

3.3.2.1. In Pakistan, the Dominant Driver Loses Tree Cover on an Annual Basis

In 2020, Pakistan had 1.37 Mha of natural forest, extending over 1.7% of its land area. In 2024, it lost 451 ha of natural forest, equivalent to 68.2 kt of CO2 emissions. Between 2001 and 2019, 7.1 percent of tree cover loss in Pakistan occurred in regions where deforestation was the primary source of loss. (GFW, 2024)

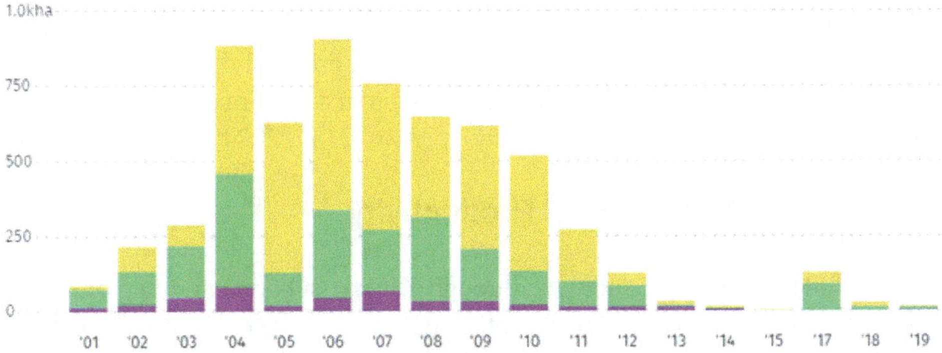

3.3.2.2. In Pakistan, the Loss of Tree Coverage

In the middle of 2001 and 2020, the highest two sections in Pakistan were liable for 90 percent of all tree cover decline. The highest tree cover loss was in the

N.W.F.P., with 4.46 hectares compared to an average of 1.21 hectares as presented in below Table 1.

Table 1.

Ranking	Regions of Pakistan	Coverage (in hectares)
1	N.W.F.P.	4.46 Kha
2	F.A.T.A.	4.25 Kha
3	Punjab	434 Kha
4	Azad Kashmir	395 Kha
5	Northern Areas	113 Kha

(GFW, 2024)

Above statistics show heavy deforestation instead of afforestation bolting climatic changes, defacing the natural beauty of planet earth. It also inculcates other factors such as pollution, smog, dust storms and flood. Human activity must be monitored for safe life considering Sustainable Development Goal (Environment).

4. IMPACTS OF SMOG

4.1 Economic Impacts: Growth, Productivity, and Public Finance

It estimates that ambient air pollution costs Pakistan roughly 6.5% of GDP annually, reflecting mortality, morbidity, reduced labor productivity, and other welfare losses. This order of magnitude makes air quality a first-rank macroeconomic issue, on par with energy security or water scarcity (Awez et al., 2024; World bank, 2023)

4.2 Firm- and Worker-Level Productivity

Smog lowers labor productivity through illness, subclinical symptoms (e.g., headaches, eye irritation), and absenteeism, and it reduces output quality in pollution-sensitive sectors (food, textiles). High-pollution days disrupt logistics (visibility-related flight and road closures) and retail and service activity. While Pakistan-specific micro-estimates vary by sector and city, South Asian evidence consistently shows meaningful productivity losses per additional 10 $\mu g/m^3$ of $PM_{2.5}$ exposure. The macro cost captures these channels collectively (Sarfraz, 2020).

4.3 Public Outlays and Fiscal Stress

Air pollution raises healthcare demand, strains frontline services during spikes, and forces ad hoc emergency spending (e.g., enforcement drives, dust suppression, and school closure protocols). Lost school days (Section 6) and disrupted commerce also erode tax bases during peak smog weeks (Ashraf et al., 2022).

5. HEALTH IMPACTS: MORTALITY, MORBIDITY, AND LIFE EXPECTANCY

Adult and children highlight a rise in patients undergoing respiratory problems along with throat and eye irritation. Four years olds are admitted for emergency care due to excessive outdoor exposure timings. Children are advised to stay inside and masks. From security guards to manual labourers working on busy roads around Lahore, they are risking permanent damage to their lungs. Bread winners of low-income households are choking on smog (Aslam, 2022). With 11 million complains of burning eyes and headache, this winter has hit five times the legal limit. Smog has single-handedly overpowered other health epidemics in Pakistan by worsening bronchial infection, heart problems, and lung damage. Children about 35% of the population in Pakistan. They are prone to asthma and other respiratory diseases that may reduce the quality of life (Sarfraz, 2020).

5.1 Mortality and Disease Burden

Long-term exposure to $PM_{2\cdot5}$ increases ischemic heart disease, stroke, COPD, lung cancer, and lower respiratory infections. Synthesizing Global Burden of Disease/ HEI assessments, independent organizations have estimated Pakistan's air pollution – attributable deaths in the six-figure range annually; Human Rights Watch, using HEI/IHME inputs, reported ~128,000 premature deaths per year (circa late-2010s). While updated absolute figures vary by methodology and year, the burden remains very high relative to population.

5.2 Life Expectancy Impacts

The Air Quality Life Index (AQLI) links chronic $PM_{2\cdot5}$ exposure to life expectancy losses. South Asia's average life expectancy shortfall from air pollution is about five years, implying large welfare gains if Pakistan could meet WHO guidelines. Pakistan's AQLI country spotlight underscores the severity and the absence (to date) of a national $PM_{2\cdot5}$ standard (GOP, 2025).

5.3 Morbidity, Vulnerable Groups, and Equity

Children, older adults, and people with cardiopulmonary conditions face disproportionate risk. Lower-income households are more likely to rely on solid fuels for cooking/heating and to live nearer to roadways or industrial clusters, compounding exposure and health costs an equity challenge that policy must address. (WHO, 2023)

6. SOCIAL IMPACTS: EDUCATION, DAILY LIFE, AND INEQUALITY

Social life is important because it helps individuals to build strong relationships, share ideas, and feel connected to others. Interacting with friends, family, and communities reduces stress, improves mental health, and provides emotional support during difficult times. A healthy social life also boosts confidence, communication skills, and overall happiness, making life more balanced and meaningful.

Social life has a direct impact on an individual's mental health because regular interaction with others provides emotional support, reduces feelings of loneliness, and lowers the risk of anxiety and depression. Healthy social connections create a sense of belonging and purpose, which helps individuals cope with stress and maintain a positive outlook on life. On the other hand, social isolation often leads to negative mental health outcomes, highlighting the strong link between social relationships and psychological well-being (American Psychological Association, 2020).

6.1 School Closures and Educational Disruption

Punjab has repeatedly closed schools during smog emergencies. Educational cut-off/online learning reduces children learning ability. Examples include November 2023 closures ordered amid hazardous AQI and November 2024 shuttering of schools and offices, with masks mandated in Lahore and nearby districts. Such closures protect health but disrupt instruction time, complicate exam schedules, and shift caregiving burdens onto families, especially those lacking digital learning alternatives. Parents expenditure increase due to purchasing of mobile phone also mothers' have to teach their children themselves or tutoring while also constant paying their fees to school (Ashraf, 2019).

6.2 Everyday Life and Social Behavior

- High-smog days alter mobility (fewer trips, car-pooling), consumption (delay of non-essential shopping), and leisure (reduced outdoor activity). Public spaces empty while mask usage and air purifier demand rise among those who can afford them, deepening exposure inequality. *It is also* closure of parks, public venues; "green lockdown" restricts social life (Khattak and Khan, 2024).

Impact on Academic Progress

Frequent closures, *shorten teaching time*, delaying syllabus completion and exam preparation. Public schools, which lack robust digital infrastructure, cannot easily switch to online learning, widening the *learning gap* between public and private school students.

Psychological Stress on Parents and Students

Parents worry about children's health and future academic performance during prolonged closures. As reported by World Health Organization (2018) that Smog creates significant psychological stress for both parents and students as it disrupts daily routines, limits outdoor activities, and poses constant health risks. Parents experience anxiety about their children's safety and well-being, especially when schools remain open despite hazardous air quality. Students, on the other hand, face stress from school closures, missed classes, and reduced physical activity, leading to academic pressure and feelings of isolation. This continuous uncertainty and health concern can contribute to increased tension, irritability, and long-term mental health challenges for families.

6.3. Lifestyle Restrictions

Limited Outdoor Activities

Smog discourages outdoor exercise, sports, and social gatherings, especially for children and young adults. Due to restricted physical activities survey-based reports of widespread health effects showed that nearly *70% of Pakistanis* suffered smog-related health issues such as coughing (68%), flu-like symptoms (66%), breathing difficulties (37%), and eye irritation (29%). Over *two million people* visited clinics in Punjab for respiratory complaints in a single month. UNICEF warned over *11 million children under five* in the province face serious health risks due to air pollution (Ali et al., 2022).

6.4. Change in Daily Routines

Daily routine; sleeping and eating pattern change due to smog. Families prefer to stay indoors, leading to less physical activity, increased screen time, and potential mental health issues like stress and anxiety. Children are restricted in social activity more prone to psychological issues. Family conflicts arise due to annoyance and irritation resulting poor mental health condition. people limit outdoor exercise, recreational activities, and social gatherings, leading to a more sedentary lifestyle and reduced physical interaction. Schools may close or shift to online classes, and parents often keep children indoors for safety, which alters family schedules. Commuting patterns also change, with some people avoiding public transport or reducing travel, ultimately affecting productivity and social engagement (Awez et al., 2024).

6.5. Gender and Equity Dimensions

Disproportionate Impact on Women

Smog has a disproportionate impact on women due to their roles and responsibilities within households and communities. Women, especially in low-income families, are often responsible for tasks such as cooking, cleaning, and caring for children, which may require them to stay indoors in poorly ventilated spaces where indoor air pollution combines with outdoor smog, worsening health risks. Pregnant women are particularly vulnerable, as exposure to polluted air increases the risk of complications such as low birth weight and respiratory issues in newborns. Additionally, women who work outdoors, such as in agriculture or informal labor, face increased exposure without adequate protective measures, making them more susceptible to both physical and psychological health problems (UNEP, 2016).

Social Inequality

Wealthier families can afford medicines, *air purifiers, N95 masks, and sealed homes*, and health strategies, while low-income families bear the brunt of exposure without protection. Poor and marginal population is more prone to environmental cruelty. It enhances *economic pressure on households especially in low-income groups. Population is already on low line health profile again burdened with Increased healthcare spending*. Families spend more on medicines for respiratory and eye problems during smog season. Children and old aged people are rich source of infestation (Anjum et al., 2021).

Loss of Income

Due to air pollution and smog seasonal labour force affected, during heavy smog transportation services limited that halt labour activity. Same in the sector of construction due to severe weather changes, contractors and owners' reduce construction sites. Most marginalized community is labour that is real victim of such ubrupt weather changes. A labour man is not a single person instead his complete family suffers as mostly he is sole bread winner of the family. Daily wage earners, farmers, hawkers, vendors and outdoor workers face reduced work opportunities because of visibility issues or health risks, *pushing vulnerable families deeper into poverty*. These patterns intersect with income and housing, reinforcing spatial inequities across urban neighborhoods (World Bank, 2019; Azhar, A., et al., 2019).

6.6. Gendered and Occupational Dimensions

Women having school going children face dual problem, school being closed but women workplace at several departments; health, security, business and laborers is open. Women, especially in low-income households, are more exposed to indoor pollutants combined with outdoor smog due to household responsibilities like cooking in poorly ventilated spaces. Pregnant women face higher health risks, including complications during pregnancy. Occupationally, outdoor workers such as farmers, traffic police, construction laborers, and street vendors are at greater risk because their jobs require prolonged exposure to polluted air. These groups often lack access to protective gear or healthcare, making them highly vulnerable to respiratory and cardiovascular illnesses caused by smog (Iram et al., 2025)

6.7. Public Events and Social Gatherings

Smog peaks in *November–December*, coinciding with the wedding season in Punjab. Outdoor functions often get canceled or shifted indoors, increasing costs for families. Social, and sports tournaments often face postponements or cancellations due to poor visibility and health risks (Sarfraz, 2020). Smog significantly affects gatherings, political activities and games by forcing cancellations or postponements to protect public health. Outdoor religious processions or mosque attendance decline on hazardous smog days. Festivals, sports matches, weddings, and cultural programs often get disrupted during severe smog periods, leading to social isolation and economic losses for event organizers. People avoid attending outdoor gatherings due to health risks, which reduces community interaction and cultural participation. This disruption not only impacts social life but also creates financial strain for businesses and individuals relying on such events for income (Iqbal et al., 2023; World Health Organization, 2018).

6.8. Increased Urban Stress and Migration

Dense cities like Lahore experience *traffic chaos*, emergency advisories, and restrictions during smog season, reducing quality of life. Some affluent families temporarily *move to hill stations* (e.g., Murree, Abbottabad) during severe smog episodes, reinforcing social stratification .

6.9. Psychological and Behavioral Changes

Constant news alerts, AQI apps, and government warnings create a sense of health insecurity create a fear and depression. While behavioral adaptation such as mask-wearing, buying air purifiers, or staying indoors becomes normalized during winter, but not everyone can afford these measures—highlighting a class divide.

7. POLICY LANDSCAPE AND PROGRESS

7.1 National Policy and Standards

Government of Pakistan approved a National Clean Air Policy (NCAP) in March and April 2023. These frameworks prioritize fuels and vehicles, industrial emissions (including kilns), waste burning, and agricultural residue management. Independent reviews approve the direction while noting implementation and monitoring gaps. Pakistan still lacks a binding national $PM_{2.5}$ standard, a gap. (GOP, 2021).

7.2 Cleaner Fuels and Vehicles

Pakistan adopted Euro-5 petrol in 2020 and Euro-5 diesel in 2021, improving sulfur content and enabling cleaner vehicle technologies, an important step to reduce transport emissions over time. Full benefits hinge on inspection/maintenance, fleet renewal, and enforcement against fuel adulteration.

7.3 Brick Kiln Modernization

Punjab has promoted zig-zag kiln retrofits and, via its Clean Air Policy, curbed conventional kiln operations. Modern kiln designs cut particulate emissions dramatically (studies report reductions exceeding 60% for PM and CO), improve fuel efficiency, and lower per-brick emissions intensity—although coverage remains incomplete and enforcement episodic. punjablaws.punjab.gov.pkpmc.ncbi.nlm.nih.gov

7.4 Emissions Inventories and Data Systems

The Lahore emissions inventory (2023) is a foundational move toward routine source quantification. Scaling this approach to other districts and linking it to compliance and permitting systems will enable evidence-based targeting and cost-effective abatement sequencing (Gov. of Punjab, 2021).

8. CASE STUDY: LAHORE'S WINTER SMOG EPISODES (2023–2024)

- November 2023: Punjab declared a health emergency; Lahore endured multi-day hazardous AQI, prompting school closures and temporary restrictions. Residents reported respiratory symptoms, and visibility fell, disturbing transport.
- November 2024: Authorities again closed schools and offices and mandated masks across Lahore division and other districts. These episodic actions reduced exposure acutely but highlight the need for upstream, year-round controls on sources.

9. STRATEGIES: A PRIORITIZED, EVIDENCE-BASED ABATEMENT PORTFOLIO

Given the sector shares (Section 3) and Pakistan's institutions, a least-cost, high-impact portfolio would include:

1. Residential Energy Transition (23% share in Punjab)

- Scale LPG/electric/biogas access and cleaner cookstoves; target poor households with subsidies/credit.
- Pair with behavioral and ventilation interventions to reduce indoor and neighborhood exposure.

2. Transport (16%)

- Implement periodic vehicle inspection & maintenance (I/M), anti-smoke enforcement, and low-sulfur Euro-5 compliance.
- Bus fleet renewal (CNG/electric hybrids), BRT expansion, and last-mile electrification for 2/3-wheelers.
- Dust management (paving shoulders, mechanical sweeping) on high-ADT corridors.

3. Industry and Kilns (16%)

- Complete zig-zag conversion, require stack emission limits, and continuous emissions monitoring for major stacks; incentivize natural gas or electric firing where feasible.

4. Agriculture (13%)

- Enforce no-burn rules with in-situ residue management, straw markets, and machine subsidies (e.g., Happy Seeder-type implements).
- Promote value chains for straw (briquettes, packaging fiber) to create farmer incentives.

5. Waste (7%)

- Ban routine open burning; accelerate door-to-dump collection coverage, landfill gas capture, and material recovery.
- Use remote sensing + hotlines for rapid enforcement.

6. Airshed Coordination

- Establish cross-border alert and enforcement compacts with Indian Punjab/Haryana on residue burning windows, leveraging satellite detection and harmonized penalties/subsidies (Gov. of Punjab, 2021).

10. GOVERNANCE, FINANCE, AND IMPLEMENTATION CAPACITY

- Institutions: Clarify mandates among the federal Ministry of Climate Change, provincial EPAs, transport authorities, and municipalities to avoid gaps and overlaps; integrate clean-air targets into transport, energy, agriculture, and industrial plans.

- Finance: Between 2015–2022, only ~0.7% of international development funding targeted ambient air pollution in South Asia. Pakistan can expand results-based financing, green bonds, and polluter-pays instruments to scale proven measures.
- Data: Expand reference-grade monitors and QA/QC protocols, link permits to emissions data, and publish near-real-time dashboards for accountability.

11. SOCIAL POLICY LINKAGES: PROTECTING PEOPLE WHILE CLEANING THE AIR

- Health system readiness: Forecast smog weeks, stock inhalers & steroids, and set triage protocols; push SMS alerts for vulnerable groups.
- Education continuity: Pre-authorize hybrid classes and smog-day packets to keep learning on track when schools close, minimizing inequities between private and public pupils.
- Just transition: Pair kiln modernization and vehicle scrappage with worker re-training and micro-credit to avoid job losses in informal sectors.

12. OUTLOOK: WHAT SUCCESS WOULD LOOK LIKE BY 2030

If Pakistan implements the portfolio above and coordinates regionally on the airshed, several outcomes are plausible by 2030:

- Double-digit percentage reductions in annual $PM_{2.5}$ in Punjab's cities, with winter peaks moderated by residue-burning controls.
- Fewer emergency closures, steadier commerce, and measurable productivity gains.
- Life-expectancy improvements (closing the AQLI gap) and lower cardiopulmonary admissions during smog season.
- Institutionalized monitoring and compliance embedded across energy, transport, agriculture, and industry.

REFERENCES

Ali, M., Siddique, I., & Abbas, S. (2022). Characterizing air pollution and its association with emission sources in Lahore: A guide to adaptation action plans to control pollution and smog. *Applied Sciences (Basel, Switzerland)*, *12*(10), 5102. DOI: 10.3390/app12105102

Allen, J. C., & Barnes, F. D. (1985). The causes of deforestation in developing countries. *Annals of the Association of American Geographers*, *75*(2), 163–184. DOI: 10.1111/j.1467-8306.1985.tb00079.x

American Psychological Association. (2020). The importance of social connection. https://www.apa.org/news/press/releases/2020/05/importance-social-connection

Anjum, M. S., Ali, S. M., Subhani, M. A., Anwar, M. N., Nizami, A.-S., Ashraf, U., & Khokhar, M. F. (2021). An emerged challenge of air pollution and ever-increasing particulate matter in Pakistan; a critical review. *Journal of Hazardous Materials*, *402*, 123943. DOI: 10.1016/j.jhazmat.2020.123943 PMID: 33254830

Ashraf, F., (2019). Smog analysis and its effect on reported ocular surface diseases: A case study of 2016 smog event of Lahore, Atmos. Environ.

Ashraf, I., Ahmad, R. U., & Tareen, H. K. (2022). Worsening situation of smog in Pakistan: A tale of three cities. *Annals of Medicine and Surgery (London)*, *79*. Advance online publication. DOI: 10.1016/j.amsu.2022.103947 PMID: 35860153

Aslam, A. 2022. Impact of the built environment on climate change risk perception and psychological distancing: empirical evidence from Islamabad, Pakistan. Environ. Sci. Pol.

Awez, J., Khan, T., & Ali, M. (2024). Assessing air pollution crisis in South Asia: The health, economic, and environmental implications through case studies of Lahore, Delhi, and Dhaka. Pakistan Languages and Humanities Review, 8(2), 224–235.

Azhar, A.. (2019). The economic burden of outdoor air pollution-related morbidity and mortality in the cities of Lahore and Karachi, Pakistan. *Journal of Environmental Management*, *251*, 109557.

FAO. 2010. Global forest resources assessment- (2010). FAO forestry paper No. 163. UN Food and Agriculture Organization, Rome. pp 85-146. https://www.fao.org/3/a-i1757e.pdf

GFW. 2024. Global Forest Watch https://www.globalforestwatch.org/dashboards/country/PAK/?category=forest-change&location=WyJjb3VudHJ5IiwiUEFLIl0%3D

Government of Pakistan. (2021). Pakistan Clean Air Action Plan 2021. Ministry of Climate Change. https://www.mocc.gov.pk

Government of Punjab. (2021). *GOVERNMENT OF THE PUNJAB, ENVIRONMENT PROTECTION DEPARTMENT POLICY NUMBER – No. SOG/EPD/7-26/2013(P-II)*. Government of Punjab.

Iqbal, M., Khan, A., Rauf, A., & Ahmad, Z. (2023). Air pollution and its socio-economic impacts in South Asia: Challenges and strategies. *Journal of Environmental Research and Public Health*, *20*(4), 215–228. DOI: 10.3390/ijerph2004215 PMID: 36833854

Iram, S., Qaisar, I., Shabbir, R., Pomee, M. S., Schmidt, M., & Hertig, E. (2025). Impact of air pollution and smog on human health in Pakistan: A systematic review. *Environments (Basel, Switzerland)*, *12*(2), 46. DOI: 10.3390/environments12020046

Khattak, M., & Khan, S. (2024). Smog and Health in Punjab: Analyzing Awareness, Protective Behaviors, and Public Support for Air Quality Policies. *Annals of Human and Social Sciences*, (05), 293–305.

Reuters. (2024). IQAir World Air Quality Report 2024 (Pakistan 73.7 μg/m^3 in 2023).

Sarfraz. Z. (2020). The Social and Economic Burden of Smog in Pakistan, Editorial;, Pakistan Journal of Surgery & Medicine.

United Nations Environment Programme. (2011). UNEP billion tree campaign reaches 12 billionmilestone.https://www.unep.org/newscentre/Default.aspx?DocumentID=2659&ArticleID=8930

United Nations Environment Programme. (2016). Gender and environment: Air pollution impacts on women. https://www.unep.org

Urban Unit. (2024). https://urbanunit.gov.pk/

WHO. (2021). Global Air Quality Guidelines. iris.who.int

WHO. (2023). https://www.who.int/emergencies/situation-report

World Bank. (2019). The cost of air pollution: Strengthening the economic case for action. https://www.worldbank.org/en/topic/environment/publication/the-cost-of-air-pollution

World Bank. (2019). The cost of air pollution: Strengthening the economic case for action. https://www.worldbank.org/en/topic/environment/publication/the-cost-of-air-pollution

World Bank. (2023). Air Pollution and Public Health in South Asia / airshed analysis._lib.icimod.org

World Bank, (2025). Pakistan: Country Climate and Development Report (air pollution ≈ 6.5% of GDP).

World Health Organization. (2018). Air pollution and child health: Prescribing clean air. https://www.who.int/publications/i/item/air-pollution-and-child-health

Zulfiqar, A., Ghaffar, M. M., Shahzad, M., Weis, C., Malik, M. I., Shafait, F., & Wehn, N. (2021). AI-Forest Watch: Semantic segmentation based end-to-end framework for forest estimation and change detection using multi-spectral remote sensing imagery. *Journal of Applied Remote Sensing*, *15*(2), 024518. DOI: 10.1117/1.JRS.15.024518

Chapter 6

Economic Dynamics of Air Pollution in Türkiye and Pakistan: An Empirical Assessment of the Environmental Kuznets Curve and Pollution-Led Growth

Asli Tasbasi
https://orcid.org/0000-0002-9640-8582
Işık University, Turkey

Maham Akhtar
https://orcid.org/0009-0009-3413-1010
Işık University, Turkey

ABSTRACT

Türkiye and Pakistan, despite differing levels of economic development, face similar macroeconomic challenges such as income inequality, inflation and debt. Both countries also experience environmental pressures from industrialization and rapid urbanization, with air pollution emerging as a critical concern affecting economic productivity and sustainable development. This study conducts a comparative analysis of air pollution in Türkiye and Pakistan from 1980 to 2023, using the Autoregressive Distributed Lag (ARDL) bounds testing approach to examine the short and long run relationships between air pollution, urbanization, industrialization, energy consumption and macroeconomic policies. The analysis tests the Environmental Kuznets Curve (EKC) for Türkiye and the pollution-led growth hypothesis for

DOI: 10.4018/979-8-3373-2469-2.ch006

Pakistan. Findings reject the EKC for Türkiye but confirm pollution-led growth in Pakistan, offering insights for effective environmental regulation and sustainable development strategies.

1. INTRODUCTION

Air pollution is one of the most important environmental problems today, due to the prevalence and depth of its economic and social consequences, particularly for emerging and developing economies. These countries, during their economic growth and industrialization processes, are facing severe levels of air pollution due to their reliance on fossil fuel consumption, energy-intensive production methods and relatively weak environmental regulations. As a result, they are struggling to mitigate the adverse effects of worsening air quality.

In the context of emerging economies, rapid urbanization and industry-driven growth are among the primary causes of air pollution. In these countries, high levels of air pollution increase socio-economic costs, place an additional burden on health systems and in the long run, threaten sustainable development. Due to income inequality, low-income households are often concentrated in areas closer to polluting industries, exposing them to greater environmental risks.

On the other hand, developing economies face serious constraints in implementing effective policies and practices to reduce air pollution, due to technological and financial limitations and institutional weaknesses. Compliance with international environmental norms and standards often imposes high adaptation costs on these economies, leading to conflicts between macroeconomic priorities and environmental sustainability objectives (UNDP, 2024). Moreover, the health problems caused by air pollution reduce labor productivity, limiting these countries' growth potential and negatively impact on agricultural production and the efficiency of natural resource use (FAO, 2023). In developing economies too, low-income households bear disproportionately the burden of economic and social problems caused by air pollution.

In this context, addressing air pollution requires not only environmental policies but also the redesign of economic development strategies. As is the case globally, for emerging and developing economies, green growth strategies, investments in clean energy and a transition to low-carbon production technologies should be embraced as key components of sustainable development. However, realizing such a transformation necessitates multi-dimensional cooperation, including technology transfer, international financing mechanisms and capacity-building programs (IPCC, 2023).

Mainstream perspectives on global carbon emissions and air pollution often overlook historical responsibilities, framing today's emerging and developing economies as the primary contributors to the climate crisis and environmental degradation.

However, the developed economies, which industrialized early and are historically responsible for the majority of greenhouse gas accumulation in the atmosphere, continue to rank among the highest in per capita emissions and consumption levels (IPCC, 2023). Despite this, a global approach that fails to adequately account for the historical emission debts of advanced economies continues to dominate the design of environmental policies and the allocation of financial responsibilities. As a result, emerging and developing economies face an unfair financial and regulatory burden, further complicating the already difficult task of balancing development priorities with environmental sustainability goals (Puaschunder, 2020).

This study adopts an approach grounded in the principle of Common but Differentiated Responsibilities (CBDR), which emphasizes a historically contextualized, fair and equitable distribution of environmental responsibilities. This principle is formalized in Rio Earth Summit of United Nations in 1992. It argues that, in sharing obligations to address environmental problems all economies are equal; yet, in consideration of the historical contributions of developed economies as major polluters, the disproportionate financial and regulatory burden on emerging and developing economies should be alleviated. In this regard, Türkiye and Pakistan provide a suitable basis for a comparative analysis aimed at understanding the impacts of industrialization and urbanization processes on air pollution in two economies at different stages of development.

Türkiye, classified as an emerging economy, has experienced significant increases in energy demand and fossil fuel consumption in recent decades, driven by accelerated industrialization and urbanization processes. Consequently, substantial deterioration in air quality has been observed. In major metropolitan areas such as Istanbul and Ankara, levels of particulate matter (PM2.5) and nitrogen dioxide (NO_2) frequently exceed the World Health Organization (WHO) limits. Emissions concentrated in industrial zones pose serious risks to environmental and public health.

Pakistan, on the other hand, as a developing economy, is confronted with chronic air pollution problems due to rapid and often unregulated urbanization, coupled with its dependence on coal and inefficient biofuel use for energy production. In major cities such as Lahore and Karachi, deteriorating air quality is further exacerbated by practices in the agriculture and transportation sectors.

Analyzing the air pollution dynamics of these two countries considering their industrialization and urbanization indicators offers valuable insights for developing policy recommendations on how growing economies can balance their development strategies with environmental sustainability goals. This study aims to fill a gap in the literature by contributing to an understanding of how fair transition approaches based on historical responsibility can support low-carbon development models in countries such as Türkiye and Pakistan.

In the study, a two-dimensional analytical framework has been adopted using data from the years 1980-2023. While Türkiye is evaluated within the scope of the Environmental Kuznets Curve (EKC) hypothesis, the analysis questions whether Pakistan exhibits a pollution-led growth (PLG) pattern. These theoretical frameworks align with the respective levels of economic development of both countries. The empirical findings from the related analyses do not confirm the Environmental Kuznets Curve hypothesis for Türkiye while they support the pollution led growth in Pakistan.

The rest of the study is structured as follows. The second section provides a review of the relevant literature. The third section presents a general assessment of the economic performance of Türkiye and Pakistan by analyzing key macroeconomic indicators and sets the nexus between air pollution and economic dynamics in the two countries. The fourth section introduces the data and methodology used in the study, accompanied by a discussion of the analysis results. The fifth section outlines the main findings and the existing environmental policy frameworks in both countries. The sixth section includes the conclusion and policy implications.

2. LITERATURE REVIEW

The relationship between economic growth and environmental degradation has been widely explored and a substantial body of empirical literature investigates this nexus across various nations; however, there remains a void in the related literature in terms of addressing the comparative economic dynamics of emerging and developing economies. Findings derived from studies on advanced economies may not fully capture the environmental-economic interactions prevalent in these economies since emerging and developing economies often exhibit unique structural characteristics such as rapid industrialization, demographic pressures and institutional constraints that distinguish them from their developed counterparts.

Kuznets Curve was developed by Simon Kuznets in 1955 to illustrate the relationship between per capita income and income inequality as an inverted-U pattern. According to Kuznet's theory, as per capita income rises, income inequality initially increases but eventually starts to decrease after reaching a certain threshold. Essentially, in the early stages of economic growth, income distribution tends to become more unequal, but as development progresses, income tends to be shared more evenly (Kuznets, 1955). Decades later, researchers began to apply a similar framework to environmental issues and empirically proved that environmental degradation and per capita income often exhibit the same inverted-U relationship. This gave rise to what is now referred to as the Environmental Kuznets Curve (EKC), first introduced by Panayotou (1993), describing the connection between economic growth and environmental quality (Dinda, 2004).

Development of the EKC model shifted the discussions on the environment's limited capacity to absorb pollution and waste toward the idea that increased economic growth might be the key to addressing environmental challenges. The model has been constructed around the pattern that developing economies often endure more acute pollution and environmental harm during their growth phases, due to technological lags and limited innovation capacity compared to developed nations. This leads to an almost automatic acceptance of substantial environmental degradation as an inevitable trade-off, in the pursuit of rapid economic growth at any cost. As a result, this approach has contributed to escalating levels of urban and industrial waste, deteriorating air, land and water quality, loss of biodiversity and the intensification of the climate crisis. The idea is still controversial, with empirical evidence proving contradictory cases.

Numerous empirical studies have investigated the EKC hypothesis within the context of Türkiye too, which can be grouped according to their methodological approach, the time period analyzed and their specific focus areas. Focusing on the studies from the last two and a half decades, early wave of research mainly examined the direct relationship between economic growth and CO_2 emissions using time-series econometric methods. Başar and Temurlenk (2007) tested the EKC hypothesis for Türkiye over the period 1968–2003. Using cointegration and causality analyses, they found evidence of an inverted U-shaped relationship between per capita income and CO_2 emissions, supporting the traditional EKC hypothesis.

Two years later, Akbostancı, Türüt-Aşık and Tunç (2009) conducted both time-series and panel data analyses to explore the EKC for CO_2 emissions at the national level (1968–2003) and for sulfur dioxide (SO_2) at the province level (1992–2001). Their results indicated a monotonically increasing relationship at the national level, contradicting the EKC hypothesis, while the province-level analysis showed some support for an inverted U-shape for SO_2 emissions.

A second group of studies expanded the EKC analysis by including energy consumption, renewable energy and sector-specific variables. Bölük and Mert (2015) examined the effects of renewable and non-renewable energy consumption on CO_2 emissions in Türkiye over the period 1971–2010. The authors found that renewable energy consumption helps reduce emissions, supporting the EKC hypothesis, while non-renewable energy worsens environmental quality.

Among research with advanced econometric techniques, Çetin, Ecevit and Yücel (2018) tested the EKC hypothesis over 1960–2013 by incorporating structural breaks into their analysis. Using the Zivot-Andrews and Lumsdaine-Papell unit root tests along with the Gregory-Hansen cointegration test, they confirmed the EKC hypothesis, identifying critical points where policy interventions might have altered the growth-emissions trajectory.

Özataç, Gökmenoğlu and Taşpınar (2017) explored the dynamic relationship among energy consumption, CO_2 emissions and economic growth in Türkiye for the period 1960–2013. Utilizing the ARDL approach with structural break considerations, they found long-run cointegration, supporting the EKC hypothesis, but also stressed the importance of renewable energy use.

Karahasan and Pınar (2022) adopted a regional analysis, focusing on subnational disparities in environmental degradation and economic growth. They employed spatial econometric techniques to show that regional income levels are linked to environmental outcomes in ways consistent with the EKC, but with significant spatial heterogeneity.

A final group of studies approached the EKC debate from a broader sustainability perspective, incorporating multi-dimensional indicators and governance considerations. Tirgil, Acar and Özgür (2021) explored the interactions among ecological footprint, human capital, energy consumption and economic growth from 1971 to 2017. Using the autoregressive distributed lag (ARDL) method, they found no evidence of the EKC hypothesis when broader sustainability indicators were used, suggesting the need to rethink the growth-environment nexus in Türkiye.

Acaroğlu, Kartal and García Márquez (2023) presented one of the most recent contributions, examining the impact of innovation and technological progress on environmental degradation in Türkiye. Their findings indicated that innovation-driven growth can decouple economic development from environmental harm, offering conditional support for a revised EKC framework when green technologies are effectively deployed.

Looking at the pioneering work of the last two and a half decades, while numerous studies examine the validity of the EKC hypothesis for Türkiye, they have not discussed the unique conditions stemming from its status as an emerging economy. By focusing on Türkiye, this chapter aims to provide a more tailored analytical approach for emerging economies, evaluating factors such as energy dependence on fossil fuels, urbanization trends and policy implementation capacities that are critical to emerging economy context.

Contrary to existing research that has primarily focused on the direct relationship between economic growth and CO_2 emissions, this study offers a contribution by focusing on Türkiye's industrialization and rapid urbanization as critical explanatory variables in testing the EKC hypothesis. By incorporating indicators of industrial output and urban population growth, the study seeks to provide a more nuanced understanding of the environmental impacts of development in an emerging market context. This approach acknowledges that the economic and environmental dynamics of emerging economies like Türkiye differ substantially from those of developed nations, where most EKC models have traditionally been applied.

Similarly, the pollution-led growth or "grow now, clean later" perspective, which suggests that developing nations prioritize economic expansion at the expense of environmental quality, has yet to be comprehensively tested in the context of Pakistan. This gap is significant for several reasons. Pakistan, as the world's fifth most populous nation, partner of the strategic China Pakistan Economic Corridor(CPEC) and heavily dependent on fossil fuel energy, plays an increasingly pivotal role in the global economy. Its rapid urbanization, industrial expansion and energy consumption patterns contribute to severe air quality challenges. Yet, few studies have systematically examined how Pakistan's growth trajectory intertwines with environmental degradation, specifically air pollution. Understanding this dynamic is crucial, not only for informing Pakistan's environmental policy but also for offering insights applicable to other densely populated, developing nations facing similar trade-offs between economic development and environmental sustainability.

By focusing on Pakistan, this research aims to contribute to the literature by filling the empirical and analytical gaps regarding the economic drivers of air pollution in developing economies. Such insights can support more nuanced policy design, both domestically and in comparable contexts globally.

Moreover, unlike previous studies, which typically focus on a single-country analysis, this research adopts a comparative perspective by juxtaposing Türkiye's results with those from Pakistan—a developing economy experiencing similar economic and environmental pressures.

Furthermore, the study's methodological framework, which integrates industrialization and urbanization as central variables, extends the scope of EKC analysis beyond conventional economic indicators, offering more policy-relevant insights for managing growth-environment trade-offs in emerging economies.

3. MACROECONOMIC AND AIR POLLUTION CONTEXTS OF TÜRKIYE AND PAKISTAN

As of 2023, the Turkish economy has reached approximately 1.1 trillion USD in terms of nominal GDP, while the Pakistani economy is approximately 375 billion USD as shown in Figure 1 (The World Bank, World Development Indicators 2024). While per capita income is approximately 10,600 USD in Türkiye, this figure is 1,600 USD in Pakistan. In Türkiye, the services and industry sectors constitute 58.8% and 27% of GDP, respectively, while in Pakistan, the share of agriculture is 23%, the share of industry is 24% and the share of services is 53% (World Bank, 2024). These indicators reveal that Türkiye is a more industrialized and integrated economy with foreign trade, while Pakistan is more dependent on agriculture and has a lower income level. However, public debt sustainability and high inflation stand out as common and chronic macroeconomic problems in both countries. Pakistan achieved an average annual growth rate of around 6.5% in the 1980s with privatization, deregulation and openness policies. However; in the 1990s, increasing external debt burden, weak institutional structure and international sanctions limited growth performance, and the public debt/GDP ratio reached 102% as of 1999 (The World Bank, 2022). Although short-term growth momentum was seen in the 2000s, structural fragilities, disruptions in energy supply and political uncertainties, especially in the post-2008 period, negatively affected economic stability.

Figure 1. GDP in Türkiye and Pakistan (1980-2023)

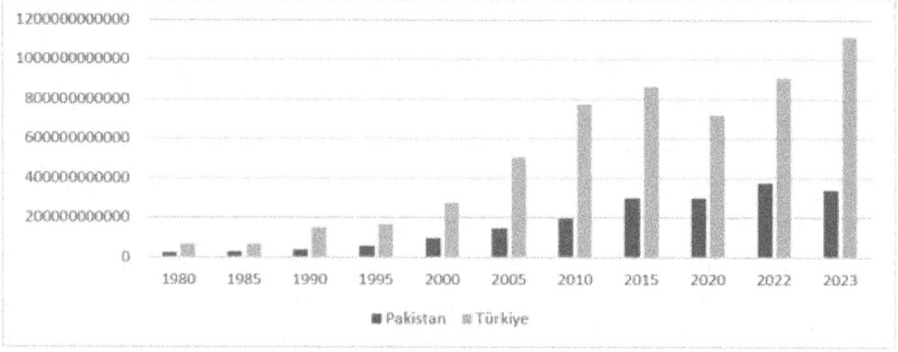

Source: The World Bank, World Development Indicators 2024

In the 1980s, Türkiye underwent a structural transformation in its economic structure by switching from an import-substitution industrialization strategy to an export-based growth model. During this period, significant increases were observed in foreign trade volume and industrial production. In the 1990s, high public deficits, political instability and fiscal indiscipline led to serious macroeconomic imbalances in the economy. The annual inflation rate based on the consumer price index ranged between 60% and 80% during this period and exceeded 100% in the 1994 crisis (TÜİK, 2025). The ratio of public debt to GDP also increased in a similar manner (TCMB, 2025). In the post-2001 period, thanks to structural reforms and fiscal consolidation policies implemented within the framework of IMF-supported programs, inflation declined and growth accelerated again. However, the heterodox monetary policies and low-interest rate strategy preferred in the 2010s increased inflationary pressures and led to exchange rate instability.

Figure 2. Inflation Rates (%) in Türkiye and Pakistan (1980-2023)

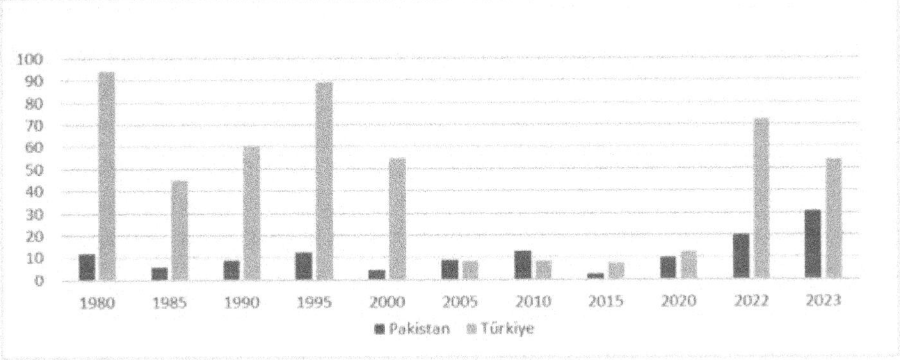

Source: The World Bank, World Development Indicators 2024

According to OECD (2023) data, annual consumer inflation in Türkiye was 72.3% at the end of 2022 and 53.8% as of 2023 (see Figure 2). According to the same institution's data, the public debt/GDP ratio, which was 75.5% in 2001, decreased to 25.4% as of 2023 (see Figure 3).

Figure 3. Central government debt, total (% of GDP) in Türkiye and Pakistan (1980-2023)

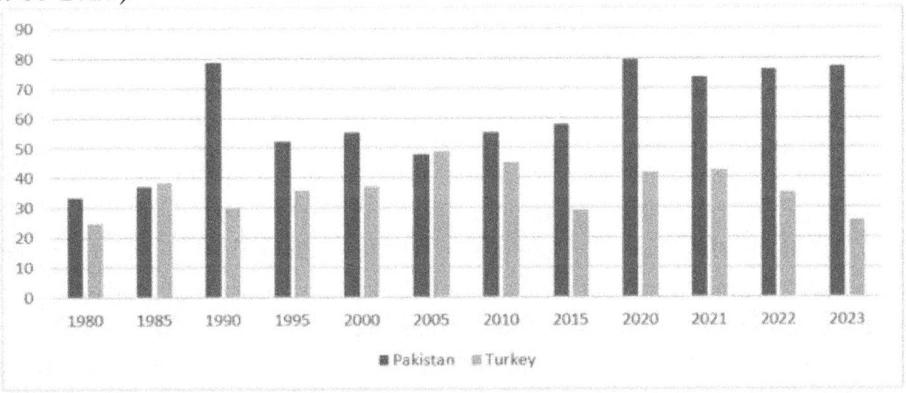

Source: The World Bank, World Development Indicators 2024

Türkiye is a country that meets more than 70% of its energy demand through imports. As of 2023, approximately one-third of its total electricity production will come from coal. According to a 2024 report by Reuters based on Ember data, Türkiye rose to first place in coal-fired electricity production in Europe with 36 TWh of coal-fired electricity production in the first four months of 2024 (Reuters, 2024). Although renewable energy investments have accelerated in recent years, the share of fossil fuels in energy production continues to remain high. This poses a risk in terms of environmental sustainability as well as energy supply security. Similarly, energy policies in Pakistan are strongly dependent on fossil fuels. As of 2023, approximately 59% of the country's electricity production will come from fossil fuels. Hydroelectricity accounts for 24%, nuclear energy for 8.8% and renewable sources (including solar and wind) for only 2.7% (Ember, 2023). As part of the energy infrastructure projects carried out with China (especially within the scope of CPEC), investments in coal-fired power plants have increased, which has brought about externalities such as air pollution and carbon emissions.

Figure 4. Fossil Fuel Consumption, % of total in Türkiye and Pakistan (1980-2015)

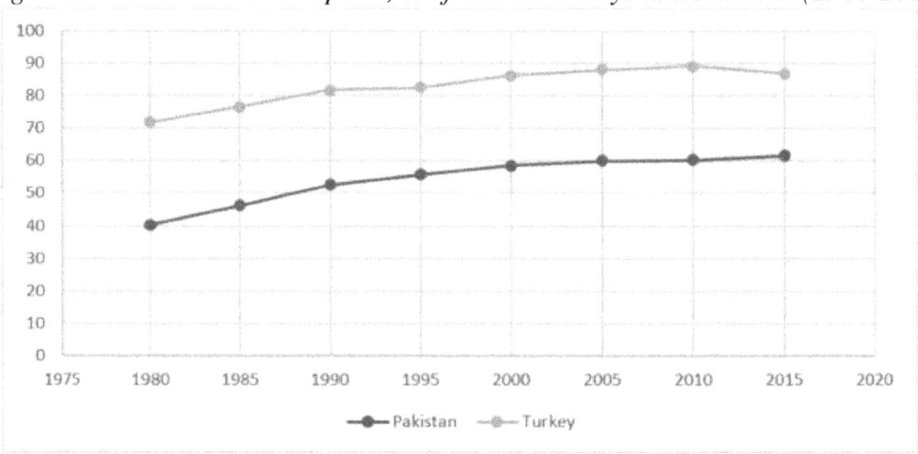

Source: The World Bank, World Development Indicators 2024

In Türkiye, the urbanization rate, which was 44% in the 1980s, has reached 77% as of 2023. During this process, industrialization and urbanization have progressed in parallel and the increasing number of vehicles and intensive industrial activities, especially in metropolitan cities, have been the main causes of air pollution. Coal-fired power plants in Türkiye cause approximately 1 ton of CO_2 emissions per capita per year (IEA, 2023). It is known that particulate matter (PM10 and PM2.5) concentrations in cities such as Istanbul, Ankara and Kocaeli exceed WHO threshold values.

Table 1. Key Air Quality Indicators for Türkiye and Pakistan (2024)

Indicator	Türkiye	Pakistan
Country Ranking	Not in top 10 most polluted	3rd most polluted
National Average PM2 (WHO Guideline= 5 μg/m³)	26 ug/m³	73.7 ug/m³
Air Quality Classification	Moderate Pollution	Unhealthy
Times above WHO Guidelines	5.2x	14.7x

Source: WHO Global Air Quality Database, IQAir 2024, national statistics

In Pakistan, the urbanization rate is 39% as of 2023 (Pakistan Bureau of Statistics, 2023). However, rapid and irregular urbanization, combined with infrastructure deficiencies, further aggravates environmental degradation. Industrialization processes have progressed unplanned and energy-intensive industrial clusters in large cities (e.g. Karachi, Lahore, Faisalabad) have seriously affected air quality. According to the World Air Quality Report 2023, Lahore ranks among the most polluted cities in the world (IQAir, 2024). As seen in the examples of Türkiye and Pakistan, economic growth and the increase in energy demand are largely met by fossil fuels, which negatively affects environmental quality. Both countries face similar challenges in balancing energy security with environmental sustainability. In order for economic development goals to be sustainable in the long term, the energy basket should be shifted to non-carbon intensive sources, environmentally friendly policies should be adopted in urban planning and environmental externalities should be internalized.

Table 2. Primary Sectoral Contributors to Air Pollution in Türkiye and Pakistan

Sector	Türkiye	Pakistan
Transportation	25 million vehicles, 50% >10 years old, high diesel use	Aging fleet, rickshaws with 2-stroke engines, limited Euro 5 fuel
Energy & Industry	Coal-fired power (36 TWh, 2024); cement, steel, and chemical sectors dominant	Brick kilns, cement, fertilizer, and textile industries emit high PM2.5
Agriculture	Limited field burning	69 million tons of crop residue burned annually
Waste	Limited data	20% of emissions from open burning of plastics
Residential Heating	High coal use in low-income areas	Biomass and wood burning in rural areas

4. DATA, METHODOLOGY AND FINDINGS

This study employs a comparative perspective based on empirical evidence to examine the economic dynamics of air pollution in Türkiye and Pakistan. In line with the development stages of the two countries, the Environmental Kuznets Curve (EKC) hypothesis is tested for Türkiye, whereas the Pollution-Led Growth (PLG) hypothesis is explored for Pakistan. The data set covers the period from 1980 to 2023, and all variables are compiled from the World Bank Development Indicators database except the coal consumption. The coal consumption data are collected from the International Energy Agency (IEA). All data, variable definitions, and descriptive statistics are consolidated in Table 3 below.

Table 3. Data Definitions and Sources

Variable	Definition	Unit	Source
Emissions	CO_2 from fossil fuels and cement production	Metric tons per capita	WDI (2024)
GDP per capita	Real GDP per capita	Constant USD	WDI (2024)
Industrialization	Industry value-added (% of GDP)	Percent	WDI (2024)
Urbanization	Urban population (% of total)	Percent	WDI (2024)
Coal Consumption (Pakistan)	Coal use (million short tons)	Million tons	IEA (2024)

Industrialization and urbanization are central to the analysis because they capture the structural and demographic factors driving emissions in rapidly transforming economies. Industrial value-added measures production intensity and technology level, while urbanization reflects population-driven energy and transport pressures—both crucial for EKC and PLG testing.

4.1. The EKC Analysis for Türkiye

The EKC hypothesis posits an inverted-*U* relationship between income and environmental degradation—i.e., as the economy grows, pollution (or emissions) initially increases, peaks and then starts to decline beyond a certain income level. Accordingly, in the model specification, GDP per capita should have a positive coefficient while its square is expected to be negative. In the emerging economies context, we adopt the following model in order to include industrialization and urbanization effects:

$$\varepsilon_{EKC}=\beta_0+\beta_1(\text{GDP per capita})+\beta_2(\text{GDP per capita})^2+\beta_3(\text{industrialization})+\beta_4(\text{urbanization})+\varepsilon$$

where ε_{EKC} stands for emissions in the EKC context.

To test the proposed relationships, the Autoregressive Distributed Lag (ARDL) bounds testing approach is applied. The ARDL method is particularly suitable for small sample sizes and for datasets with a mix of stationary I(0) and first-difference stationary I(1) variables. It also allows for simultaneous estimation of short-run and long-run dynamics.

For stationarity testing, The Augmented Dickey-Fuller (ADF) test is used to assess the order of integration of all variables. According to the results, urbanization is stationary at level (I(0)), while all other variables are stationary at first difference (I(1)).

The optimal lag structure is determined using the Akaike Information Criterion (AIC). The Pesaran *et al.* (2001) bounds testing procedure is applied to examine the existence of a long-run equilibrium relationship. If the calculated F-statistic exceeds the upper bound of critical values, cointegration is confirmed. The Error Correction Term (ECT) is used to verify the speed of adjustment toward equilibrium.

Descriptive statistics for the Turkish dataset reveal notable variation in emissions and economic indicators. Summary statistics and estimation results are presented below (see Tables 4, 5 and 6 respectively).

Table 4. Descriptive Statistics

Statistical Values	Mean	Maximum	Minimum	Std. Dev.	Observations
Emissions	.2902093	.489	.17	.0811916	43
GDP per Capita	5825.351	12578.19	1209.051	3984.074	43
GDP Per Capita²	4.94e+07	1.58e+08	1461806	5.23e+07	43
Industrialization	28.22899	32.97471	23.49444	2.681022	43
Urbanization	64.36149	77.022	43.78	9.059644	43

According to the empirical results from the ARDL bounds test, the Environmental Kuznets Curve hypothesis for Türkiye is not confirmed. Despite a statistically significant long-run relationship between air pollution and the selected economic indicators, the signs and statistical significance of GDP and its square contradict the expected inverted-U pattern and thus do not align with the theoretical expectations of the EKC. These results suggest that economic growth has not yet reached the turning point expected by the EKC framework or that other structural factors, such as energy mix, policy interventions or shifts in sectoral composition, may obscure the traditional pattern.

Table 5. ARDL Long Term Results

Variables	Coefficients	Std. Errors	t-Statistics	P>|t|
Emissions	-1.171932	0.2258049	-5.19	0
GDP per capita	-0.000021	.0000107	-1.96	0.061
(GDP per capita)^2	1.01e-09	6.28e-10	1.60	0.122
Industrialization	-.0040839	.001844	-2.21	0.037
Urbanization	-.0051298	.0017705	-2.90	0.008

Short run dynamics exhibit limited statistical significance and the error correction term indicates a strong adjustment toward equilibrium. Overall, the negative effects of industrialization and urbanization on emissions in the long run indicate that policy-driven structural changes may have contributed more to environmental improvements than income growth alone.

Table 6. Short term results

Variables	Coefficients	Std. Errors	T-stats	P>|t|
Emissions	-.3459776	.1855976	-1.86	0.075
GDP per capita	-7.42e-06	.0000124	-0.60	0.554
(GDP per capita)^2	-6.06e-11	7.64e-10	-0.08	0.937
Industrialization	.0003814	.0029775	0.13	0.899
Urbanization	-.0138566	.0385194	-0.36	0.722
CointEq (-1) *	.925617	.1889568	4.90	0.000

4.2. Pollution-Led Growth Analysis for Pakistan

The pollution led growth hypothesis posits pollution to be a driver of economic growth, especially in developing economies dependent on resource intensive industrialization. Testing the pollution-led growth hypothesis using an ARDL bounds approach for Pakistan over the period 1980–2023, our model includes total CO_2 emissions (in metric tons), GDP per capita, industrialization (including construction, % of GDP) and coal consumption (in million short tons):

$$\varepsilon_{PLG}=\beta_0+\beta_1(\text{GDP per capita})+\beta_2(\text{industrialization})+\beta_3(\text{coal consumption})+\varepsilon$$

where ε_{PLG} stands for emissions in the pollution led growth context.

According to the results for short-run dynamics; CO_2 emissions are positively and significantly associated with GDP, supporting the pollution-led growth hypothesis (see Table 7). GDP per capita has a negative and significant coefficient, implying that immediate increases in per capita income are associated with a reduction in emissions. This could reflect increased environmental awareness or the adoption of cleaner technologies during short-term growth episodes. Industry value-added shows a marginally significant positive effect, indicating its role in short-term growth. Coal consumption has a negative and highly significant short-run coefficient, which appears counterintuitive. Biomass substitution in the energy mix may be the explanation to this. The error correction term is large and significant at the 10% level, indicating strong convergence toward long-run equilibrium.

Table 7. Short term results

Variables	Coefficients	Std. Errors	T-stats	P>|t|
Emissions	1.05979	.2563596	4.13	0.000
GDP per capita	-.0165513	.0047589	-3.48	0.002
Industrialization	.5110135	.2684899	1.90	0.070
Coal Consumption	-.0018963	.000406	-4.67	0.000
CointEq (-1) *	-8.642899	4.647459	-1.86	0.076

As for the long-run results, CO_2 emissions continue to have a significant and positive impact on GDP in the long run, further validating the pollution-led growth hypothesis (see Table 8). GDP per capita shows an insignificant effect in the long run. Industrialization shows a negative long-run relationship with GDP, which may be due to low productivity sectors or negative externalities-i.e. environmental damage from the existing industrial base. Coal consumption becomes positive and significant in the long run, implying a structural dependency of economic growth on fossil fuels.

Table 8. ARDL Long Term Results

Variables	Coefficients	Std. Errors	t-Statistics	P>|t|
Emissions	.268735	.1041773	2.58	0.017
GDP per capita	.0039397	.0089895	0.44	0.665
Industrialization	-1.87528	.927144	-2.02	0.055
Coal Consumption	.0046372	.0014305	3.24	0.004

The findings support the pollution led growth hypothesis, suggesting that Pakistan's economic growth trajectory is still heavily reliant on pollution-intensive activities. These results underline the importance of reevaluating energy policy in the country, focusing on a green growth trajectory while acknowledging the current growth dependence on fossil fuels.

5. CONCLUDING REMARKS

The rejection of the EKC in Türkiye reflects persistent structural dependence on fossil fuels, limited green technology diffusion, and inconsistent enforcement of environmental standards. Despite long-run cointegration, the economy's reliance on imported coal and energy-intensive industries prevents a decoupling of growth from emissions. These findings highlight the need for proactive environmental policy, particularly in promoting cleaner production technologies and enhancing regulatory enforcement.

The validation of the PLG hypothesis highlights Pakistan's structural reliance on pollution-intensive sectors. The negative short-run coal coefficient may reflect temporary substitution toward biomass and hydropower, while the positive long-run coefficient indicates entrenched fossil fuel dependence. Institutional weakness and fragmented environmental governance exacerbate this trend.

Türkiye's environmental governance is anchored in the National Air Quality Strategy and Action Plan (2018–2028), Zero Waste Initiative (2017), and the Climate Law Draft (2023), implemented by the Ministry of Environment, Urbanization and Climate Change. Pakistan operates under the Clean Air Action Plan (2023), National Climate Change Policy (2021), and Punjab Smog Policy (2017), coordinated by the Ministry of Climate Change & Environmental Coordination. These frameworks aim to reduce pollution through monitoring, emissions control, and renewable energy adoption.

However, the effectiveness of these policy frameworks remains constrained by weak regulatory enforcement, institutional fragmentation, and limited fiscal capacity. In Türkiye, frequent policy shifts, overlapping jurisdictional mandates, and inadequate monitoring mechanisms have hindered the translation of strategic goals into measurable environmental outcomes. Similarly, in Pakistan, enforcement gaps at the provincial level, outdated industrial standards, and insufficient coordination between federal and local agencies dilute the impact of environmental legislation. Both countries face challenges in ensuring policy continuity, integrating environmental considerations into economic planning, and mobilizing private investment for green technologies. Strengthening institutional capacity, transparency, and compliance mechanisms is therefore essential to transform existing frameworks from aspirational plans into instruments of effective environmental governance.

6. CONCLUSION AND POLICY IMPLICATIONS

This study explored the economic dynamics of air pollution in Türkiye as an emerging economy and Pakistan as a developing economy through the Environmental Kuznets Curve and the Pollution-Led Growth hypotheses respectively. Using ARDL bounds testing, we provided empirical insights based on data from 1980 to 2023.

For Türkiye, the EKC hypothesis was not confirmed. Although a long-run relationship between emissions and economic variables is found, the results suggest that the country is yet to reach the income threshold required for emissions to decline. The implications are that economic growth alone may not yield environmental improvements without targeted policy interventions that prioritize sustainable practices and technological innovation.

In contrast, the analysis supported the Pollution-Led Growth hypothesis in Pakistan. According to the empirical findings, emissions have a statistically significant positive relationship with GDP in both the short and long run, indicating a reliance on environmentally harmful growth mechanisms. Coal consumption plays a dual role—detrimental in the short run but positively associated with GDP in the long run—highlighting the urgent need for cleaner alternatives.

Based on the empirical findings, it is clear that Türkiye should seek technological and structural transformation to decouple growth from emissions. Therefore, policy recommendations that force technological innovation in industry; may include targeted interventions such as a "technology-forcing" industrial policy that compels firms to integrate cleaner production processes and adopt energy-efficient technologies, expanding carbon taxes and emissions trading systems—aligned with the EU Carbon Border Adjustment Mechanism (CBAM) and channeling public and private capital toward renewable energy manufacturing, circular economy initiatives and sustainable urban infrastructure.

Pakistan, on the other hand, should focus on energy diversification strategies and just transition frameworks. These may include expanding renewable energy capacity beyond hydropower; prioritizing solar, wind, and biogas, establishing a just transition framework to manage the socio-economic impacts of phasing out coal and fossil fuel employment, and strengthening environmental governance to monitor emissions.

To sum up, for both Turkish and Pakistani cases, the results underscore the critical role of environmental governance, green investment and international cooperation. While Türkiye should strengthen its transition policies to break away from pollution-intensive growth, Pakistan must prioritize energy reform and equitable development to mitigate the socio-economic consequences of environmental degradation. Both economies must enhance regulatory frameworks, support technological upgrades and engage in global climate initiatives. Both economies require coherent, multi-level strategies integrating technological innovation, equitable energy transition, and international cooperation to achieve sustainable growth aligned with global climate commitments. Only through such multidimensional strategies can they achieve a fair and sustainable growth trajectory aligned with global climate commitments.

REFERENCES

Acaroğlu, E., Kartal, C., & Fausto, P. G. M. (2023, May). Testing the environmental Kuznets curve hypothesis in terms of ecological footprint and CO_2 emissions through energy diversification for Turkey. *Environmental Science and Pollution Research International*, *30*(22), 63289–63304. DOI: 10.1007/s11356-023-26278-w

Akbostancı, E., Türüt-Aşık, S., & Tunç, G. İ. (2009). The Relationship Between Income and Environment in Turkey: Is There an Environmental Kuznets Curve? *Energy Policy*, *37*(3), 861–867. DOI: 10.1016/j.enpol.2008.09.088

Başar, S., & Sinan Temurlenk, M. (2007). Çevreye Uyarlanmış Kuznets Eğrisi: Türkiye Üzerine Bir Uygulama. *Atatürk Üniversitesi İktisadi Ve İdari Bilimler Dergisi*, *21*(1), 1–12.

Bölük, G., & Mert, M. (2015). The Renewable Energy, Growth and Environmental Kuznets Curve in Turkey: An ARDL Approach. *Renewable & Sustainable Energy Reviews*, *52*(C), 587–595. DOI: 10.1016/j.rser.2015.07.138

Central Bank of the Republic of Turkey (TCMB). (2025). "Public Finance – Debt Stock (% of GDP)" table. Retrieved [02.04.2025] from https://www.tcmb.gov.tr/wps/wcm/connect/a26d8980-d468-4d79-beaf-b1b02c7fe4e3/6c11-1.pdf

Çetin, M., Eyyup, E., & Yucel, E. (2018). The impact of economic growth, energy consumption, trade openness, and financial development on carbon emissions: Empirical evidence from Turkey. *Environmental Science and Pollution Research International*, *25*(36), 36589–36603. DOI: 10.1007/s11356-018-3526-5

Dinda, S. (2004). Environmental Kuznets Curve Hypothesis: A Survey. *Ecological Economics*, *49*(4), 431–455. DOI: 10.1016/j.ecolecon.2004.02.011

FAO. 2023. The State of the Food and Agriculture: Revealing the True Cost of Food to Transform Agrifood Systems. Rome. https://doi.org/.DOI: 10.4060/cc7724en

IPCC. (2023). Summary for Policymakers. In *Climate Change 2023: Synthesis Report. Contribution of Working Groups I, II and III to the Sixth Assessment Report of the Intergovernmental Panel on Climate Change* [Core Writing Team, H. Lee and J. Romero (eds.)]. (pp. 1–34). IPCC., DOI: 10.59327/IPCC/AR6-9789291691647

IQAir. (2024). *2023 World Air Quality Report: Highlighting global air quality performance*. Geneva: IQAir. Retrieved from https://www.greenpeace.org/static/planet4-india-stateless/2024/03/44a856c8-2023_world_air_quality_report.pdf

Kuznets, S. (1955). Economic Growth and Income Inequality. *The American Economic Review*, *XLV*(1), •••.

Organisation for Economic Co-operation and Development (OECD). *OECD Economic Surveys: Türkiye 2023*. Paris: OECD Publishing. (2023, February). Retrieved from https://www.oecd.org/en/publications/2023/02/oecd-economic-surveys-turkiye-2023_816ce823.html

Özatac, N., Gökmenoğlu, S., & Taşpınar, N. (2017). Testing the EKC Hypothesis by Considering Trade Openness, Urbanization and Financial Development: The Case of Turkey. *Environmental Science and Pollution Research International*, *24*(20), 16690–16701. DOI: 10.1007/s11356-017-9317-6

Pakistan Bureau of Statistics. (2023). *7th Population & Housing Census 2023 – Final Results*. Islamabad: Pakistan Bureau of Statistics. Retrieved from https://www.pbs.gov.pk/wp-content/uploads/2020/07/National-Census-Report-2023.pdf

Panayotou, T. (1993). *Empirical tests and policy analysis of environmental degradation at different stages of economicdevelopment (No. 992927783402676)*. International Labour Organization.

Puaschunder, J. (2020). *Governance & Climate Justice: Global South & Developing Nations*. Palgrave Macmillan Cham. DOI: 10.1007/978-3-319-63281-0

The World Bank. (2022). *Pakistan Country Climate and Development Report*. World Bank.

The World Bank. (2024). World Development Indicators. https://databank.worldbank.org/source/world-development-indicators

Tirgil, Abdullah, Yasin Acar and Önder Özgür. 2021. "Revisiting the environmental Kuznets curve: evidence from Turkey." *Environment, Development and Sustainability: A Multidisciplinary Approach to the Theory and Practice of Sustainable Development*, Springer, vol. 23(10), pages 14585-14604, October.

Turkish Statistical Institute-TÜİK. (2025). "Inflation & Price" statistics (Consumer Price Index). Retrieved from TÜİK database. Retrieved [19.01.2025] from https://data.tuik.gov.tr/Kategori/GetKategori?dil=2&p=enflasyon-ve-fiyat-106

UNDP. 2024. The Sustainable Development Goals Report 2024. United Nations. https://unstats.un.org/sdgs/report/2024/The-Sustainable-Development-Goals-Report-2024.pdf

Chapter 7
Sustainable Practices in Urban Health:
Addressing Pollution and Inequality

Richa Joshi
https://orcid.org/0009-0006-7286-6585
Eternal University, India

Jagneet Kour
https://orcid.org/0009-0003-0341-3636
Eternal University, India

Raino Bhatia
https://orcid.org/0000-0002-5379-3003
Eternal University, India

ABSTRACT

This chapter will investigate sustainable practices that can effectively address these intertwined challenges. By emphasizing green infrastructure—such as urban forests, green roofs, permeable pavements, and community gardens—we can identify strategies that improve urban environments. Green infrastructure not only enhances air and water quality but also creates vital spaces for recreation and social interaction, thereby fostering community well-being. Moreover, this chapter will examine the necessity of integrating environmental policies with public health initiatives. Sustainable urban planning can lead to the creation of resilient communities, where health considerations are embedded in urban design. By promoting equitable access to green spaces, enhancing public transportation, and investing in clean energy solutions, cities can cultivate healthier living environments that benefit all residents.

DOI: 10.4018/979-8-3373-2469-2.ch007

INTRODUCTION

As urban populations continue to grow, cities are grappling with a range of significant challenges that threaten the health and well-being of their residents. Rapid urbanization has resulted in an intricate mix of issues, such as rising pollution levels, increasing social inequality, and insufficient urban infrastructure to meet the demands of growing populations. These factors collectively undermine public health, contributing to a rise in non-communicable diseases, mental health disorders, and a general decline in quality of life. Understanding the interconnection of environmental conditions, social disparities, and urban infrastructure is critical to addressing these challenges effectively. This chapter explores the current state of urban health in light of these growing issues and advocates for sustainable solutions aimed at alleviating the detrimental effects of urbanization. By focusing on integrated strategies that improve environmental health, promote social equity, and enhance infrastructure, cities can be transformed into healthier, more resilient, and inclusive environments. Through such initiatives, urban areas can foster communities where health and sustainability go hand in hand, mitigating the risks posed by pollution and inequality. While the connection between urban environments and health is well-established, this chapter seeks to provide a synthesized analysis of the integrated strategies that simultaneously address environmental, infrastructural, and social determinants of health. This chapter aims to explore the critical role of sustainable urban planning in addressing health challenges in cities. By reviewing key strategies, case studies, and best practices, the chapter will provide readers with a comprehensive understanding of how sustainable practices can create healthier, more equitable urban environments.

The Devastating Impact of Pollution on Urban Health

Pollution is one of the most significant health threats faced by urban populations. It manifests in various forms of air, water, and soil contamination each of which has profound and far-reaching health consequences for city dwellers (World Health Organization, 2020). Poor air quality, caused by emissions from vehicles, industrial activities, and construction, has been linked to respiratory issues, cardiovascular diseases, and neurological problems (Zeffer, 2024). Water contamination, often resulting from untreated sewage and industrial waste, leads to gastrointestinal disorders, increases cancer risks, and disproportionately impacts vulnerable populations (Allaq et al., 2023). Additionally, soil pollution, driven by the deposition of hazardous chemicals and waste, contributes to health issues such as heavy metal poisoning, developmental disorders, and other chronic conditions (Hanif & Benitez, 2025). The public health implications of pollution are not limited to physical ailments. Exposure to environmental hazards is also closely tied to mental health issues, particularly among marginalized communities. The pervasive nature of pollution contributes to stress, anxiety, and depression, further exacerbating the health challenges faced by urban populations (Ventriglio et al., 2021). As such, tackling urban pollution is essential not only for improving physical health outcomes but also for promoting mental well-being and social equity.

Research Objectives

- To provide an in-depth analysis of how environmental factors, such as air quality, green spaces, waste management, and transportation systems, directly influence urban public health outcomes. This objective will establish the foundation for understanding the link between sustainability and health in cities.
- To explore the benefits of green infrastructure such as urban forests, green roofs, and community gardens in mitigating urban heat, improving air quality, reducing stress, and promoting physical activity. This section will focus on the multifaceted health benefits of integrating nature into urban design.
- To analyze how sustainable transportation options like cycling, walking, and public transit can reduce urban pollution, encourage physical activity, and promote social equity. The chapter will look at examples of cities that have successfully implemented these solutions and the health impacts they have achieved.
- To assess the importance of sustainable waste management practices such as recycling, composting, and waste-to-energy technologies in reducing environmental pollution and safeguarding public health. The chapter will highlight how circular economy approaches contribute to both environmental sustainability and public health.
- To provide real-world examples from around the world, particularly from India, of cities that have implemented successful urban planning projects aimed at improving health outcomes. This includes examining projects that integrate green infrastructure, promote sustainable transportation, and adopt innovative waste management practices (Kumar et al., 2018)
- To discuss the key barriers cities face in adopting sustainable urban practices, including political, financial, and social challenges. Additionally, the chapter will explore the opportunities that lie in overcoming these barriers to achieve healthier, more resilient urban environments.
- To emphasize the importance of ensuring that sustainable urban health practices are inclusive, equitable, and accessible to all segments of the popula-

tion, especially marginalized and vulnerable groups. The chapter will highlight strategies for integrating social equity into urban planning for health.

- To offer evidence-based policy recommendations for urban planners, government officials, and stakeholders to adopt sustainable practices that improve urban health. These recommendations will focus on practical, scalable solutions that can be implemented in both developed and developing urban areas.

REVIEW OF LITERATURE: SUSTAINABLE PRACTICES IN URBAN HEALTH

The rapid urbanization experienced over the past few decades has resulted in a complex array of challenges that significantly affect the health of urban residents, including pollution, social inequality, and insufficient infrastructure. Understanding the interconnections among these issues is crucial for developing sustainable urban health practices that cater to the varied needs of city populations. An increasing body of literature has investigated the impacts of pollution on urban health, the function of green infrastructure in alleviating these impacts, the effects of social inequality, and the necessity for policy reforms that align environmental sustainability with public health objectives.

Mage et al. (1996) emphasized that urban air, water, and soil pollution are widely acknowledged as major determinants of public health. Air pollution, predominantly stemming from vehicular emissions, industrial operations, and energy generation, ranks among the most urgent environmental health threats in urban settings. Research indicates that exposure to air pollution is associated with a variety of non-communicable diseases (NCDs), including respiratory conditions such as asthma and chronic obstructive pulmonary disease (COPD), as well as cardiovascular issues and cognitive decline. Ompad et al. (2007) noted that communities with lower incomes are disproportionately affected by environmental hazards like air and water pollution and often lack access to green spaces and quality healthcare services. Oberndorfer et al. (2007) demonstrated that urban forests play a crucial role in decreasing air pollution by absorbing particulate matter and serving as carbon sinks, while green roofs contribute to reducing the urban heat island effect, enhancing energy efficiency in buildings. Nelly et al. (2011) highlighted how socio-economic status exacerbates health disparities in urban environments. Low-income neighborhoods are frequently located near industrial areas or major traffic routes, resulting in higher pollution levels and associated health risks. These communities, therefore, suffer not only from higher exposure to environmental hazards but also from reduced access to health-promoting resources like clean air, green spaces, and healthcare services. Social inequality also has serious repercussions for mental health. Coutts (2016) emphasized that community gardens represent another essential aspect of green infrastructure. They not only improve air quality but also promote social cohesion, enhance food security, and support both physical and mental health. Wallace et al. (2016) highlighted that green infrastructure including urban forests, green roofs, community gardens, and permeable pavements has gained recognition for its potential to enhance urban health by mitigating environmental pollution and improving the urban ecosystem. Mundorf et al. (2018) have also argued about health disparities in urban settings are heavily influenced by social determinants, including income, education, housing, and healthcare access. Their systematic review outlines the numerous health benefits associated with green spaces, such as better air quality, reduced heat stress, improved mental well-being, and increased opportunities for physical activity. According to the World Health Organization (2020), urban air pollution contributes significantly to premature deaths globally, particularly in cities with high traffic congestion and industrial activities. Water quality is another vital component influencing public health in urban areas, especially in economically disadvantaged communities. Ventriglio et al. (2021) found that prolonged exposure to fine particulate matter (PM2.5) has been correlated with elevated mortality and morbidity rates among urban populations. For example, Feng et al. (2021) found that community gardens provide urban residents, particularly those in low-income neighborhoods, with access to fresh produce, thereby addressing food insecurity while also encouraging physical activity and fostering mental health through community involvement. Furthermore, sustainable transportation infrastructure such as bike lanes and pedestrian-friendly streets offers significant health benefits by reducing air pollution and promoting physical activity. Flèche et al. (2021) demonstrated that research on active transportation has shown that cycling and walking can substantially lower the risk of cardiovascular diseases, obesity, and mental health disorders. In addition to enhancing physical health, these transportation methods contribute to decreased carbon emissions, aligning environmental sustainability with public health goals, as Nguyen et al. (2021) observed that residing in areas with poor environmental conditions characterized by high air pollution, limited green space, and inadequate healthcare access has been linked to elevated rates of stress, anxiety, and depression. Additionally, social isolation, often worsened by substandard housing and a lack of community engagement, further intensifies these mental health challenges. Social isolation in urban environments can be exacerbated by inadequate urban planning, which fails to provide accessible community spaces for social interaction and physical activity. Allaq et al. (2023) conducted research highlighting the disproportionate impact of waterborne diseases on marginalized populations, stressing the urgent need for improved sanitation and access to clean water in urban environments. Contaminated water sources can lead to waterborne illnesses such as cholera, dysentery, and hepatitis, which predominantly affect low-income urban dwellers who often lack access to proper water treatment systems. Soil pollution, often a byproduct of industrial waste, agricultural runoff, and improper disposal of hazardous materials, also presents significant health risks. Hanif & Benitez (2025) indicated that exposure to contaminated soils, particularly heavy metals like lead and arsenic, can cause neurological damage and developmental issues, particularly in children. Urban areas with poorly managed waste disposal practices are especially susceptible to such contamination, which further exacerbates health inequities in urban environments.

The literature reviewed underscores the critical need for integrated policies that address both environmental and social factors affecting urban health. As urbanization continues to increase, understanding the multifaceted relationships between pollution, green infrastructure, and social inequality will be essential for developing sustainable urban health strategies. Policymakers must prioritize investments in green infrastructure, sustainable transportation, and equitable healthcare access to reduce the health disparities faced by urban populations, particularly those in low-income areas. By addressing these challenges in a holistic manner, cities can become more resilient, healthier, and more equitable spaces for all their inhabitants.

Social Inequality: The Silent Killer of Urban Health

Social inequality is another major factor exacerbating urban health issues. Low-income communities and marginalized groups are often the most affected by environmental hazards, as they tend to reside in areas with poor air quality, limited access to clean water, and inadequate infrastructure (Ompad et al., 2021). These populations also face higher barriers to accessing healthcare services, education, and other social determinants of health, further contributing to health disparities (Nelly et al., 2011).The inequitable distribution of resources in urban environments creates a vicious cycle in which disadvantaged groups are more likely to suffer from chronic health conditions, limited access to healthy food, and insufficient opportunities for physical activity (Díaz et al., 2021). This cycle is reinforced by political and economic marginalization, which often excludes these communities from decision-making processes related to urban development and resource allocation (Caniglia, 2016). Addressing these disparities requires comprehensive strategies that ensure all urban residents, regardless of their socioeconomic status, have access to essential services and resources that promote good health.

Inadequate Infrastructure: A Recipe for Disaster

In many urban areas, inadequate infrastructure significantly contributes to the worsening of public health. Outdated transportation systems, inefficient waste management, and lack of green spaces all have detrimental effects on health outcomes (Nix et al., 2020). Insufficient access to public transportation forces many urban residents to rely on private vehicles, exacerbating air pollution and contributing to sedentary lifestyles, which in turn increase the risk of chronic diseases such as obesity and diabetes (Mundorf et al., 2018). Poor waste management systems often result in the accumulation of hazardous materials, contaminating water sources and increasing the likelihood of disease outbreaks (Giusti, 2009; Velasco Perez et al., 2021). Furthermore, the lack of green spaces, which help reduce urban heat islands and promote mental well-being, significantly reduces the quality of life in densely populated areas (Nguyen et al., 2021). The absence of these restorative environments is directly linked to higher rates of psychological stress and poorer overall mental health in urban populations (Bratman et al., 2019).

Integrating Environmental Policies with Public Health Initiatives: A Call to Action

To effectively address urban health challenges, it is crucial to integrate environmental policies with public health initiatives. Sustainable urban planning that prioritizes health outcomes can create resilient communities that are better equipped to cope with the challenges of climate change, pollution, and social inequality. This includes policies that promote equitable access to green spaces, invest in clean energy solutions, and encourage sustainable transportation systems. A key mechanism for this integration is the Health in All Policies (HiAP) approach, which systematically considers the health implications of decisions across all sectors of government to improve population health and health equity (Leppo et al., 2013). By aligning environmental and health goals, cities can create environments that not only mitigate health risks but also foster social equity and economic opportunity.

I. IDENTIFY KEY URBAN HEALTH THREATS

1.1 Assess Pollution's Impact on Non-Communicable Diseases

Pollution continues to be one of the most severe threats to urban health, influencing a wide array of non-communicable diseases (NCDs) and contributing to an overall decline in the quality of life for urban dwellers. Air pollution, in particular, is a major concern, with sources such as vehicular emissions, industrial activities, and energy production releasing harmful pollutants into the atmosphere. Fine particulate matter (PM2.5), a significant pollutant found in urban air, has been linked to a wide range of chronic respiratory and cardiovascular diseases, including asthma, chronic obstructive pulmonary disease (COPD), lung cancer, and heart disease (Mage et al., 1996). For vulnerable populations such as children, the elderly, and those with pre-existing health conditions—long-term exposure to PM2.5 and other pollutants can lead to premature death and a reduced quality of life. Water pollution remains another critical challenge, particularly in densely populated urban areas where infrastructure may be inadequate to handle waste and water treatment. Low-income communities, especially in developing nations, are disproportionately affected by contaminated drinking water. Contaminants such as heavy metals, pesticides, and pathogens from untreated sewage often find their way into water supplies, resulting in the spread of waterborne diseases such as cholera, dysentery, and hepatitis. Allaq et al. (2023) highlight that these diseases are especially pervasive in slums and informal settlements where water and sanitation infrastructure is often lacking or substandard. Similarly, soil contamination, which is often a consequence of industrial waste, chemical runoff, and improper waste disposal, poses a substantial health risk. Urban soil contamination with toxic substances such as heavy metals (e.g., lead, arsenic, cadmium) has been associated with a range of health complications, particularly in children who are more vulnerable to these toxins. Long-term exposure to contaminated soil can lead to neurological disorders, developmental delays, and increased risks for cancers (Hanif & Benitez, 2025). This environmental degradation is especially prevalent in industrial and marginalized urban areas, further exacerbating health disparities.

1.2 Analyze Urban Living Conditions and Health Correlations

Urban living conditions are among the most critical determinants of health, and they are inextricably linked to social, economic, and environmental factors. In many cities, overcrowded housing, poor sanitation, and limited access to healthcare are pervasive, resulting in a multitude of health risks. In high-density areas where urbanization has outpaced infrastructure development, substandard living conditions lead to the proliferation of infectious diseases, respiratory illnesses, and other health issues. Crowded housing often characterized by inadequate ventilation, lack of insulation, and moisture buildup can increase the spread of airborne diseases like tuberculosis and exacerbate asthma and allergies. Overcrowding also places significant stress on sanitation systems, leading to contamination of drinking water and the spread of gastrointestinal diseases (Nelly et al., 2011).

The lack of green spaces in urban areas is a particularly pressing issue, especially in low-income neighborhoods. Research has shown that green spaces such as parks, gardens, and forests play a vital role in promoting physical and mental health. The absence of such spaces leads to increased levels of stress, anxiety, depression, and obesity, as residents lack access to recreational areas and opportunities for physical activity (Ompad et al., 2007). Additionally, noise pollution, often a byproduct of dense traffic and construction activities in urban areas, has been linked to a range of health problems, including sleep disturbances, hypertension, and cardiovascular diseases. For vulnerable populations, including the elderly and children, such noise exposure can exacerbate existing health conditions and increase the likelihood of chronic disease.

Furthermore, poor access to healthcare remains a significant barrier to urban health. In many cities, particularly in informal settlements and slums, there is limited access to affordable or quality healthcare services, making it difficult for residents to seek timely treatment for both chronic and acute conditions. Without accessible medical care, preventable diseases such as diabetes, hypertension, and mental health disorders remain untreated, leading to severe health outcomes and a lower quality of life for individuals in these communities.

1.3 Evaluate the Effects of Climate-Related Events on Health Risks

The impacts of climate change are rapidly becoming one of the most significant drivers of health risks in urban environments, with climate-related events such as heatwaves, flooding, storm surges, and wildfires becoming increasingly frequent and intense in cities worldwide. These events not only exacerbate existing health problems but also introduce new risks to urban populations.

Heatwaves, for example, are associated with an increase in heat-related illnesses such as heatstroke, dehydration, and cardiovascular stress. Urban areas are particularly vulnerable to heat due to the urban heat island effect, where cities experience higher temperatures than surrounding rural areas due to the concentration of buildings, roads, and other heat-absorbing materials. Vulnerable populations, including the elderly, children, and those with pre-existing conditions like cardiovascular disease or respiratory disorders, are at heightened risk during periods of extreme heat (WHO, 2020). Heat stress can also exacerbate mental health issues such as anxiety and depression, which are already prevalent in densely populated, low-income neighborhoods with limited access to cooling resources.

Flooding, often linked to increased rainfall and sea level rise, poses both direct and indirect health risks. In addition to the physical risks associated with drowning, flooding can lead to displacement, loss of property, and mental trauma. The disruption of essential services like healthcare and sanitation during floods also exacerbates the spread of waterborne diseases such as cholera, typhoid, and hepatitis A (Mundorf et al., 2018). Moreover, stagnant floodwaters are often breeding grounds for disease-carrying mosquitoes, increasing the risk of vector-borne diseases such as malaria and dengue fever. In addition to flooding, extreme weather events like storms and hurricanes can lead to widespread property damage, injuries, and mental health challenges. Following such events, urban areas often experience disruptions in infrastructure, such as water supply contamination, electrical outages, and health-care service interruptions, which hinder recovery efforts and heighten public health risks. Additionally, mental health impacts from natural disasters are profound, with affected individuals facing post-traumatic stress disorder (PTSD), depression, and anxiety due to loss of life, displacement, and ongoing economic hardship.

Finally, the long-term effects of climate change, such as rising temperatures and extreme weather patterns, can contribute to the spread of infectious diseases by altering the geographic distribution of pathogens and disease vectors. The spread of vector-borne diseases, such as malaria, Zika virus, and dengue, is expected to increase as warmer temperatures extend the range of mosquitoes and other vectors into new urban areas (WHO, 2020).

SUSTAINABLE PRACTICES: THE KEY TO UNLOCKING HEALTHIER URBAN ENVIRONMENTS

To address the intertwined challenges of pollution, inequality, and infrastructure deficiencies, sustainable practices offer a promising pathway. Integrating green infrastructure such as urban forests, green roofs, and permeable pavements can significantly improve environmental conditions in cities. These practices help mitigate the urban heat island effect, improve air and water quality, and provide much-needed spaces for recreation and social interaction (Coutts et al., 2022). For instance, urban forests have been shown to improve air quality, reduce stress, and promote physical activity, all of which contribute to better public health. Similarly, green roofs help reduce storm water runoff, improve energy efficiency, and create habitats for urban wildlife, all while offering residents opportunities for relaxation and recreation (Oberndorfer et al., 2007). Moreover, sustainable transportation systems such as pedestrian-friendly infrastructure, bike lanes, and expanded public transit are essential for reducing pollution levels and promoting physical activity. However, the successful implementation of these systems often faces challenges related to existing urban sprawl, car-centric culture, and significant upfront investment costs, which must be strategically overcome through phased planning and strong political will (Litman, 2022).Encouraging walking and cycling can help reduce air pollution, improve cardiovascular health, and foster a sense of community (Flèche et al., 2021). In addition, innovative waste management strategies that prioritize recycling and waste-to-energy systems can significantly reduce the environmental impact of urban living, contributing to cleaner cities and better health outcomes (Giusti, 2009; Velasco Perez et al., 2021; Soni et al., 2025).

Sustainable urban design goes beyond simply mitigating environmental damage; it fosters healthier, more livable urban spaces that contribute to the well-being of all residents, especially vulnerable populations. Key strategies such as green infrastructure, sustainable transportation, and innovative waste management not only reduce pollution but also create inclusive urban environments that enhance quality of life. The integration of sustainability into urban health strategies is a pivotal step toward creating cities that are resilient in the face of climate change, more equitable in access to resources, and healthier for all their residents. As we explore these sustainable practices, we will see how they can transform urban health challenges into opportunities for positive change.

Green Infrastructure: A Solution for Pollution and Mental Well-Being

Green infrastructure refers to natural or semi-natural systems that deliver environmental, economic, and social benefits. Examples include urban forests, green roofs, and community gardens, which can mitigate pollution, reduce heat, and improve mental health outcomes. Urban forestry, for instance, provides numerous health benefits by improving air quality, reducing the urban heat island effect, and offering spaces for physical and mental rejuvenation. Furthermore, plants help capture and store rainwater, reducing the risk of urban flooding while also filtering out harmful substances before they reach local water systems (Liu & Jensen, 2018)

Air Quality and Pollution Mitigation

Urban forests, for example, act as natural air filters. Trees absorb pollutants such as carbon dioxide (CO2), particulate matter (PM), and other harmful substances, improving overall air quality. Studies show that areas with higher tree canopy cover have lower levels of air pollution, leading to reduced respiratory conditions like asthma and bronchitis. Furthermore, plants help capture and store rainwater, reducing the risk of urban flooding while also filtering out harmful substances before they reach local water systems.

Waste Management Innovations: A Circular Approach to Health

Effective waste management is essential for maintaining urban health. Sustainable waste practices such as recycling, composting, and the implementation of waste-to-energy technologies can reduce pollution, conserve resources, and decrease landfill waste. By adopting a circular economy approach, cities can ensure that waste is minimized and reused, reducing the burden on landfills and the associated health risks. However, these technologies must be carefully managed to avoid harmful emissions and ensure that they do not contribute to air pollution. Critics argue that WTE can disincentives waste reduction and recycling at the source, and that emissions of dioxins and heavy metals remain a concern without advanced filtration systems (Khandelwal et al., 2019). Therefore, WTE should be considered only after maximizing reduction, reuse, and recycling efforts (Giusti, 2009; Velasco Perez et al., 2021).

Reducing the Urban Heat Island Effect

Urban areas, particularly those with dense concrete infrastructure, are prone to the urban heat island (UHI) effect, where cities experience higher temperatures than surrounding rural areas due to the absorption of heat by buildings, roads, and other impervious surfaces. This elevated temperature can cause heat-related illnesses, increase energy consumption, and exacerbate air pollution. Green infrastructure, such as green roofs and urban parks, provides cooling benefits by absorbing sunlight and releasing moisture into the atmosphere through evapotranspiration. This process reduces ambient temperatures, making cities more comfortable and less prone to heat stress.

Reducing Pollution through Active Transportation

Encouraging walking and cycling not only reduces the reliance on fossil-fuel-based vehicles but also lowers the overall carbon footprint of cities. Walking and cycling infrastructure, such as pedestrian-friendly streets, protected bike lanes, and safe crossings, helps create an environment conducive to active transportation. This reduces traffic congestion, lowers emissions of harmful pollutants, and improves air quality, which has immediate health benefits, including a reduction in respiratory and cardiovascular diseases.

Sustainable Transportation: Building Healthier Cities on Foot and Bike

Sustainable transportation options, such as walking, cycling, and public transit, are critical to reducing urban pollution. By promoting bike-friendly streets and pedestrian walkways, cities can lower vehicle emissions and encourage physical activity. Public transportation systems can also reduce traffic congestion, lower pollution levels, and improve access to essential services.

Public Health and Environmental Policy Integration

Policymakers must work across sectors to integrate health goals with urban development plans. This involves creating policies that reduce environmental hazards while simultaneously addressing health inequities. For example, zoning laws can be designed to limit industrial emissions in residential areas, while public health campaigns can promote physical activity through green space use.

Promoting Equity through Environmental and Health Policies

To ensure that all residents benefit from these policies, particular attention must be paid to marginalized communities. This requires targeted interventions that address the specific needs of vulnerable populations, including access to healthcare, safe housing, and clean environments. Health equity can be achieved by ensuring that sustainable practices benefit everyone, regardless of income, race, or geographic location

Improving Mental Health and Social Cohesion

Beyond environmental and physical health benefits, green infrastructure has significant positive effects on mental well-being. Research has consistently shown that access to green spaces reduces stress, anxiety, and depression, while also enhancing cognitive function. Urban parks, community gardens, and tree-lined streets provide opportunities for physical activity, social interaction, and relaxation, which are essential for mental health. These green spaces also promote social cohesion by providing common areas where people from diverse backgrounds can come together, fostering a sense of community and belonging.

Improved Physical Activity and Health

Sustainable transportation modes walking, cycling, and public transit also promote physical activity. Studies have shown that people who engage in active transportation are more likely to meet the recommended daily physical activity guidelines, which can reduce the risk of obesity, diabetes, and heart disease. Active transportation fosters healthier populations and can contribute to a reduction in the burden of chronic diseases that strain healthcare systems.

Equitable Access to Transportation

Sustainable transportation options can also foster greater equity in cities. In many urban areas, low-income communities face significant transportation barriers, with limited access to reliable public transit and safe walking or cycling routes. By prioritizing public transit systems that are affordable, accessible, and inclusive, cities can ensure that all residents have equitable access to jobs, healthcare, and other essential services. Additionally, enhancing public transit options reduces the reliance on private cars, which can alleviate traffic congestion and make urban areas more navigable for everyone.

Recycling and Composting: Reducing Landfill Burden

Recycling and composting are fundamental components of a circular economy that reduces the volume of waste sent to landfills and minimizes pollution. By recycling materials like paper, plastics, metals, and glass, cities can reduce the environmental footprint of waste disposal. Composting organic waste not only diverts food waste from landfills but also creates nutrient-rich soil that can be used in urban agriculture or green infrastructure projects. These practices help lower greenhouse gas emissions and reduce contamination of air, soil, and water.

Waste-to-Energy Technologies

Another innovative waste management solution is waste-to-energy (WTE) technology, which involves converting non-recyclable waste into energy through processes like incineration, anaerobic digestion, or gasification. WTE systems can help cities reduce waste volumes while generating renewable energy. However, these technologies must be carefully managed to avoid harmful emissions and ensure that they do not contribute to air pollution. When implemented correctly, WTE can be a sustainable solution for cities with high waste generation rates, especially in urban areas where landfill space is limited.

Reducing Environmental Impact and Promoting Public Health

In addition to reducing waste, sustainable waste management practices also have direct health benefits. Proper waste disposal and management reduce the risk of water contamination and prevent the spread of diseases caused by pests, such as rodents and mosquitoes, which thrive in poorly managed waste environments. Furthermore, effective waste management systems can help keep public spaces cleaner, creating healthier and more pleasant environments for residents.

Creating Resilient Communities

Sustainable urban planning can help create resilient communities where health considerations are central to the design of the built environment. Cities should focus on providing equitable access to green spaces, promoting clean transportation, and building healthy housing that minimizes health risks. By embedding public health into urban planning, cities can create environments that nurture both people and nature.

HOLISTIC APPROACHES: INTEGRATING SUSTAINABILITY ACROSS URBAN SYSTEMS

While green infrastructure, sustainable transportation, and waste management are individually important, their true impact on urban health can be maximized when they are integrated into a holistic urban planning strategy. Sustainable urban systems work together to create a mutually reinforcing cycle of environmental health, economic opportunity, and social well-being (Giusti, 2009; Velasco Perez et al., 2021).

Integrated Urban Planning: Urban planning must incorporate sustainability across all sectors housing, transportation, infrastructure, and public health. The use of sustainability assessment frameworks, such as BREEAM for Communities or LEED for Cities, can provide a structured methodology for evaluating and guiding such integrated planning (Berardi et al., 2025). By designing neighborhoods that promote walking, cycling, and access to green spaces, cities can reduce pollution, improve health outcomes, and increase residents' overall quality of life.

Building Resilient and Inclusive Communities: Urban sustainability should not only focus on environmental improvements but also on promoting social equity. A sustainable urban environment must ensure that all residents, regardless of income, race, or social status, have access to the same opportunities for a healthy life. This includes access to clean air, water, food, safe housing, and transportation. Community engagement is essential to creating inclusive, resilient cities. Local residents should have a voice in the planning and design of their neighborhoods to ensure that sustainable practices meet their unique needs.

Collaborative Efforts for Urban Health: Lastly, cities must embrace cross-sectoral collaboration to address the complex challenges of urban health. This requires cooperation between urban planners, environmental experts, public health professionals, and community leaders. By working together, these stakeholders can create integrated policies that support both environmental sustainability and public health, fostering cities that are not only green and resource-efficient but also healthy, equitable, and resilient.

URBAN PLANNING CASE STUDIES: TACKLING HEALTH CHALLENGES WITH SUSTAINABLE PRACTICES

To provide a practical understanding of how sustainable urban planning strategies can address health challenges, this chapter reviews case studies from cities around the world that have successfully integrated green infrastructure, sustainable transportation, and community involvement into their urban planning efforts. These case studies demonstrate that combining environmental sustainability with health-focused urban design can lead to improved public health outcomes, enhanced resilience, and more equitable urban environments.

1. The Green Streets Initiative - Portland, Oregon, USA

Portland, known for its environmental leadership, has implemented a comprehensive green infrastructure program to address water pollution, storm water runoff, and the urban heat island effect. The Green Streets initiative is a key element of the city's broader approach to sustainable urban planning and climate adaptation.

- **Green Streets:** The initiative uses green infrastructure elements, including rain gardens, bioswales, and permeable pavements, to capture and filter storm water before it enters the sewer system. These systems reduce the risk of flooding and improve water quality by removing pollutants from rainwater.
- **Community Involvement:** The city has involved local communities in designing and maintaining green spaces, fostering a sense of ownership and encouraging residents to take part in environmental stewardship.
- **Health Benefits:** By reducing urban heat islands and improving air and water quality, the Green Streets program helps mitigate respiratory and heat-related illnesses, particularly in low-income neighborhoods, which are often more vulnerable to the effects of pollution and climate change.

Results

- Significant reduction in storm water runoff and improved water quality in local rivers and streams.
- Increased urban greenery, which has led to a decrease in the urban heat island effect, improving overall livability.
- Enhanced mental and physical health outcomes for residents due to the increased availability of green spaces.

2. The High Line - New York City, USA

The High Line is a 1.45-mile-long elevated park built on an old freight rail line on Manhattan's West Side. What began as an abandoned piece of infrastructure has been transformed into a vibrant public space that integrates green infrastructure and fosters social cohesion.

- **Green Infrastructure:** The High Line includes a variety of vegetation, including grasses, shrubs, and trees that help mitigate the urban heat island effect, improve air quality, and increase biodiversity.
- **Sustainable Design:** The Park incorporates sustainable design features such as rainwater harvesting systems and energy-efficient lighting. The landscaping helps manage storm water, preventing runoff and improving water quality.
- **Public Health and Well-Being:** The High Line provides a green space for walking, cycling, and socializing, encouraging physical activity and promoting mental well-being through access to nature. The park also hosts community events, art installations, and cultural activities, contributing to social interaction and community engagement.

Results

- Positive effects on mental health, with studies showing that visitors to the High Line experience reduced stress and improved mood.
- Increase in local property values, which has sparked gentrification but also led to increased investment in surrounding neighborhoods.
- The High Line has also become a model for similar green infrastructure projects globally, demonstrating the potential for reimagining urban spaces for both environmental and public health benefits.

3. Vauban District - Freiburg, Germany

Vauban is a sustainable district in Freiburg, Germany, designed as an eco-neighborhood focused on reducing energy consumption, promoting sustainable transportation, and creating a community-oriented living environment. It is considered one of the best examples of green urban planning in Europe.

- **Sustainable Transport:** Vauban has prioritized walking, cycling, and public transportation over car use. There are no parking spaces in front of homes, and the streets are designed to discourage car traffic, reducing air pollution and promoting physical activity.

- **Energy-Efficient Homes:** The homes in Vauban are energy-efficient, using solar panels, passive heating, and high-performance insulation to minimize energy consumption and reduce greenhouse gas emissions.
- **Community Involvement:** Residents were involved in the planning and design process, which helped ensure that the development met the needs of the community. A focus on shared spaces, community gardens, and social interaction fosters a sense of connection among neighbors.
- **Health Outcomes:** Vauban promotes a healthy lifestyle by encouraging active transportation, reducing pollution, and providing spaces for outdoor activities. The neighborhood's low-carbon footprint and energy-efficient buildings also contribute to a cleaner environment.

Results

- A dramatic reduction in car use, with more than 50% of trips made by walking, cycling, or public transport.
- The eco-friendly design has led to high levels of energy efficiency, with many homes achieving passive house standards.
- High resident satisfaction and community engagement, with an active neighborhood association that promotes health and social cohesion.

4. Bogotá's Ciclovía - Bogotá, Colombia

Bogotá's Ciclovía is a pioneering initiative that temporarily closes certain streets to motor vehicles every Sunday and holiday, transforming them into car-free spaces for pedestrians, cyclists, and other non-motorized forms of transport. It began in 1974 as an effort to address traffic congestion, pollution, and the lack of public spaces for physical activity.

- **Temporary Open Streets:** Every Sunday, over 75 miles of roads are closed to cars and opened up for cyclists, pedestrians, and rollerblades. This initiative is one of the most extensive open streets programs in the world.
- **Health and Recreation:** Ciclovía encourages physical activity by providing a safe, accessible environment for exercise. The program also offers health workshops, fitness classes, and recreational activities along the route.
- **Community Engagement:** The program fosters a sense of community, with people of all ages and backgrounds coming together to exercise, socialize, and enjoy their city in a car-free environment.

Results:

- Ciclovía has become a model for other cities around the world, with over 100 cities in various countries implementing similar programs.
- Increased physical activity among residents, with studies showing a significant reduction in obesity and cardiovascular diseases in participants.
- Reduced traffic-related pollution and noise, improving air quality and the overall urban environment.
- Greater social cohesion and community engagement, with many participants reporting a greater sense of connection to their city and fellow residents.

5. The Copenhagen Metro and Bicycle Infrastructure - Copenhagen, Denmark

Copenhagen is a global leader in sustainable urban transport, particularly in its focus on cycling and public transportation. The city's comprehensive metro system and extensive cycling infrastructure have been central to its efforts to reduce pollution, improve air quality, and promote public health.

- **Cycling Infrastructure:** Copenhagen has one of the highest rates of cycling in the world, with over 50% of residents commuting by bike. The city has created a network of dedicated bike lanes, bike bridges, and bicycle-friendly policies that encourage cycling as a primary mode of transport.
- **Public Transportation:** Copenhagen's metro system, which opened in 2002, is a driverless and energy-efficient network that provides fast, reliable public transportation. It is fully integrated with the city's bus and bicycle networks, making it easy for residents to choose sustainable modes of transport.
- **Health Benefits:** By reducing car traffic and promoting cycling, Copenhagen has improved air quality, decreased respiratory and cardiovascular diseases, and increased physical activity among its residents.

Results

- A reduction in traffic-related emissions, contributing to cleaner air and improved health outcomes.
- Increased physical activity, with cycling becoming the primary mode of transportation for a significant portion of the population.
- Copenhagen is consistently ranked as one of the happiest and healthiest cities in the world, with high life expectancy and low rates of chronic diseases.

6. Singapore's Garden City Initiative - Singapore

Singapore has long been known for its commitment to green urban planning, particularly through its "Garden City" initiative. The city-state integrates nature into urban spaces to improve the environment, public health, and quality of life for its residents.

- **Green Roofs and Vertical Gardens:** Singapore is home to numerous green roofs, vertical gardens, and other forms of green infrastructure that enhance biodiversity, reduce the heat island effect, and improve air quality.
- **Park Connector Network:** The city's extensive network of park connectors links parks, nature reserves, and waterfronts, making it easy for residents to access green spaces and engage in outdoor activities.
- **Smart Urban Solutions:** Singapore uses advanced technologies to monitor and manage environmental issues, such as air quality, waste management, and energy use, helping to optimize sustainability in urban spaces.

Results

- Singapore has one of the highest percentages of green cover in any major city, with over 47% of the city-state covered by greenery.
- The city has become a model for integrating green spaces into urban environments, with significant improvements in residents' mental and physical well-being due to increased access to nature.
- The city's commitment to sustainability has led to significant reductions in carbon emissions and improved air quality, contributing to better health outcomes.

Lessons Learned from Case Studies

These case studies illustrate the power of sustainable urban planning to address health challenges in cities. By integrating green infrastructure, promoting sustainable transportation, and engaging communities in the planning process, cities can create healthier, more resilient environments. The success of these initiatives demonstrates that urban health challenges are not insurmountable—innovative, inclusive, and sustainable practices can transform cities into healthier, more livable spaces for all residents.

URBAN PLANNING CASE STUDIES IN INDIA: ADDRESSING HEALTH CHALLENGES THROUGH SUSTAINABLE PRACTICES

India's urban centers face a unique set of health and environmental challenges, including rapid urbanization, air pollution, water scarcity, inadequate healthcare, and social inequality. However, several Indian cities have begun to implement innovative urban planning strategies to address these challenges, particularly by integrating green infrastructure, promoting sustainable transportation, and improving community health outcomes (Sharma & Kumar, 2022).

1. The Delhi Yamuna Riverfront Development - New Delhi, India

The Yamuna River, one of India's most important rivers, has been severely polluted and encroached upon in New Delhi, contributing to environmental degradation and health problems for millions of residents. The Yamuna Riverfront Development Project aims to restore the river's ecological health while creating green public spaces and improving air and water quality.

- **Riverfront Development:** The project focuses on creating green spaces along the Yamuna River, with parks, walking tracks, and recreational areas to provide residents with access to nature. This also includes the rejuvenation of wetlands to help filter wastewater and improve water quality.
- **Flood Mitigation and Waste Management:** The project has integrated flood control measures and waste management initiatives to prevent contamination of the river. The development aims to create sustainable systems to reduce industrial and domestic pollution that flows into the river.
- **Green Infrastructure:** The project plans to plant trees and create artificial wetlands along the riverbanks, helping to improve air quality and support biodiversity.

Results:

- **Improved Public Health:** The development has enhanced access to green spaces, contributing to better mental and physical health for urban residents. The addition of green infrastructure is expected to reduce air pollution levels in surrounding areas.
- **Water Quality Improvement:** The riverfront's wastewater treatment facilities are expected to significantly reduce the contamination of the Yamuna River, improving drinking water quality and reducing waterborne diseases.

- **Social Inclusion:** The creation of public spaces along the riverfront provides an opportunity for the city's marginalized populations to access nature and engage in community activities.

2. The Smart Cities Mission – Multiple Cities, India

Launched in 2015 by the Government of India, the Smart Cities Mission (Kumar, 2018) aims to create 100 smart cities across the country with a focus on sustainable urban development, improving infrastructure, and enhancing the quality of life for residents. These cities use technology and data-driven solutions to improve services, including health, transportation, and environmental management.

- **Data-Driven Urban Planning:** Smart cities leverage IoT (Internet of Things) and data analytics to monitor air and water quality, waste management systems, and traffic congestion. By gathering real-time data, these cities can respond proactively to urban health challenges.
- **Green Infrastructure:** Many smart cities have incorporated green spaces such as urban parks, rooftop gardens, and rainwater harvesting systems to reduce pollution and enhance the well-being of residents.
- **Sustainable Mobility:** Electric vehicle (EV) infrastructure, bike-sharing programs, and improved public transportation networks are part of the sustainable transportation initiatives in these cities.
- **Health and Hygiene Programs:** Smart cities use smart waste management systems and public health monitoring tools to prevent disease outbreaks and improve sanitation, with a special focus on marginalized areas.

Results

- **Improved Environmental Health:** The use of smart technologies to monitor air quality and waste management has led to cleaner cities with lower pollution levels, reducing respiratory illnesses and other health issues.
- **Enhanced Public Services:** Improved transportation networks and smart health monitoring systems have helped increase access to essential services, especially for vulnerable communities.
- **Community Engagement:** The mission has focused on citizen participation in decision-making and urban development, ensuring that residents' needs are addressed.

3. The Bio-Waste Management Initiative – Pune, India

Pune, one of India's fastest-growing cities, has faced significant challenges related to waste management. As urban populations grow, managing organic waste and reducing its environmental impact has become increasingly difficult. The Bio-Waste Management Initiative was launched to reduce waste and promote sustainability.

- **Segregation at Source:** Pune has implemented a citywide program for segregating waste at the source, encouraging households and businesses to separate biodegradable waste from non-biodegradable waste. This is crucial for the city's efforts to reduce landfill pressure and minimize pollution.
- **Composting and Waste-to-Energy Projects:** Organic waste is diverted to composting facilities, where it is processed into high-quality compost. In addition, some waste is used for generating biogas, which provides a sustainable energy source.
- **Community Involvement:** Pune has partnered with local communities, NGOs, and residents to promote awareness about waste segregation and composting. The city has also implemented door-to-door waste collection services to ensure efficient waste management.

Results

- **Reduced Pollution:** By diverting organic waste from landfills, the city has reduced methane emissions, which are a significant source of air pollution and contribute to climate change.
- **Improved Public Health:** Reducing waste and improving sanitation practices have had a direct impact on public health, with fewer incidents of waste-related diseases.
- **Sustainable Urban Growth:** The bio-waste management system has enabled Pune to manage its rapid urban growth without compromising environmental health.

4. The Shivaji Nagar Redevelopment Project – Bengaluru, India

Shivaji Nagar, a densely populated and underdeveloped area in Bengaluru, has long struggled with poor infrastructure, inadequate waste management, and environmental pollution. The city's efforts to redevelop this area focus on improving public health through sustainable urban practices.

- **Green Infrastructure in Urban Planning:** The redevelopment plan includes creating green spaces, improving drainage systems, and planting trees to combat the urban heat island effect. A network of green parks and recreational spaces has been incorporated to increase the area's resilience to climate change.
- **Sustainable Transport Networks:** The project encourages the use of public transport and bicycles by providing well-designed bus stops, cycle lanes, and pedestrian walkways. This helps reduce traffic congestion and lowers air pollution.
- **Affordable Housing and Health Facilities:** The redevelopment includes the construction of affordable housing units with access to basic amenities, including healthcare facilities, water supply, and sanitation, which have direct health benefits.

Results

- **Improved Health Infrastructure:** The provision of healthcare facilities and better housing conditions has improved the overall health of the residents, reducing the spread of diseases linked to overcrowding and poor sanitation.
- **Environmental Benefits:** Green spaces and sustainable transportation options have reduced pollution levels and improved air quality in a previously congested area.
- **Social Equity:** The redevelopment plan has focused on providing better living conditions for low-income groups, addressing health inequities and social exclusion.

5. The Clean Air Initiative – Delhi, India

Delhi has been grappling with some of the worst air quality levels in the world, which has a direct impact on public health, especially respiratory and cardiovascular diseases. The Clean Air Initiative, launched in 2017, aims to improve air quality through various sustainable measures, including green infrastructure and air pollution reduction strategies.

- **Improved Air Quality Monitoring:** The initiative uses air quality monitoring stations across the city to measure pollutants such as PM2.5, NOx, and SO2, and to create actionable data for policymaking.
- **Green Infrastructure Projects:** Delhi has increased its tree cover and created green corridors to absorb pollutants and provide cleaner air. Urban forests and parks have been established in areas prone to pollution.

- **Public Awareness Campaigns and Clean Transport:** The initiative promotes the use of electric vehicles, carpooling, and the expansion of the public transport system. There is also a strong push for clean energy sources and better waste management practices.

Results

- **Improved Health Outcomes:** Decreased levels of particulate matter in the air have contributed to improved respiratory health outcomes, with a reduction in asthma and other lung diseases.
- **Cleaner Air:** The green corridors and reduced vehicular emissions have significantly improved the air quality in several parts of the city.
- **Policy Influence:** The initiative has influenced national policies on air pollution and environmental health, contributing to wider reforms across India.

Lessons Learned from Case Studies

These case studies illustrate the power of sustainable urban planning to address health challenges in cities. By integrating green infrastructure, promoting sustainable transportation, and engaging communities in the planning process, cities can create healthier, more resilient environments. The success of these initiatives demonstrates that urban health challenges are not insurmountable innovative, inclusive, and sustainable practices can transform cities into healthier, more livable spaces for all residents.

Sustainable practices ranging from the implementation of green infrastructure to the promotion of sustainable transportation and the adoption of innovative waste management strategies hold immense potential in transforming urban environments into healthier, more resilient spaces. These practices are not just environmentally beneficial; they play a pivotal role in improving public health, fostering social equity, and ensuring the well-being of urban populations.

ENVIRONMENTAL SUSTAINABILITY AND HEALTH

At the heart of urban health challenges lies the persistent issue of pollution. Cities around the world, particularly in developing regions, face rapidly rising pollution levels, including poor air quality, water contamination, and soil degradation. These environmental stressors contribute significantly to non-communicable diseases (NCDs) such as respiratory diseases, cardiovascular ailments, mental health disorders, and even cancer. Green infrastructure, such as urban forests, green roofs, and community gardens, offers a direct response to these challenges. Urban greenery improves air quality by filtering pollutants, mitigates the urban heat island effect, and provides essential spaces for physical and mental rejuvenation. These natural solutions not only enhance environmental quality but also significantly reduce the burden of disease caused by pollution.

Sustainable transportation systems, such as pedestrian-friendly streets, bike lanes, and efficient public transit, contribute to both environmental and public health improvements. These systems lower vehicle emissions, reduce traffic-related air pollution, and encourage physical activity, such as walking and cycling, which have direct benefits for cardiovascular health, mental well-being, and social connectedness. The integration of these systems into urban infrastructure can help reduce the prevalence of lifestyle-related health conditions while promoting a shift towards more sustainable, low-carbon urban mobility.

Innovative waste management practices such as recycling, composting, and waste-to-energy technologies are key to reducing the environmental impacts of urbanization. Cities that implement circular economy models can minimize landfill waste, conserve resources, and reduce exposure to health hazards associated with poor waste disposal, such as vector-borne diseases and pollution from leachates. The health benefits of effective waste management are particularly evident in low-income neighborhoods, where inadequate waste infrastructure often leads to environmental and health crises.

Addressing Inequality through Sustainable Urban Practices

A critical component of urban health challenges is social inequality. Urban populations are diverse, and within cities, there are stark contrasts between affluent and marginalized communities in terms of access to clean air, water, healthcare, and green spaces. Sustainable practices must be designed to address these disparities. The equitable distribution of green spaces and access to healthcare facilities are crucial for ensuring that all residents benefit from urban health improvements. The principle of environmental justice dictates that low-income and vulnerable populations—who often bear the brunt of pollution and inadequate infrastructure should be prioritized in the development of sustainable urban solutions.

Building Resilient and Livable Cities

As urban populations continue to grow and the impacts of climate change intensify, cities must adopt resilient strategies that integrate health, environmental sustainability, and social equity into urban planning and development. Urban areas are increasingly exposed to climate-related risks such as heatwaves, flooding, and extreme weather events. Green infrastructure not only mitigates the effects of these events such as by cooling the urban environment and managing storm water but also enhances the adaptability of urban areas, helping them withstand and recover from climate shocks.

To build truly resilient cities, it is essential that urban planning integrates health considerations at all levels. Health impact assessments (HIAs) should be embedded in the urban planning process, ensuring that all new developments whether residential, commercial, or recreational are evaluated for their potential health impacts. Furthermore, policies should encourage the development of mixed-use communities that prioritize walkability, active transport, and access to green spaces.

Policy and Governance: The Path Forward

Incorporating sustainability into urban planning requires strong leadership and coordinated action across multiple sectors, including health, environment, transportation, and housing. Governments, urban planners, and local communities must collaborate to ensure that sustainable practices are implemented effectively. Policymakers should take an integrated approach to urban development, aligning environmental and health goals with social and economic priorities.

Governments can play a pivotal role by creating incentives for green infrastructure projects, supporting sustainable transportation systems, and providing funding for waste management innovations. However, successful implementation also relies on the active participation of local communities. Engaging residents in the planning process and empowering them to participate in sustainability initiatives such as community gardens, local recycling programs, or green space management ensures that these efforts are locally relevant and widely supported.

A Holistic Approach to Urban Health and Sustainability

The interconnections between environmental sustainability, health, and social equity are undeniable. Urban health challenges cannot be addressed in isolation. Rather, they require a holistic approach that integrates sustainability across all urban systems housing, transport, energy, waste management, and healthcare. Only through an interconnected strategy that prioritizes both environmental and social well-being can cities unlock the full potential of sustainable practices.

This approach offers profound long-term benefits, including reduced healthcare costs, enhanced quality of life, and improved social cohesion. By fostering healthier, more equitable cities, sustainable practices contribute not just to the well-being of current residents but also to the long-term viability of urban environments.

The Urgent Need for Action

The urgency of adopting sustainable urban health practices cannot be overstated. With cities becoming the epicenters of population growth, environmental degradation, and social inequality, the challenges we face are growing ever more complex. However, by embracing sustainable urban practices, we can tackle these challenges head-on, creating cities that are not only livable and resilient but also promote the health and well-being of all their residents. Urban sustainability, therefore, is not merely an environmental imperative it is a critical component of urban health, equity, and resilience.

CONCLUSION

This is not just a theoretical vision; it is an achievable future that demands immediate attention and action. The case studies from Portland, Vauban, Bogotá, and Indian cities like Pune and Delhi demonstrate that the co-benefits for health and sustainability are tangible. The common thread in these successes is multi-stakeholder governance, community engagement, and policies that prioritize long-term well-being over short-term convenience. Only through integrated, inclusive, and sustainable urban planning can we ensure that future generations inherit cities that are healthier, more equitable, and more sustainable.

REFERENCES

Allaq, A. A. A., Mahid, H. H., Yahya, E. B., Saleh, A. M., Sidik, N. J., Abdulsamad, M. A., Saad, N. R., Elengoe, A., & Abdul-Aziz, A. (2023). Emerging drinking water borne diseases: A review on types, sources and health precaution. *Journal of Pharmaceutical Research International*, *35*(31), 1–17. DOI: 10.9734/jpri/2023/v35i317462

Berardi, U., Civiero, P., Eicker, U., Gavalda, O., Montella, I., Ortiz, J., . . . Salom, J. (2025). Tools and protocols for cities/neighbourhood sustainability assessment. In Sustainability Certifications, Labels and Tools in the Built Environment (pp. 22-78). Routledge.

Bratman, G. N., Anderson, C. B., Berman, M. G., Cochran, B., de Vries, S., Flanders, J., Folke, C., Frumkin, H., Gross, J. J., Hartig, T., Kahn, P. H.Jr, Kuo, M., Lawler, J. J., Levin, P. S., Lindahl, T., Meyer-Lindenberg, A., Mitchell, R., Ouyang, Z., Roe, J., & Daily, G. C. (2019). Nature and mental health: An ecosystem service perspective. *Science Advances*, *5*(7), eaax0903. Advance online publication. DOI: 10.1126/sciadv.aax0903

Caniglia, B. S., Vallée, M., & Frank, B. (Eds.). (2016). *Resilience, environmental justice and the city*. Routledge. DOI: 10.4324/9781315652054

Coutts, C. (2016). *Green infrastructure and public health*. Routledge. DOI: 10.4324/9781315647623

Díaz, C., González, D., & Rodríguez, M. A. (2021). Addressing urban health inequality through integrated social and environmental policies. *International Journal of Environmental Research and Public Health*, *18*(6), 3089. DOI: 10.3390/ijerph18063089

Giusti, L. (2009). A review of waste management practices and their impact on human health. Waste management, 29(8), 2227-2239.

Hanif, M. N., & Benitez, I. B. (2025). Soil Pollution in Urban Environments: Sources, Consequences, Potential Mitigation Strategies and the Importance of Sustainable Urban Development. *Water, Air, and Soil Pollution*, *236*(14), 945. DOI: 10.1007/s11270-025-08421-0

Khandelwal, H., Thalla, A. K., Kumar, S., & Kumar, R. (2019). Life cycle assessment of municipal solid waste management options: A review. *Journal of Cleaner Production*, *209*, 630–653. DOI: 10.1016/j.jclepro.2018.10.233

Kumar, N. M., Goel, S., & Mallick, P. K. (2018). Smart cities in India: Features, policies, current status, and challenges. 2018 Technologies for Smart-City Energy Security and Power (ICSESP), 1-4.

La Flèche, M., Cuss, C. W., Noernberg, T., Shotyk, W., & Karst, J. (2021). Trace metals as indicators of tree rooting in bituminous soils. *Land Degradation & Development*, *32*(5), 1970–1980. DOI: 10.1002/ldr.3848

Leppo, K., Ollila, E., Pena, S., Wismar, M., & Cook, S. (2013). Health in all policies: seizing opportunities, implementing policies. Sosiaali-ja terveysministeriö.

Litman, T. M. (2022). Evaluating transportation equity: Guidance for incorporating distributional impacts in transport planning. Institute of Transportation Engineers. *ITE Journal*, *92*(4), 43–49.

Liu, L., & Jensen, M. B. (2018). Green infrastructure for sustainable urban water management: Practices of five forerunner cities. *Cities (London, England)*, *74*, 126–133. DOI: 10.1016/j.cities.2017.11.013

Mage, D., Ozolins, G., Peterson, P., Webster, A., Orthofer, R., Vandeweerd, V., & Gwynne, M. (1996). Urban air pollution in megacities of the world. *Atmospheric Environment*, *30*(5), 681–686. DOI: 10.1016/1352-2310(95)00219-7

Mundorf, N., Redding, C. A., & Paiva, A. L. (2018). Sustainable transportation attitudes and health behavior change: Evaluation of a brief stage-targeted video intervention. *International Journal of Environmental Research and Public Health*, *15*(1), 150. DOI: 10.3390/ijerph15010150

Nelly, V., de Snyder, S., Friel, S., Fotso, J. C., Khadr, Z., Meresman, S., & Patil-deshmukh, A. (2011). Social Conditions and Urban Health Inequities: Realities, Challenges and Opportunities to Transform the Urban Landscape through Research and Action. *Journal of Urban Health*, *88*(6), 1183–1193. DOI: 10.1007/s11524-011-9609-y

Nguyen, P. Y., Astell-Burt, T., Rahimi-Ardabili, H., & Feng, X. (2021). Green space quality and health: A systematic review. *International Journal of Environmental Research and Public Health*, *18*(21), 11028. DOI: 10.3390/ijerph182111028

Nix, E., Paulose, J., Shrubsole, C., Altamirano-Medina, H., Davies, M., Khosla, R., Belesova, K., & Wilkinson, P. (2020). Evaluating housing health hazards: Prevalence, practices and priorities in Delhi's informal settlements. *Journal of Urban Health*, *97*(4), 502–518. DOI: 10.1007/s11524-020-00442-w

Oberndorfer, E., Lundholm, J., Bass, B., Coffman, R. R., Doshi, H., Dunnett, N., Gaffin, S., Köhler, M., Liu, K. K. Y., & Rowe, B. (2007). Green roofs as urban ecosystems: Ecological structures, functions, and services. *Bioscience*, *57*(10), 823–833. DOI: 10.1641/B571005

Ompad, D. C., Galea, S., Caiaffa, W. T., & Vlahov, D. (2007). Social determinants of the health of urban populations: Methodologic considerations. *Journal of Urban Health*, *84*(S1, Suppl 1), 42–53. DOI: 10.1007/s11524-007-9168-4

Sharma, M., & Kumar, S. (2022). Analysing the spatial patterns and trends of urban growth in Rohtak city, India. *Sustainable Environment*, *8*(1), 2051268. DOI: 10.1080/27658511.2022.2051268

Soni, A., Gupta, S. K., Rajamohan, N., & Yusuf, M. (2025). Waste-to-energy technologies: A sustainable pathway for resource recovery and materials management. *Materials Advances*, *6*(14), 4598–4622. Advance online publication. DOI: 10.1039/D5MA00449G

Velasco Perez, M., Sotelo Navarro, P. X., Vazquez Morillas, A., Espinosa Valdemar, R. M., & Hermoso Lopez Araiza, J. P. (2021). Waste management and environmental impact of absorbent hygiene products: A review. *Waste Management & Research*, *39*(6), 767–783. DOI: 10.1177/0734242X20954271

Ventriglio, A., Bellomo, A., di Gioia, I., Di Sabatino, D., Favale, D., De Berardis, D., & Cianconi, P. (2021). Environmental pollution and mental health: A narrative review of literature. *CNS Spectrums*, *26*(1), 51–61. DOI: 10.1017/S1092852920001303

Wallace, R., Green, S., & Agarwal, G. (2016). Promoting the health benefits of walking and bicycling to work: A qualitative exploration of the role of healthcare providers in addressing barriers to active commuting. *Sports and Exercise Medicine–Open Journal*, *2*(2), 24–32. DOI: 10.17140/SEMOJ-2-135

World Health Organization (WHO). (2020). *Health and the environment: A global perspective*. World Health Organization. https://www.who.int/publications/i/item/health-and-environment-a-global-perspective

Zeffer, W. (2024). An Examination of Environmental Health Literacy: Capacities, Structures, and Engagement (Doctoral dissertation, University of Pittsburgh).

Chapter 8

Horticulture-Based Strategies for Air Quality Management: Global Trends in Smog Mitigation and Pollution Control Through Green Infrastructure

Shreya Singh
https://orcid.org/0000-0001-8907-9947
Chhatrapati Shahu Ji Maharaj University, Kanpur, India

Shivansu Sachan
https://orcid.org/0000-0002-9893-8018
Chhatrapati Shahu Ji Maharaj University, Kanpur, India

ABSTRACT

Air pollution remains one of the most pressing global environmental challenges, with smog and deteriorating air quality posing significant threats to human health, ecosystems, and urban sustainability. Traditional mitigation approaches, while effective to an extent, are often resource-intensive and technologically complex. In recent years, horticulture-based strategies, particularly the integration of green infrastructure, have emerged as cost-effective and sustainable solutions for smog mitigation and pollution control. This paper examines global trends in leveraging horticultural interventions such as urban forests, vertical gardens, green roofs, and phytoremediation techniques to improve air quality. Emphasis is placed on the mechanisms by which vegetation captures particulate matter ($PM_{2.5}$, PM_{10}), absorbs gaseous pollutants (NOx, SO_2, CO_2), and enhances urban microclimates.

DOI: 10.4018/979-8-3373-2469-2.ch008

The study synthesizes evidence from diverse geographical regions, highlighting successful case studies and policy initiatives that prioritize green infrastructure as a core component of air quality management

1. INTRODUCTION

Air pollution remains a dominant environmental issue worldwide, responsible for an estimated **7** *million premature deaths annually* (WHO, 2023). Rapid urbanization, industrialization, and vehicular emissions have intensified air quality degradation, particularly in developing economies. Traditional approaches to smog mitigation—including air filtration technologies, emission control policies, and cleaner fuel adoption—are essential but insufficient to address the multidimensional nature of the problem.This chapter explores *global trends in horticulture-based air quality management*, focusing on urban green infrastructure as a critical tool for pollution control. It synthesizes research findings, discusses successful international case studies, and evaluates future prospects for integrating horticulture into urban sustainability strategies. Air pollution has become one of the most pressing environmental challenges of the 21st century, affecting nearly every region of the world. The *World Health Organization (WHO, 2023)* estimates that air pollution contributes to *seven million premature deaths annually,* primarily due to cardiovascular and respiratory diseases. Rapid urbanization, industrial growth, and increased vehicular traffic have intensified air quality deterioration, particularly in developing economies. Among the many pollutants, *particulate matter ($PM_{2.5}$ and PM_{10}),* nitrogen oxides (NOx), sulfur dioxide (SO_2), carbon monoxide (CO), and volatile organic compounds (VOCs) are the most concerning due to their harmful impacts on human health and the environment.

Urban smog, a complex mixture of these pollutants, results from photochemical reactions involving NOx and VOCs in the presence of sunlight. Smog episodes are frequent in cities like Beijing, Delhi, and Los Angeles, where meteorological conditions, emissions, and population density combine to create hazardous air quality. Traditional solutions to address smog and pollution—such as *stringent emission norms, fuel quality improvements, and air filtration systems*—play a critical role but are insufficient when applied in isolation. These approaches often involve high costs, energy consumption, and limited long-term effectiveness.

Against this backdrop, *horticulture-based strategies*, particularly those integrated into *green infrastructure*, have emerged as viable, sustainable solutions for improving air quality. Plants act as natural biofilters, capturing pollutants on leaf surfaces, absorbing gases through stomata, and improving urban microclimates by reducing heat islands. Green infrastructure includes components such as *urban forests, green roofs, vertical gardens, roadside plantations, and bio-walls*, which not only purify air but also enhance biodiversity, sequester carbon, and provide psychological benefits to urban residents.

This chapter critically examines horticulture-based strategies for air quality management, focusing on global trends, underlying scientific mechanisms, and practical applications. It also highlights policy frameworks, case studies, and future directions to integrate green solutions into urban sustainability planning.

2. LITERATURE REVIEW

2.1 Air Pollution and Its Impacts

Air pollution is a complex environmental problem that has significant implications for human health, ecosystems, and the global economy. It is primarily characterized by the presence of harmful substances in the atmosphere, including *particulate matter (PM), nitrogen oxides (NO_X), sulfur dioxide (SO_2), carbon monoxide (CO), volatile organic compounds (VOCs), and ground-level ozone (O_3)*. These pollutants are generated from various anthropogenic sources such as *vehicular emissions, industrial activities, construction dust, biomass burning, and energy production* (Gurjar et al., 2016). Additionally, natural sources such as wildfires and dust storms exacerbate air quality deterioration in many regions.

Particulate Matter (PM), particularly $PM_{2.5}$ (particles with aerodynamic diameters less than 2.5 micrometers), poses the greatest risk to human health because of its ability to penetrate deep into the respiratory tract and enter the bloodstream. Long-term exposure to $PM_{2.5}$ is associated with *chronic bronchitis, lung cancer, cardiovascular diseases, and premature mortality* (WHO, 2022). *Nitrogen oxides (NO_X)*, produced largely by motor vehicles and industrial combustion, contribute to respiratory problems, ground-level ozone formation, and acid rain. Similarly, *sulfur dioxide (SO_2)*, a byproduct of coal combustion, aggravates asthma and causes environmental acidification. VOCs, commonly released from fuel combustion and industrial solvents, are precursors to tropospheric ozone—a major component of photochemical smog (Jacob, 1999).

From an economic perspective, air pollution leads to *massive financial losses*, including *increased healthcare expenditures*, *reduced labor productivity*, damage to infrastructure, *and lower agricultural yields*. According to the *OECD (2021)*, outdoor air pollution costs the global economy over *USD 5 trillion annually*, a figure that is expected to rise without significant mitigation strategies. Moreover, the *World Bank (2020)* reported that welfare losses attributable to air pollution account for more than *4.8% of global GDP*.

Environmental consequences of poor air quality extend to *soil degradation, water contamination, and biodiversity loss*. Acid deposition resulting from SO_2 and NO_X emissions alters soil chemistry, reducing nutrient availability and impairing crop yields. Aquatic ecosystems suffer from reduced pH levels, leading to species extinction and disruption of food chains. Additionally, persistent organic pollutants (POPs) and heavy metals adhere to airborne particles, contaminating terrestrial and aquatic habitats.

Urban smog episodes illustrate the severe impact of uncontrolled air pollution. The *Great Smog of London in 1952* caused over *12,000 premature deaths* and catalyzed the enactment of the UK Clean Air Act (Bell & Davis, 2001). In recent decades, Asian megacities such as *Beijing, Delhi, and Jakarta* have experienced recurrent smog events, with AQI levels frequently exceeding hazardous thresholds. These occurrences underscore the urgent need for *innovative air quality management strategies* that complement conventional regulatory measures such as emission standards and industrial controls.

2.2 Green Infrastructure and Air Quality

Green infrastructure (GI) refers to strategically planned networks of *natural and semi-natural elements* such as parks, urban forests, wetlands, green roofs, and vegetated corridors designed to deliver *ecosystem services* in urban landscapes (Tzoulas et al., 2007). GI is increasingly recognized as an effective tool for *air quality improvement* because of its capacity to *filter pollutants, regulate microclimates, and provide carbon sequestration* (Escobedo et al., 2011).

Plants within green infrastructure systems improve atmospheric conditions through multiple mechanisms:

- **Particulate Matter Deposition**: Leaves capture PM via *physical interception* and *electrostatic adhesion*. Leaf morphology—such as the presence of trichomes (hair-like structures) and rough surfaces—enhances pollutant capture efficiency (Nowak et al., 2014).
- **Gaseous Pollutant Absorption**: Through *stomatal uptake*, plants absorb NO_2, SO_2, O_3, and CO_2. These gases are either incorporated into metabolic pathways or stored in plant tissues.
- **Oxygen Generation and Carbon Sequestration**: Photosynthesis reduces CO_2 concentrations while releasing oxygen, improving ambient air quality.
- **Urban Heat Island (UHI) Mitigation**: By reducing surface and air temperatures, green spaces decrease *ozone formation*, which is temperature-dependent.

The air purification potential of urban vegetation is significant. For instance, *Nowak et al. (2014)* estimated that urban trees in the United States remove approximately *711,000 metric tons of air pollutants annually*, valued at *USD 3.8 billion in public health benefits*. Similarly, a study by *Escobedo et al. (2011)* found that urban forests in Santiago, Chile, reduce PM_{10} concentrations by up to *9%*, contributing to improved public health outcomes.

Beyond outdoor benefits, *green roofs and vertical greenery systems (VGS)* enhance *building energy efficiency and indoor air quality*. They act as thermal insulators, reducing energy consumption and greenhouse gas emissions, while also filtering airborne pollutants near the source of human exposure. Consequently, *green infrastructure represents a multi-functional solution* addressing not only air quality but also climate resilience, stormwater management, and urban aesthetics.

2.3 Role of Horticulture in Smog Mitigation

Horticultural strategies constitute a core component of green infrastructure for *smog and pollution control*. Unlike traditional gray infrastructure, which relies on mechanical filtration and chemical scrubbing, horticultural systems leverage *biological processes* for pollutant removal, making them cost-effective and sustainable.

Plants act as *natural biofilters*, intercepting dust particles on their surfaces and adsorbing gaseous pollutants via stomata. Certain species exhibit *phytoremediation capabilities*, metabolizing or immobilizing toxic substances such as heavy metals and persistent organic compounds (Zhou et al., 2020). The *NASA Clean Air Study (1989)* demonstrated that indoor plants such as *Areca palm (Dypsis lutescens)*, *Ficus benjamina*, and *Sansevieria trifasciata* effectively remove *benzene, formaldehyde, and trichloroethylene*, which are common VOCs in indoor environments.

Species selection is critical for optimizing pollutant removal efficiency. Characteristics such as *leaf surface roughness, stomatal density, leaf area index (LAI), and tolerance to pollution stress* determine plant effectiveness. For instance:

- *Ficus elastica* (Rubber plant) and *Azadirachta indica* (Neem) have high LAI, enabling significant PM capture.
- *Polyalthia longifolia* (Ashoka tree) and *Delonix regia* (Gulmohar) thrive in polluted environments, making them suitable for roadside plantations.
- Indoor species such as *Epipremnum aureum* (Money plant) and *Chlorophytum comosum* (Spider plant) are effective in VOC absorption.

Indoor horticulture solutions, including *green walls, living walls, and bio-walls*, are gaining traction in corporate and residential buildings, particularly *in high-rise urban environments with poor natural ventilation*. These installations not only improve indoor air quality but also enhance psychological well-being, reduce stress, and improve cognitive performance, as evidenced by studies in environmental psychology (Lohr, 2010).

3.1 Asia: Tackling Severe Smog through Urban Greening

Asia, home to some of the fastest-growing economies and most densely populated cities, faces severe air pollution challenges that threaten public health, economic productivity, and environmental sustainability. According to the World Health Organization (WHO) and IQAir (2022), 22 of the 30 most polluted cities globally are located in Asia, with India and China dominating the list. The primary drivers of this crisis include rapid industrialization, overreliance on fossil fuels, urban traffic congestion, biomass burning, and uncontrolled construction activities. Seasonal factors, such as winter inversions and crop residue burning, exacerbate smog episodes, particularly in regions like the Indo-Gangetic Plain.

Against this backdrop, horticulture-based strategies—comprising large-scale afforestation, urban forestry, vertical greenery systems (VGS), green roofs, and community gardening—have emerged as critical components of sustainable air quality management. These interventions not only address particulate matter ($PM_{2.5}$ and PM_{10}) and gaseous pollutants (NO_2, SO_2, O_3, and VOCs) but also mitigate urban heat islands, improve aesthetics, and enhance biodiversity.

3.1.1 The Scope of Air Pollution in Asia and the Need for Green Solutions

Asia's urbanization trajectory has been unprecedented. Cities like Beijing, Delhi, and Jakarta have witnessed explosive growth in population and vehicular density, leading to escalating emissions from transportation and industry. For example, Delhi often experiences AQI levels exceeding 400, categorized as "severe," during winter months (CPCB, 2020). Beijing, despite stringent emission control measures, continues to battle high $PM_{2.5}$ concentrations during smog episodes. These conditions demand interventions that complement traditional emission reduction strategies with nature-based solutions capable of delivering long-term benefits.

Horticulture-based approaches are particularly suitable for Asia due to their adaptability, cost-effectiveness, and potential for community engagement. Urban greening initiatives can intercept dust particles, absorb harmful gases, and reduce ambient temperatures, thereby indirectly curbing ground-level ozone formation. Moreover, horticulture projects align with global sustainability agendas, including the United Nations Sustainable Development Goals (SDGs) and the Paris Climate Agreement.

3.1.2 China: Pioneering Large-Scale Afforestation and Urban Greening

China stands out as a global leader in implementing horticulture-based air quality interventions, primarily through its large-scale afforestation programs and urban greening policies. Recognizing the detrimental effects of desertification and dust storms on air quality, the Chinese government launched the *Great Green Wall*, officially known as the *Three-North Shelter Forest Program*, in 1978.

The Great Green Wall: Scope and Impact

The Great Green Wall is an ambitious ecological engineering project stretching over 4,500 kilometers across northern China, designed to serve as a barrier against encroaching deserts and to mitigate dust storms originating from the Gobi Desert. By 2020, the project had planted over 66 billion trees, covering an estimated 35 million hectares of land (Cao et al., 2011). The benefits of this initiative extend beyond desertification control. Satellite imagery and air quality monitoring indicate a reduction in dust concentrations by approximately 20% in regions adjacent to afforestation belts (NASA, 2021). Additionally, urban areas within the influence zone of these shelterbelts have recorded $PM_{2.5}$ reductions of up to 12%, highlighting the direct link between horticulture and air pollution mitigation.

Urban Greening Initiatives in Beijing and Shanghai

In addition to rural afforestation, Chinese megacities have embraced urban greenery to counteract vehicular and industrial emissions. Beijing, for example, has significantly expanded its green cover through the development of urban parks, green corridors, and roadside plantations. The city also promotes *vertical greenery systems (VGS)* and *green roofs* on commercial and residential buildings to maximize greenery in space-constrained environments. Studies reveal that these green installations contribute to localized cooling effects, reducing ground-level ozone formation and trapping particulate matter on leaf surfaces.

Shanghai has adopted similar measures, integrating *green infrastructure into zoning regulations* and incentivizing developers to incorporate vegetated façades and rooftop gardens. These efforts not only improve air quality but also enhance the city's resilience against urban heat islands.

Policy and Governance Framework

China's success in horticulture-based air quality management is underpinned by strong policy support and institutional mechanisms. The inclusion of *"Ecological Civilization"* in the country's national development agenda emphasizes the integration of green infrastructure into urban planning. Moreover, substantial financial investments in research, species selection, and maintenance have ensured the sustainability of these programs. However, challenges such as monoculture plantations, water consumption concerns, and biodiversity trade-offs necessitate adaptive management strategies.

3.1.3 India: Urban Forestry and Vertical Gardens as Pollution Mitigation Tools

India faces some of the most severe air pollution episodes globally, with cities like Delhi, Kanpur, and Lucknow frequently recording hazardous AQI levels. Vehicular emissions, industrial discharges, biomass burning, and construction dust are the dominant contributors. Recognizing the urgency of the situation, the Government of India launched the *National Clean Air Programme (NCAP)* in 2019, targeting a 20–30% reduction in PM concentrations by 2024 (MoEFCC, 2019). Horticulture-based solutions constitute a critical component of this strategy.

Delhi Metro Vertical Gardens

One of the most innovative interventions in India is the installation of vertical gardens on metro pillars and flyover walls in Delhi. These structures feature ornamental and air-purifying species such as *Philodendron, Syngonium, and Money Plant*, arranged in modular panels equipped with automated irrigation systems. The Central Pollution Control Board (CPCB) reported that areas adjacent to these green walls exhibited a *14–17% reduction in suspended dust particles* and a *10% decrease in* NO_2 *concentrations* (CPCB, 2020). These localized improvements, though modest on a city-wide scale, underscore the potential of vertical greening in high-traffic corridors.

Tree Transplantation and Urban Forestry

In addition to vertical gardens, Delhi and other cities have initiated *tree transplantation projects* to preserve mature trees during infrastructure development. Furthermore, urban forestry programs under NCAP encourage municipalities to plant native and pollution-tolerant species such as *Azadirachta indica (Neem)*, *Polyalthia longifolia*, and *Ficus religiosa* along roadsides, medians, and institutional campuses.

Community Engagement and Challenges

While these interventions are commendable, their scalability is hindered by factors such as *water scarcity*, *high installation and maintenance costs*, and limited public participation. The success of horticultural strategies depends on sustained community involvement, which remains weak in many Indian cities. To overcome this barrier, NGOs and citizen-led groups are promoting micro-greening initiatives, including balcony gardens and community parks, which complement official programs.

3.1.4 Regional Examples: South and Southeast Asia

Beyond China and India, several other Asian nations are incorporating horticultural strategies into their urban sustainability agendas:

- *Singapore* has institutionalized green infrastructure through the *Green Plan 2030*, promoting vertical gardens, green roofs, and biodiversity corridors. The *Skyrise Greenery Incentive Scheme (SGIS)* offers financial support to building owners for installing vegetated façades, significantly improving local air quality and reducing ambient temperatures by 2–3°C (Tan et al., 2013).
- *Bangkok, Thailand* faces persistent smog episodes, particularly during the dry season. The city government has launched initiatives to create *green belts and rooftop gardens*, prioritizing native species that require minimal water and maintenance.
- *Jakarta, Indonesia* is combating vehicular emissions by developing *urban forests and pocket parks* in residential and commercial districts. These measures are integrated into climate resilience plans to address both air quality and flooding risks.

3.1.5 Key Lessons from Asia's Experience

Asia's experiments with horticulture-based air quality management offer several important lessons for global replication:

1. Policy Integration is Crucial

Programs such as China's Ecological Civilization and India's NCAP demonstrate that horticultural interventions succeed when embedded in broader policy frameworks.

2. Technology Enhances Efficiency

IoT-enabled irrigation systems, modular green wall designs, and drone-based monitoring have improved the scalability and maintenance of urban greening projects.

3. Community Engagement Ensures Sustainability

Citizen participation, as seen in Singapore's community gardening programs, is vital for ensuring long-term success and reducing maintenance burdens on municipal authorities.

4. Species Selection Matters

Pollution-tolerant, low-water-demand species ensure both ecological and economic sustainability. Native species are generally preferred for their adaptability and biodiversity support.

3.1.6 Persistent Challenges and Future Directions

Despite notable progress, Asia faces several obstacles in mainstreaming horticulture-based air quality solutions:

- *Land Constraints* in densely populated urban cores necessitate innovative approaches like vertical greening and rooftop gardens.
- *Water Scarcity* in arid regions, particularly northern India and northern China, challenges the maintenance of green infrastructure.
- *High Initial Costs* deter widespread adoption of technologies such as modular green walls and automated irrigation systems.
- *Institutional Fragmentation* between municipal bodies, environmental agencies, and urban developers often delays implementation.

To overcome these barriers, governments must *strengthen regulatory frameworks, expand financial incentives, and invest in research* to develop climate-resilient horticultural solutions. Furthermore, *regional cooperation platforms* can facilitate knowledge sharing, allowing cities to replicate successful models and avoid common pitfalls.

3.2 Europe: Integrating Green Roofs and Vertical Gardens

European cities have institutionalized green infrastructure through *policy mandates and incentives*. For example:

- *Paris* enforces legislation requiring *green roofs or solar panels* on all new commercial buildings, contributing to *air quality improvement and heat island mitigation*.

- *Berlin* implements a *Green Roof Strategy*, offering subsidies for vegetated rooftops, while *London* integrates urban tree planting under its *Clean Air Strategy*.
- Scandinavian countries such as *Sweden and Denmark* have adopted *eco-district planning*, embedding green corridors and vertical gardens into urban design.

Empirical studies indicate that these measures contribute to *reductions in PM and NO_2 concentrations*, while enhancing *urban biodiversity and aesthetic appeal*.

3.3 North America: Green Infrastructure as Policy

In the *United States*, green infrastructure adoption is driven by a combination of *environmental policies, civic engagement, and economic incentives*:

- *Chicago* boasts over *500 green roofs* covering more than *5.5 million square feet*, improving *local air quality and building energy efficiency*.
- *New York's MillionTreesNYC program* successfully planted over *one million trees*, providing pollutant removal and carbon sequestration benefits.
- *California's Urban Forestry Act* promotes large-scale tree planting to reduce *ozone concentrations in smog-prone areas* such as Los Angeles.

Canada also supports urban greening through municipal programs in *Toronto and Vancouver*, emphasizing *green roofs, street trees, and sustainable landscaping*.

3.4 Middle East and Africa: Adapting to Harsh Climates

Regions with *arid climates* face unique challenges in implementing horticultural solutions due to *water scarcity and extreme temperatures*. Nonetheless, innovative strategies are emerging:

- *United Arab Emirates (UAE)* promotes *indoor vertical farming* and *green wall installations* in commercial and residential complexes. These initiatives integrate *hydroponic and aeroponic systems* to minimize water use.
- *Dubai's Sustainable City* incorporates *rooftop gardens and greenhouses*, reducing indoor air pollution and cooling energy demand.
- In *Africa*, Nairobi's *Green Belt Movement* champions afforestation to combat *air pollution, deforestation, and soil erosion*. However, limitations such as *limited funding, inadequate policy frameworks, and lack of technical expertise* remain significant obstacles.

4. HORTICULTURAL TECHNIQUES FOR SMOG AND POLLUTION CONTROL

Urban Forestry

Urban forests are among the most effective strategies for air quality improvement. Trees reduce PM by up to *85 g per tree annually*, depending on species and canopy size. *Neem (Azadirachta indica)* and *Peepal (Ficus religiosa)* are highly effective in capturing dust and gases in tropical climates.

Green Roofs

Green roofs consist of vegetation layers installed on rooftops. They reduce dust deposition, absorb gaseous pollutants, and moderate building temperatures. Studies in Germany show that *green roofs can reduce* $PM_{2.5}$ *by up to 0.2* g/m^2 *annually*, contributing to localized air quality improvement.

Vertical Gardens

Vertical greenery systems (VGS) offer space-efficient solutions for dense urban environments. These systems combine ornamental and pollutant-tolerant plants to maximize air filtration. In Delhi, vertical gardens along highways reduced dust levels by *up to 20%* within six months of installation.

Phytoremediation

This technique employs plants to remove heavy metals, VOCs, and gaseous pollutants from the atmosphere. Species like *Chrysopogon zizanioides (Vetiver grass)* absorb lead and arsenic, while *Epipremnum aureum (Money plant)* reduces indoor VOCs.

Bio-Walls and Indoor Plants

Indoor green walls enhance air quality in office spaces and residential buildings. *NASA's research* identifies species such as *Aloe vera, Peace Lily, and Bamboo Palm* as effective in VOC reduction, making them ideal for polluted indoor environments.

5. CASE STUDIES

5.1 Beijing, China: The Great Green Wall Project

Background

Beijing, the capital of China, has long been synonymous with severe air pollution, particularly during winter months when *temperature inversions trap pollutants close to the ground*. The rapid industrialization of China during the late 20th century, coupled with heavy reliance on *coal-based energy*, vehicular emissions, and massive construction projects, led to *$PM_{2.5}$ concentrations exceeding 500 μg/m^3*, nearly *20 times higher than WHO safe limits*. Such conditions resulted in *hazardous AQI levels*, frequent smog episodes, and a public health crisis marked by rising cases of respiratory and cardiovascular diseases. Studies by the Chinese Ministry of Ecology and Environment highlighted that air pollution-related illnesses accounted for *over 1.6 million premature deaths annually* in China (HEI, 2020).

To counter this environmental degradation, the Chinese government initiated multiple *green infrastructure interventions*, of which the *Great Green Wall Project*, officially named the *Three-North Shelter Forest Program*, remains the most ambitious. Initially launched to prevent desertification and dust storms originating from the Gobi Desert, the project evolved into a *multi-functional ecological restoration initiative* that includes *air pollution control* as a core objective.

Project Scale and Implementation

The Great Green Wall Project began in *1978* and spans a projected *4,500 km* across northern China. The initiative involves *planting more than 35 billion trees*, creating a *massive ecological barrier* that serves as a defense against dust storms while enhancing urban air quality. The project is divided into *phases spanning several decades*, with completion expected by *2050*.

Key features of the implementation include:

- **Species Selection**: Tree species such *as Populus spp., Salix spp., and Pinus spp.* were chosen for their fast growth rates, *dense canopy structures*, and high tolerance to arid conditions.
- **Urban Integration**: Alongside rural afforestation, Beijing introduced *green belts, urban forests, and roadside plantations*, forming a *green buffer* around high-emission zones.

- **Technological Integration**: The use of *remote sensing, GIS mapping, and IoT-enabled irrigation systems* ensures efficient monitoring and adaptive management.

Impact on Air Quality

The impact of the Great Green Wall and associated urban greening efforts is well documented:

- **Reduction in $PM_{2.5}$ Levels**: NASA (2021) reported that *$PM_{2.5}$ concentrations dropped by 12%* in areas surrounding afforestation zones.
- **Localized Dust Reduction**: Chen et al. (2020) observed that urban shelterbelts reduced *dust concentrations by 15–20%* during high-pollution episodes.
- **Improved AQI During Dust Storms**: The frequency and intensity of dust storms decreased significantly, reducing particulate matter carried into Beijing.

Socio-Economic and Ecological Benefits

The project generated *millions of rural jobs*, providing income stability in impoverished regions. Ecologically, it *increased carbon sequestration capacity*, improved *biodiversity*, and *stabilized soil*. Economically, reductions in air pollution-related healthcare costs and productivity losses have provided *long-term financial benefits*.

Limitations

Despite its achievements, the project faces criticism:

- **Monoculture Plantations**: Over-reliance on single species poses *biodiversity risks* and vulnerability to pests.
- **Water Stress**: Certain species require *high water input*, raising sustainability concerns in arid regions.
- **Soil Degradation**: Poor site selection and inappropriate species have led to localized soil salinization.

5.2 Delhi, India: Vertical Garden Initiatives

Background

Delhi consistently ranks among the *world's most polluted cities*, with AQI levels often exceeding *400 (hazardous)* during winter due to a combination of *vehicular emissions, crop residue burning, industrial pollution, and unfavorable meteorology*. Public health studies indicate that *air pollution accounts for over 30,000 premature deaths annually in Delhi* (CPCB, 2020). The city's dense population and limited land availability necessitate *innovative space-efficient solutions* for air quality improvement.

Implementation

Delhi's *vertical garden program* emerged as a *localized horticultural intervention* to mitigate dust and gaseous pollutants in congested urban corridors. Key components include:

- **Roadside Green Walls**: Installed on *flyover pillars, metro station walls, and high-traffic corridors*.
- **Plant Selection**: Shade-tolerant and hardy species such as *Philodendron, Syngonium, Money Plant (Epipremnum aureum)*, and *Ficus species* dominate installations due to their *high dust retention and gas absorption capacity*.
- **Infrastructure Support**: Automated *drip irrigation systems*, nutrient supply mechanisms, and *modular planting panels* ensure sustainability.

Impact on Air Quality

Empirical data from the *Central Pollution Control Board (2020)* revealed:

- **Suspended Dust Reduction**: Dust levels decreased by *14–17%* in areas adjacent to vertical gardens.
- **NO_2 Concentration Drop**: Localized reductions of *10%* in nitrogen dioxide levels were recorded during high traffic hours.
- **Temperature Moderation**: Microclimatic cooling effects improved pedestrian comfort and reduced ground-level ozone formation.

Cost and Maintenance

The installation cost ranges between *INR 2,500–3,000 per square meter*, with additional expenses for *water and nutrient management*. Without government subsidies or corporate sponsorship, large-scale deployment remains financially challenging.

Citizen Participation

Non-governmental organizations and *resident welfare associations* are driving *community greening initiatives*, including *balcony gardens, micro-vertical walls*, and *rooftop plantations*, complementing municipal efforts.

5.3 Singapore: City in a Garden

Background

Singapore exemplifies *policy-driven urban greening*, positioning itself as a *"City in a Garden"*. Constrained by *land scarcity* and high *urban density*, Singapore integrates *green infrastructure with smart technologies* to achieve *air quality goals* and *urban livability*.

The Green Plan 2030

Singapore's *Green Plan 2030* mandates:

- **Expansion of Urban Greenery**: Increase in *urban tree cover* and *creation of nature parks*.
- **Vertical Greening Incentives**: Through the *Skyrise Greenery Incentive Scheme (SGIS)*, developers receive *funding subsidies* for green roofs and living walls.
- **Integration with Building Codes**: Mandatory incorporation of green elements in *commercial and residential projects*.

Impact on Air Quality and Urban Climate

Research shows:

- **$PM_{2.5}$ Reduction**: Vertical greenery and rooftop gardens contribute to *6–8% reductions* in particulate matter.

- **Temperature Reduction**: Surface temperatures decrease by *2–3°C*, mitigating heat island effects and reducing ozone formation.
- **Psychological Benefits**: Green infrastructure improves *mental well-being*, supporting Singapore's holistic sustainability agenda.

Integration of Smart Technologies

Singapore employs:

- **IoT-enabled Irrigation**: Sensors monitor *soil moisture and nutrient levels*, minimizing water use.
- **AI-driven Plant Health Systems**: Predictive algorithms identify stress indicators, optimizing maintenance.

Lessons for Other Nations

Singapore demonstrates that *policy incentives*, combined with *technological innovation* and *public-private partnerships*, are crucial for *scaling green infrastructure* in dense urban landscapes.

5.4 New York City, USA: MillionTreesNYC Program

Launched in *2007*, the *MillionTreesNYC initiative* sought to plant *one million trees by 2017*, a goal achieved ahead of schedule. The program's success was driven by *community engagement*, volunteer participation, and *inter-agency collaboration*. Environmental benefits include:

- **Pollutant Removal**: NYC trees remove *1,821 metric tons of pollutants annually*, including PM_{10}, ozone, and NO_2 (Nowak et al., 2018).
- **Public Health Benefits**: Air quality improvements reduce respiratory illnesses and associated healthcare costs.
- **Climate Co-benefits**: Increased carbon sequestration supports *climate mitigation goals*.

5.5 Paris, France: Green Roof Legislation

Paris *enforces legislation requiring green roofs or solar panels* on *new commercial buildings*. The city also:

- **Creates Green Corridors**: Linking *parks, riversides, and urban plazas* for improved ventilation.
- **Pollutant Reduction**: Studies reveal green roofs reduce *particulate matter by 5–7%*, particularly in *traffic-heavy areas.*
- **Heat Island Mitigation**: Vegetated roofs lower ambient temperatures by *1–2°C*, reducing *ozone formation.*

5.6 Nairobi, Kenya: Urban Green Belt Initiative

Background

Nairobi faces *dual challenges* of vehicular emissions and *dust storms*, compounded by *rapid urbanization*. Air pollution levels often exceed national standards, contributing to rising respiratory ailments.

Implementation

The city's *Urban Green Belt Initiative* involves:

- **Roadside Plantations**: Using *native drought-resistant species* such as *Acacia and Jacaranda.*
- **Community Afforestation Drives**: Led by the *Green Belt Movement*, mobilizing *local communities* for tree planting.

Impact

Preliminary data indicate:

- **Dust Deposition Reduction**: *12–15% decrease* in dust levels along major corridors.
- **Enhanced Biodiversity**: Increased *urban canopy cover* provides habitat for birds and pollinators.
- **Public Awareness**: The initiative fosters *citizen engagement in environmental stewardship.*

6. CHALLENGES AND LIMITATIONS

Despite the proven benefits of horticulture-based strategies for air quality management, several challenges persist:

6.1 Space Constraints

High-density urban areas often lack sufficient space for large-scale tree planting. Innovative solutions such as *vertical gardens and rooftop greenery* require structural modifications, limiting scalability.

6.2 High Maintenance Costs

Vertical gardens, green roofs, and indoor green walls demand *regular irrigation, fertilization, and pest management.* The operational costs can deter widespread adoption, especially in low-income regions.

6.3 Species Selection and Climate Adaptability

Not all plant species are equally effective in pollutant absorption. For example, broad-leaved species typically capture more particulate matter than needle-leaved species. However, species selection must also account for *climate adaptability*, water requirements, and resilience to pests.

6.4 Water Scarcity

In arid and semi-arid regions, the irrigation needs of large-scale green projects can *strain local water resources*, creating sustainability issues.

6.5 Policy Gaps

Many urban development plans lack *mandatory provisions for green infrastructure*. Even where policies exist, **enforcement is weak**, and funding is often insufficient.

6.6 Social Equity Concerns

Access to green infrastructure is often uneven, favoring affluent neighborhoods. This creates disparities in exposure to clean air and the associated health benefits.

7. FUTURE DIRECTIONS AND POLICY RECOMMENDATIONS

The integration of horticulture-based strategies for air quality management represents an innovative, sustainable, and long-term approach to combating urban smog and pollution. However, to make these strategies scalable, impactful, and equitable across different geographies, future efforts must focus on systemic planning, institutional reforms, public participation, and technological innovation. The following subsections outline key policy recommendations and future directions to mainstream horticulture-based interventions in urban ecosystems globally.

7.1 Green Building Codes and Regulatory Frameworks

One of the most effective ways to institutionalize horticulture-based air quality solutions is through *mandatory green building codes*. Governments should legislate building regulations that make the inclusion of green roofs, vertical gardens, and rooftop vegetation compulsory for new residential, commercial, and institutional developments. Such codes have already proven effective in cities like *Paris*, where legislation requires all new commercial buildings to incorporate either solar panels or green roofs.

Key Recommendations:

- **Mandatory Provisions:** National and municipal building authorities should revise construction norms to require a minimum percentage of green space coverage on rooftops and vertical façades.
- **Retrofitting Guidelines:** Policies should include provisions for retrofitting existing buildings with green walls and rooftop gardens, supported by structural audits to ensure safety.
- **Tax Incentives and Rebates:** Governments can offer tax reductions, property tax rebates, and expedited building permits for developers who adopt green infrastructure.
- **Standardization:** Development of *national and international standards for green infrastructure design, plant selection, and maintenance* to ensure uniformity and quality across projects.

Global Examples:

- **Singapore's Skyrise Greenery Incentive Scheme (SGIS):** Provides subsidies up to 50% of installation costs for green roofs and vertical gardens.
- **Toronto Green Roof Bylaw:** Mandates green roofs for buildings larger than 2,000 square meters.

By embedding such regulations into the legal framework, urban greening becomes not just a voluntary choice but a **compliance-driven environmental obligation**.

7.2 Community-Based Urban Forestry and Citizen Engagement

Public participation plays a pivotal role in sustaining horticultural projects. *Community-based urban forestry programs* encourage citizens to actively contribute to planting, maintaining, and monitoring trees in their neighborhoods. This participatory approach reduces the burden on municipal authorities while fostering environmental stewardship among urban residents.

Strategies for Implementation:

- **Tree Adoption Programs:** Residents, businesses, and schools can adopt trees or vertical garden units for maintenance.
- **Awareness Campaigns:** Use social media, community events, and educational workshops to inform people about the health benefits of green infrastructure.
- **Gamification:** Introduce digital platforms where citizens can track their contributions to air quality improvements, earn points, and receive rewards or recognition.
- **Collaborations with NGOs:** Partnering with environmental NGOs can amplify outreach and mobilization efforts for large-scale planting drives.

Case Example:

- **MillionTreesNYC Program:** Successfully combined government initiatives with citizen participation to plant one million trees, improving air quality and enhancing social cohesion.

Empowering communities ensures *long-term sustainability and accountability* of horticulture-based projects, making urban greening a shared responsibility rather than an isolated government task.

7.3 Development of Species-Specific Guidelines

The effectiveness of horticultural interventions depends heavily on *selecting plant species that are best suited for local environmental conditions and pollution challenges*. Plants differ in their ability to capture particulate matter, absorb gaseous pollutants, and adapt to stress factors like heat, drought, or pests.

Actionable Measures:

- **Pollution Absorption Index (PAI):** Develop indices to rank plant species based on their pollutant absorption capacity, canopy density, and surface morphology.
- **Climate-Specific Suitability Maps:** Research institutions should create region-wise plant suitability charts for rooftop gardens, roadside plantations, and vertical greenery.
- **Low-Maintenance and Native Species:** Prioritize native and drought-resistant species to minimize irrigation needs and ensure ecological harmony.
- **Indoor Plant Guidelines:** For indoor air quality improvement, identify species effective in absorbing VOCs (e.g., Areca Palm, Peace Lily, and Snake Plant).

Benefits of Species Optimization:

- **Enhanced Air Purification:** Maximizing pollutant uptake through targeted species selection.
- **Resource Efficiency:** Reducing water and nutrient requirements.
- **Biodiversity Conservation:** Avoiding monoculture practices that can degrade soil and attract pests.

Collaboration between *botanists, urban planners, and environmental scientists* is critical to developing scientifically validated and widely disseminated species-specific guidelines.

7.4 Integration of Digital Technologies and Smart Monitoring

Technological innovation can transform horticulture-based pollution control from a static intervention to a *dynamic, data-driven system*. The integration of *IoT (Internet of Things), AI (Artificial Intelligence), and Big Data analytics* allows for real-time monitoring and predictive modeling of both plant health and air quality improvements.

Key Technological Interventions:

- **AI-Powered Air Quality Monitoring:** Deploy sensors to track pollutant levels and link them with vegetation performance metrics.
- **IoT-Enabled Irrigation Systems:** Automate watering schedules based on soil moisture, temperature, and humidity data, reducing water wastage.

- **Drone-Assisted Monitoring:** Use drones for large-scale green belt inspections, pest detection, and vegetation health analysis.
- **Mobile Applications:** Develop apps that provide live dashboards for urban residents to monitor neighborhood air quality and greenery coverage.

Benefits:

- **Cost Efficiency:** Automated systems reduce labor costs associated with maintenance.
- **Precision Management:** Real-time insights enable rapid response to plant stress or irrigation failures.
- **Transparency:** Publicly accessible data enhances accountability and citizen trust.

Singapore's use of *AI-driven irrigation systems* and plant health diagnostics serves as a global model for how smart technologies can support green infrastructure management.

7.5 Public-Private Partnerships (PPPs) and Corporate Engagement

Given the high costs associated with green roofs, vertical gardens, and large-scale afforestation projects, *public-private partnerships (PPPs)* are essential for financing and sustaining these initiatives. Corporations can incorporate urban greening projects under their *Corporate Social Responsibility (CSR)* mandates, leveraging private capital for public environmental benefits.

Models of Collaboration:

- **CSR-Based Funding:** Encourage corporations to fund green walls in public spaces, schools, and hospitals as part of sustainability initiatives.
- **Green Credits:** Companies that invest in urban greening could receive *green credits*, which can offset their carbon footprint obligations.
- **Advertising Rights:** Allow businesses to sponsor green installations in exchange for branding opportunities on-site.
- **PPP Maintenance Contracts:** Assign long-term maintenance responsibilities to private firms under government supervision.

Success Example:

- **Delhi Metro Vertical Garden Initiative:** Funded by private entities under CSR, reducing project costs for the government while promoting corporate sustainability branding.

PPPs can create a *win-win model* where both the environment and businesses benefit, accelerating the pace of green infrastructure development globally.

7.6 Incentives, Funding Mechanisms, and Financial Innovation

Urban greening and horticulture-based pollution control projects require substantial capital investment, particularly in the initial phases. To ensure scalability, *innovative financial mechanisms* must be integrated into urban planning frameworks.

Proposed Mechanisms:

- **Green Bonds:** Issue municipal and corporate green bonds dedicated to financing urban forestry, green roofs, and vertical gardens.
- **Carbon Offset Programs:** Develop frameworks where companies can purchase carbon credits by investing in horticulture-based initiatives.
- **Urban Greening Credits:** Introduce credits for developers who exceed green infrastructure compliance levels, allowing them to trade these credits in the market.
- **International Climate Finance:** Leverage funding from organizations such as the *Green Climate Fund (GCF)* and the *Global Environment Facility (GEF)* for large-scale afforestation in developing countries.
- **Microfinance for Community Projects:** Provide small-scale loans to households and local communities for balcony gardens and small green walls.

Why Financial Incentives Matter:

- Lower upfront costs make green projects feasible for both governments and private actors.
- Financial innovation attracts global investors seeking *ESG (Environmental, Social, Governance)*-compliant portfolios.
- Ensures *long-term sustainability* by creating revenue streams for maintenance.

7.7 Policy Integration and Multi-Sectoral Coordination

For horticulture-based air quality management to succeed, policies must *be integrated across sectors* such as urban planning, transportation, water management, and health.

Policy Integration Steps:

- **Cross-Ministerial Committees:** Establish committees involving environment, urban development, housing, and health ministries to align objectives.
- **Air Quality Action Plans:** Incorporate horticulture as a primary mitigation measure alongside emission reduction strategies.
- **International Collaboration:** Facilitate knowledge-sharing platforms for cities to exchange best practices in green infrastructure.

CONCLUSION

Horticulture-based strategies offer an effective, sustainable, and multi-functional approach to improving air quality in urban areas. While challenges such as space limitations, maintenance costs, and policy gaps exist, innovative solutions, community engagement, and technology integration can help overcome these barriers. Global case studies demonstrate that cities embracing *green infrastructure* experience significant improvements in *air quality, climate resilience, and urban livability*. To achieve scalable impact, governments, private sectors, and communities must collaborate to embed green infrastructure in the core of urban development policies. The rapid urbanization and industrialization of the past few decades have led to unprecedented levels of air pollution, posing significant threats to human health, ecosystem stability, and overall quality of life. The emergence of smog as a persistent urban phenomenon, particularly in densely populated cities such as Beijing, Delhi, and Los Angeles, underscores the urgent need for sustainable and innovative mitigation strategies. Traditional approaches to air pollution control—such as stricter emission standards, technological upgrades in industries, and the promotion of clean energy—remain crucial, but they often overlook the complementary role of natural systems in restoring atmospheric balance. In this context, horticulture-based strategies, which integrate plants and green infrastructure into urban environments, offer a promising, nature-based solution for combating air pollution while delivering multiple co-benefits.Green infrastructure is more than an aesthetic addition to city landscapes; it functions as a dynamic system capable of delivering essential ecosystem services, including pollutant removal, carbon sequestration, microclimate regulation, and biodiversity enhancement. Urban forestry, vertical gardens, green roofs, bio-walls, and community green belts have been shown to significantly reduce concentrations of particulate matter ($PM_{2.5}$ and PM_{10}), nitrogen oxides (NO_X), sulfur dioxide (SO_2), volatile organic compounds (VOCs), and ground-level ozone. By capturing pollutants on leaf surfaces, absorbing harmful gases through stomata, and facilitating the deposition of airborne particles, vegetation serves as a natural biofilter in urban settings.

Case studies from around the world reaffirm the tangible benefits of horticulture-based interventions. Beijing's Great Green Wall project has not only reduced $PM_{2.5}$ concentrations in urban and peri-urban zones but also created ecological corridors that enhance resilience against desertification. Delhi's vertical garden initiatives, though localized, have demonstrated measurable reductions in suspended dust particles and NO_2 levels near busy intersections and metro pillars. Similarly, Singapore's "City in a Garden" approach has transformed green infrastructure into a core urban planning principle, reducing smog formation and mitigating the urban heat island effect. These examples validate the role of horticultural strategies as a critical component of integrated air quality management frameworks.The advantages of green infrastructure extend well beyond air pollution control. They contribute to climate change mitigation by sequestering carbon dioxide, provide shading that reduces urban heat, improve stormwater management, and enhance mental well-being by creating restorative urban spaces. Furthermore, green projects create employment opportunities in nursery development, landscaping, and maintenance, thus linking environmental sustainability with socio-economic development.From a public health perspective, the impact of horticulture-based interventions is particularly significant. Reducing air pollutants such as fine particulate matter and ozone lowers the risk of respiratory illnesses, cardiovascular disorders, and premature mortality. By improving air quality in schools, hospitals, and residential areas, green infrastructure directly contributes to reducing the healthcare burden associated with pollution-related diseases. These outcomes reinforce the argument that horticulture is not merely an environmental strategy but a public health imperative.Despite the proven benefits, several challenges hinder the large-scale implementation of horticulture-based strategies. Space constraints in densely populated cities often limit the feasibility of urban forestry projects, necessitating alternative solutions such as vertical gardens and rooftop greenery. Maintenance costs, particularly for high-tech installations like vertical gardens, pose financial challenges, especially in low- and middle-income countries. Additionally, species selection remains a critical determinant of success; inappropriate plant choices can lead to high water demand, invasive behavior, or vulnerability to pests.Another limitation lies in policy fragmentation and governance gaps. In many regions, air quality management policies operate in isolation from urban planning and landscape development frameworks, resulting in missed opportunities for integrated solutions. Furthermore, the absence of clear guidelines on plant species suitability, maintenance protocols, and performance monitoring hinders consistency in implementation and evaluation. Addressing these gaps requires coordinated efforts across government agencies, private stakeholders, research institutions, and civil society organizations.

Looking ahead, the fusion of horticulture with advanced technologies presents a transformative opportunity. Internet of Things (IoT)-enabled irrigation systems, AI-driven plant health monitoring, and drone-assisted maintenance can optimize resource use while ensuring high survival rates for planted species. Smart monitoring systems can link pollutant removal data with vegetation performance, creating feedback loops for adaptive management. Digital platforms and mobile applications can further enhance citizen engagement, allowing urban residents to participate in tree-planting drives, report maintenance issues, and monitor local air quality improvements in real time.

The success of these technological interventions, however, depends on the availability of infrastructure, financial resources, and technical expertise. Ensuring equitable access to such innovations is crucial for avoiding a digital divide in environmental solutions between developed and developing nations.For horticulture-based air quality solutions to achieve their full potential, enabling policy frameworks and international cooperation are essential. Governments must integrate green infrastructure into building codes, urban design regulations, and air quality management plans. Financial incentives—such as green bonds, tax rebates, and carbon credits—should be institutionalized to encourage private investment in horticultural projects. Public-private partnerships (PPPs) offer a viable model for scaling green infrastructure initiatives, leveraging corporate social responsibility (CSR) contributions and fostering long-term sustainability.

At the international level, knowledge exchange platforms can facilitate the sharing of best practices, technical guidelines, and research findings. Global initiatives such as the *UN Decade on Ecosystem Restoration (2021–2030)* provide an ideal framework for embedding horticultural strategies into climate and pollution mitigation agendas. Aligning these efforts with the *Sustainable Development Goals (SDGs)*—particularly SDG 11 (Sustainable Cities and Communities), SDG 13 (Climate Action), and SDG 15 (Life on Land)—can amplify their impact on human well-being and planetary health.The urgency of the air pollution crisis demands solutions that are not only scientifically sound but also socially inclusive, economically viable, and environmentally regenerative. Horticulture-based strategies meet all these criteria by offering a *triple win*—improving air quality, enhancing urban resilience, and delivering socio-economic benefits. As cities continue to expand, the integration of vegetation into built environments must shift from being a symbolic gesture to a *strategic necessity* embedded in urban planning.

Future research should focus on refining species selection for different climatic zones, developing cost-effective vertical garden technologies, and quantifying the long-term economic benefits of green infrastructure to strengthen the business case for policymakers and investors. Similarly, education and awareness campaigns should aim to foster a culture of green responsibility among urban residents, making community participation a cornerstone of environmental governance.In conclusion, horticulture-based air quality management represents a paradigm shift in addressing urban pollution—a shift from reactive technological fixes to proactive, nature-based solutions that harmonize human activity with ecological processes. While challenges remain, the path forward is clear: by embracing green infrastructure as an integral component of urban ecosystems, we can build healthier cities, mitigate the effects of climate change, and create a sustainable future for generations to come. The success of this endeavor hinges on *collective action, visionary policies, and a shared commitment to reimagining our cities as living, breathing systems rather than concrete jungles.*

REFERENCES

Bell, M. L., & Davis, D. L. (2001). Reassessment of the lethal London fog of 1952: Novel indicators of acute and chronic consequences of acute exposure to air pollution. *Environmental Health Perspectives, 109*(Suppl 3), 389–394. DOI: 10.1289/ehp.01109s3389 PMID: 11427388

Cao, S., Chen, L., Shankman, D., Wang, C., Wang, X., & Zhang, H. (2011). Excessive reliance on afforestation in China's arid and semi-arid regions: Lessons in ecological restoration. *Earth-Science Reviews, 104*(4), 240–245. DOI: 10.1016/j.earscirev.2010.11.002

Central Pollution Control Board (CPCB). (2020). *National Ambient Air Quality Status and Trends Report.* Ministry of Environment, Forest and Climate Change, Government of India.

Chen, B., Chen, Y., & Liu, X. (2020). Evaluating the effect of urban shelterbelts on dust storm mitigation in northern China. *Environmental Monitoring and Assessment, 192*(5), 302. DOI: 10.1007/s10661-015-4536-0 PMID: 32322989

Escobedo, F. J., Kroeger, T., & Wagner, J. E. (2011). Urban forests and pollution mitigation: Analyzing ecosystem services and disservices. *Environmental Pollution, 159*(8–9), 2078–2087. DOI: 10.1016/j.envpol.2011.01.010 PMID: 21316130

Gurjar, B. R., Jain, A., Sharma, A., Agarwal, A., Gupta, P., Nagpure, A. S., & Lelieveld, J. (2016). Human health risks in megacities due to air pollution. *Atmospheric Environment, 128*, 233–243. DOI: 10.1016/j.atmosenv.2015.12.039

Health Effects Institute (HEI). (2020). *State of Global Air 2020.* https://www.stateofglobalair.org/

Jacob, D. J. (1999). *Introduction to atmospheric chemistry.* Princeton University Press. DOI: 10.2307/j.ctt7t8hg

Lohr, V. I. (2010). Benefits of nature: What we are learning about why people respond to nature. *Journal of Physiological Anthropology, 29*(3), 111–117. DOI: 10.2114/jpa2.29.111 PMID: 17435348

Ministry of Environment. Forest and Climate Change (MoEFCC). (2019). *National Clean Air Programme (NCAP).* Government of India.

NASA. (2021). *Monitoring the impact of the Great Green Wall on dust and vegetation using satellite data.* National Aeronautics and Space Administration. https://earthobservatory.nasa.gov/

Nowak, D. J., Crane, D. E., & Stevens, J. C. (2014). Air pollution removal by urban trees and shrubs in the United States. *Urban Forestry & Urban Greening*, *4*(3–4), 115–123. DOI: 10.1016/j.ufug.2006.01.007

Nowak, D. J., Hirabayashi, S., Bodine, A., & Hoehn, R. (2018). Tree and forest effects on air quality and human health in the United States. *Environmental Pollution*, *193*, 119–129. DOI: 10.1016/j.envpol.2014.05.028 PMID: 25016465

Organisation for Economic Co-operation and Development (OECD). (2021). *The economic consequences of outdoor air pollution*. OECD Publishing., DOI: 10.1787/9789264257474-

Tan, P. Y., Wang, J., & Sia, A. (2013). Perspectives on five decades of the Singapore Green Plan: From "Garden City" to "City in a Garden". *Cities (London, England)*, *32*, 24–32. DOI: 10.1016/j.cities.2013.02.001

Tzoulas, K., Korpela, K., Venn, S., Yli-Pelkonen, V., Kaźmierczak, A., Niemela, J., & James, P. (2007). Promoting ecosystem and human health in urban areas using green infrastructure: A literature review. *Landscape and Urban Planning*, *81*(3), 167–178. DOI: 10.1016/j.landurbplan.2007.02.001

World Bank. (2020). *The cost of air pollution: Strengthening the economic case for action*. World Bank Group.

World Health Organization (WHO). (2022). *Ambient (outdoor) air pollution*. https://www.who.int/news-room/fact-sheets/detail/ambient-(outdoor)-air-pollution

World Health Organization (WHO). (2023). *Air pollution*. https://www.who.int/health-topics/air-pollution

Zhou, X., Cao, Y., & Ma, Y. (2020). Phytoremediation of air pollutants: A review. *Environmental Science and Pollution Research International*, *27*(4), 4489–4509. DOI: 10.1007/s11356-019-07053-6 PMID: 31832956

Chapter 9
Green Entrepreneurship for Clean Air:
A Bibliometric Approach to Environmental Solutions

Murat Avci
https://orcid.org/0000-0002-2355-1396
Kyrgyz-Turkish Manas University, Kyrgyzstan

Azamat Maksudunov
https://orcid.org/0000-0002-5010-513X
Kyrgyz-Turkish Manas University, Kyrgyzstan

ABSTRACT

Air pollution severely threatens environmental sustainability, public health, and economic development. While technological innovation and regulatory frameworks have been widely studied as mechanisms for addressing air quality issues, the entrepreneurial ecosystem, particularly green entrepreneurship, has emerged as a powerful but underexplored driver of sustainable solutions. This book chapter aims to provide a comprehensive overview of the literature on green entrepreneurship by analyzing scientific publications from the Web of Science. A bibliometric analysis approach has been employed for data analysis, and visualization has been conducted using R programming tools. The most relevant sources, authors, affiliations, countries, and the most cited documents were identified and discussed. The potential research trends and other findings presented in this research serve as a roadmap for future research.

DOI: 10.4018/979-8-3373-2469-2.ch009

INTRODUCTION

Environmental sustainability has emerged as one of the most critical global priorities of the twenty-first century. Rapid industrialization, urban expansion, population growth, and an ongoing dependence on fossil fuels have significantly deteriorated air quality, posing substantial risks to human health, biodiversity, and climate stability. The World Health Organization (2023) estimates that air pollution causes nearly seven million premature deaths annually, underscoring the urgency of finding effective mitigation strategies. Addressing these escalating environmental risks requires not only regulatory and policy interventions but also the mobilization of entrepreneurial innovation capable of delivering practical, scalable solutions.

Within this context, green entrepreneurship has gained prominence as a pathway that integrates environmental responsibility with economic innovation. In contrast to conventional entrepreneurship, which primarily emphasizes profit maximization, green entrepreneurship pursues dual objectives—creating economic value while reducing ecological harm (GIZ, 2014; Schaltegger & Wagner, 2011). Green entrepreneurs contribute to the transition toward low-carbon and circular economies through the development of eco-innovative products, services, and technologies (Mehta et al., 2024; Sobko et al., 2024). Empirical studies (Guo et al., 2025; Wei et al., 2024) highlight that such ventures are no longer peripheral; they have become central pillars of sustainable urban development. Initiatives such as electric mobility systems, renewable energy startups, and air quality monitoring technologies demonstrate how green enterprises actively reduce particulate matter ($PM_{2.5}$) and enhance urban environmental health (Cojoianu et al., 2024).

Despite its growing relevance, green entrepreneurship faces persistent challenges. Prior research identifies a range of barriers—limited access to green finance, inadequate sustainability-oriented training, and fragmented regulatory systems (Chandel, 2022; GIZ, 2014). In developing and transition economies, these issues are often compounded by weak institutional support and low environmental awareness among consumers. Studies (Abdelwahed et al., 2023; Chu et al., 2021) further reveal sociocultural and market-based obstacles, including inconsistent environmental policies, consumer skepticism toward green products, and the absence of sustained mechanisms for supporting early-stage clean-tech ventures. To overcome these constraints, scholars (Nadhifa et al., 2024; Yi, 2020) emphasize the importance of public–private partnerships, mentoring initiatives, and impact investment as key drivers of scalability.

Recent evidence also positions green entrepreneurship as a catalyst for broader development outcomes, including economic growth and human well-being. Neumann (2022) reports that economies with a higher prevalence of green enterprises tend to achieve superior per capita income levels and Human Development Index

(HDI) scores. Although macro-level environmental outcomes, such as reductions in CO_2 emissions, remain less visible, this may reflect the relatively small scale and early-stage nature of green ventures. Conversely, Cojoianu et al. (2024) found that cities with dense concentrations of green startups exhibit significant improvements in $PM_{2.5}$ levels, particularly when innovation ecosystems are well established. These findings suggest that green entrepreneurship can serve as both an environmental and developmental policy lever.

Nevertheless, a distinct gap remains in the literature regarding the explicit relationship between green entrepreneurship and air pollution mitigation. Much of the research to date (Chandel, 2022; Albort-Morant et al., 2017) focuses on renewable energy, waste management, and sustainable agriculture, while the specific mechanisms through which green enterprises contribute to cleaner air are rarely analyzed. This absence highlights the need for a comprehensive examination of the scholarly landscape connecting green entrepreneurship to air quality improvement. Indeed, although the link between entrepreneurship and sustainability is well documented, its intersection with air pollution and public health remains underexplored (Neumann, 2022; Wang et al., 2021).

To address this gap, this book chapter conducts a bibliometric analysis of the scientific literature on green entrepreneurship, emphasizing its intersection with clean air and environmental sustainability agendas. Using data from the Web of Science Core Collection, the study employs bibliometric tools implemented in R software to map the structure, trends, and intellectual evolution of the field. It identifies influential authors, journals, institutions, and thematic clusters to provide a comprehensive understanding of how green entrepreneurship research has developed over time.

The contribution of this chapter is twofold. First, it synthesizes existing conceptual and theoretical frameworks, clarifying the multidimensional nature of green entrepreneurship as it aligns with the Sustainable Development Goals (SDGs). Second, it offers data-driven insights through bibliometric mapping, enabling scholars and policymakers to identify research gaps, emerging topics, and collaborative patterns shaping the field's future trajectory. In particular, it illuminates the intersections among concepts such as green entrepreneurial orientation (GEO), ecological innovation, and clean air strategies, highlighting their relevance to both climate action and public health (Criado-Gomis, 2017; Hameed et al., 2021).

The chapter proceeds as follows: Section 2 provides a conceptual review of green entrepreneurship and its theoretical foundations; Section 3 examines its role in air pollution mitigation; Section 4 presents the methodological framework of the bibliometric analysis; Section 5 discusses the findings; and the final section offers implications for research and policy. Through this interdisciplinary approach, the chapter contributes to a more integrated understanding of how green entrepreneurship functions not only as an economic mechanism but also as an environmental tool for achieving cleaner air and sustainable development.

THE CONCEPTUALIZATION AND EVOLUTION OF GREEN ENTREPRENEURSHIP

Green entrepreneurship has emerged as a compelling response to the multifaceted challenges posed by environmental degradation, climate change, and the need for sustainable economic growth. Unlike traditional entrepreneurial models that primarily prioritize profit maximization, green entrepreneurship embodies a more holistic vision of value creation, integrating economic viability with environmental sustainability and social well-being. It is not only a mechanism for environmental protection but also a transformative force that redefines how business and ecological stewardship intersect in the modern world (Chandel, 2022; Schaltegger & Wagner, 2011).

The roots of green entrepreneurship are grounded in the broader discourse on sustainable development. The 1987 Brundtland Report by the World Commission on Environment and Development marked a critical turning point by introducing sustainability as a guiding principle for development: meeting the needs of the present without compromising the ability of future generations (WCED, 1987). This framework laid the intellectual groundwork for new approaches to business that move beyond economic efficiency alone. As environmental concerns intensified globally, from rising greenhouse gas emissions to loss of biodiversity, the entrepreneurial landscape began to shift toward more responsible and environmentally conscious business practices.

Green entrepreneurship emerged in this context as a response to both environmental urgency and the opportunity to develop innovative, sustainable solutions. It is generally defined as the process of identifying, developing, and implementing business ideas that contribute positively to environmental sustainability, either by offering green products and services or by operating through environmentally friendly processes (Ansari, 2025; GIZ, 2014). Dean and McMullen (2007) describe it as the exploitation of market imperfections related to environmental degradation, whereby entrepreneurs can address ecological problems while generating economic value. In essence, green entrepreneurship functions as a bridge between ecological responsibility and economic opportunity.

A key feature that distinguishes green entrepreneurship is its alignment with the "3P" model—People, Planet, and Profit (Elkington, 1997). This framework, articulated by Chandel (2022), emphasizes a balanced integration of social inclusion, environmental protection, and economic sustainability. Green entrepreneurs are motivated not solely by financial returns but by a commitment to generate broader societal benefits, such as reducing carbon emissions, conserving natural resources, and creating green jobs in underserved communities. This makes GE not only a business model but also a vehicle for environmental and social transformation.

Over time, academic interest in green entrepreneurship has grown substantially, accompanied by more nuanced conceptual frameworks. The literature now recognizes multiple forms and levels of green entrepreneurship, ranging from individual ecopreneurs who initiate grassroots green innovations to large-scale institutional entrepreneurs who integrate sustainability into corporate strategy (Mehta et al., 2024). GIZ (2014) proposes a typology distinguishing between "core green entrepreneurs," whose business models are intrinsically environmental, and "peripheral green entrepreneurs," who embed green practices into otherwise conventional business operations. This distinction is analytically useful for understanding the depth of commitment and strategic orientation within different green enterprises.

Besides, technological innovation is central to the development of green entrepreneurship. Across sectors, green entrepreneurs leverage advancements in renewable energy, smart agriculture, circular product design, and digital infrastructure to create solutions that are both economically and ecologically viable (Sobko et al., 2024). For example, the integration of digital technologies such as the Internet of Things (IoT), blockchain, and artificial intelligence has enabled more efficient monitoring of resource use, waste reduction, and transparent environmental reporting. These tools not only increase operational efficiency but also enhance trust and accountability in green markets.

Recent empirical studies (Ansari, 2025; Mehta et al., 2024; Tien et. al., 2023; Onnoghen et al., 2024) also show that green entrepreneurship can be a significant driver of clean air initiatives, particularly through innovations in transportation, energy efficiency, and urban infrastructure. Enterprises focused on clean mobility solutions, decentralized renewable energy systems, and pollution monitoring technologies exemplify how entrepreneurship can directly contribute to public health and environmental quality. Furthermore, the shift toward circular economy principles–designing out waste and keeping materials in use–has become a foundational strategy within green business models, offering both environmental benefits and long-term economic resilience (Cojoianu et al., 2024; Odeyemi et al., 2024).

The evolution of green entrepreneurship has also been accompanied by increasing policy and institutional support. International frameworks such as the Paris Agreement and the UN Sustainable Development Goals (SDGs) have reinforced the need for private-sector innovation in achieving sustainability targets. National governments and regional bodies have responded with green subsidies, tax incentives, and funding programs aimed at promoting green start-ups and sustainable innovation ecosystems (Mehta et al., 2024; Sobko et al., 2024). For instance, the European Union's expansion of green bond mechanisms and the development of digital platforms for knowledge transfer have provided critical infrastructure for GE growth (Huang et al., 2024; European Environment Agency, 2024).

Bibliometric and systematic literature reviews (Duran, 2024; Karadayı et al., 2025; Mondal & Singh, 2025; Muo & Azeez, 2019) indicate a sharp increase in scholarly output on green entrepreneurship, particularly after 2015, with focal areas including climate innovation, low-carbon transition, and circular economy practices. This growing academic attention underscores both the theoretical richness and the practical urgency of green entrepreneurship as a field of study. Nevertheless, gaps remain in our understanding of how green entrepreneurship functions across different institutional contexts, cultural settings, and stages of economic development (Karadayı et al., 2025; Mondal & Singh, 2025; Neumann, 2022).

In conclusion, green entrepreneurship represents a critical shift in how business activities are conceived and implemented in the face of escalating environmental challenges. It offers a framework through which entrepreneurs can create meaningful ecological and social impact while sustaining economic viability. As the field continues to mature, further research is needed to refine its theoretical underpinnings, evaluate its real-world outcomes, and identify pathways to scale its transformative potential across geographies and sectors. By embedding environmental consciousness into the core of business practice, green entrepreneurship stands as a vital strategy in the global pursuit of sustainability and clean air.

THEORETICAL APPROACHES TO GREEN ENTREPRENEURSHIP

Green entrepreneurship has developed into a distinct branch of entrepreneurship research, emphasizing the simultaneous creation of economic value and the pursuit of environmental and social objectives. Unlike traditional entrepreneurs, green entrepreneurs prioritize sustainability in their business models, integrating environmental concerns into core strategic decisions. Accordingly, to understand the motivations, behaviors, and systems surrounding green entrepreneurship, several theoretical frameworks have been proposed. Below is an in-depth review of these approaches.

Ecological Modernization Theory

Ecological Modernization Theory (EMT) emerged in the 1980s within the context of German and Dutch environmental sociology. EMT posits that environmental degradation is a systemic consequence of industrialization but can be mitigated through internal reforms within capitalist economies using technological and institutional innovation (Mol & Sonnenfeld, 2000).

From this perspective, green entrepreneurs function as catalysts of systemic reform, utilizing innovation and market mechanisms to minimize environmental degradation. EMT emphasizes the role of collaborative governance involving state regulations, corporate initiatives, and civil society in promoting sustainable practices (Buttel, 2000). Green startups engaged in renewable energy, circular economy models, or waste reduction embody this logic, acting as catalysts for eco-modernization within existing industrial systems (Kachynska et al., 2022).

Triple Bottom Line (TBL) Framework

Developed by John Elkington (1997), the Triple Bottom Line (TBL) framework argues that businesses should measure success not only in terms of financial performance but also environmental sustainability and social equity. The three dimensions People, Planet, and Profit form a holistic lens through which green entrepreneurship is evaluated.

TBL is especially relevant in explaining how green entrepreneurs manage trade-offs between economic viability and ecological responsibility. Startups today increasingly cater to eco-conscious consumers who evaluate products based on carbon footprint, ethical sourcing, and social responsibility, not just price and quality (Gu et al., 2022; Savitz, 2013). In this framework, green entrepreneurship is more than just an ethical imperative—it is a strategic advantage aligned with shifting consumer preferences and regulatory expectations (Cho, 2024).

Sustainable Entrepreneurship Model

The Sustainable Entrepreneurship Model integrates economic, environmental, and social objectives into a unified entrepreneurial strategy (Belz & Binder, 2017). Schaltegger and Wagner (2011: 223) define sustainable entrepreneurs as those who proactively recognize sustainability-related market opportunities and embed them within their business operations. This model conceptualizes green entrepreneurs as individuals who pursue "transformational innovation" that aligns with sustainable development goals (Hall et al., 2010). Unlike traditional entrepreneurs, they are driven by an imperative to solve societal and ecological problems through profitable ventures (Yasir et al., 2023).

The sustainable entrepreneurship model is particularly aligned with the United Nations Sustainable Development Goals (SDGs). Green entrepreneurs play a critical role in advancing goals such as affordable clean energy, responsible consumption, and climate action (Iqbal et al., 2020; United Nations, 2015).

Institutional Theory and Legitimacy

Institutional Theory offers a sociological explanation for how organizations conform to societal norms and expectations to gain legitimacy. According to DiMaggio and Powell (1983), businesses adopt environmental practices not only for efficiency but also to gain acceptance and legitimacy within institutional environments (Fernando & Lawrence, 2014). In green entrepreneurship, this is reflected in the growing adoption of certifications like ISO 14001 or EMAS. These standards signal a commitment to environmental stewardship and enhance credibility in the eyes of stakeholders (Bansal & Clelland, 2004; Gao et al., 2018).

Neo-institutional approaches further suggest that internal values, such as environmental ethics, can also drive green initiatives, reinforcing the role of organizational culture and leadership (Jennings & Zandbergen, 1995). This approach highlights that internal environmental values and leadership play a critical role in shaping the direction of green entrepreneurship.

Value-Based Entrepreneurship

Another emerging perspective is Value-Based Entrepreneurship, which argues that entrepreneurial activity is increasingly shaped by personal values and ethical beliefs. In this model, green entrepreneurs are not merely opportunity-driven but are motivated by a desire to contribute positively to society and the planet (Shepherd & Patzelt, 2011).

These entrepreneurs often adopt activist roles, using their ventures to advocate for environmental justice, biodiversity, and climate responsibility. This approach is particularly prevalent among younger generations, who view entrepreneurship as a platform for systemic change rather than just financial gain (Esqueda, 2019).

Systems Thinking and Complexity Theory

Systems Thinking offers an integrated framework for analyzing the interconnections among environmental, economic, and social systems. It argues that sustainable solutions require coordinated, multi-level interventions rather than isolated innovations (Meadows, 2008). This perspective underscores the need to move beyond linear cause-and-effect models and embrace dynamic interactions that influence long-term sustainability outcomes.

In line with this approach, green entrepreneurs operating in areas such as clean energy, sustainable agriculture, or supply chain circularity must navigate complex interdependencies. Systems thinking becomes an essential competency for understanding and addressing root causes rather than symptoms of environmental degradation (Diale et al., 2019). In this view, green entrepreneurship is not only a strategic activity but also a form of systems leadership that seeks to redesign broken structures for long-term sustainability.

Green entrepreneurship is a multi-dimensional phenomenon best understood through a variety of theoretical lenses. Ecological Modernization Theory, Triple Bottom Line, Sustainable Entrepreneurship, Institutional Theory, Value-Based Entrepreneurship, and Systems Thinking each contribute unique insights into the motivations and operations of environmentally conscious entrepreneurs.

Together, these frameworks suggest that green entrepreneurship is not a singular practice but an evolving constellation of strategies, values, and institutional interactions. As environmental challenges intensify and societal expectations shift, these theories will continue to evolve, offering new ways to understand and support green entrepreneurial ecosystems.

GREEN ENTREPRENEURSHIP AS A TOOL FOR CLEAN AIR AND ENVIRONMENTAL SUSTAINABILITY

Green entrepreneurship has become an essential driver in addressing air pollution and advancing environmental sustainability. While traditional environmental policy has relied heavily on regulatory instruments and technological advancement, the entrepreneurial domain offers a complementary and often more agile pathway for developing and deploying innovative solutions to air quality degradation. By promoting market-oriented innovation and sustainability-centered value creation, green entrepreneurship contributes effectively to resolving key environmental challenges, particularly those linked to urbanization, industrialization, and fossil fuel dependency (Ansari, 2025; Sobko et al., 2024).

Air pollution remains a global concern, contributing to an estimated seven million premature deaths annually and affecting nearly every continent through rising levels of particulate matter (PM2.5), nitrogen oxides (NOx), carbon monoxide (CO), and volatile organic compounds (VOCs) (WHO, 2023). In response, green entrepreneurs are leveraging sustainable technologies to mitigate these harmful emissions. For example, startups working on electric mobility, air purification systems, clean cookstove designs, and low-emission building materials are directly contributing to cleaner urban environments (Chandel, 2022; Mehta et al., 2024). These initiatives not only help reduce pollutants but also encourage behavior change among consumers, governments, and corporations alike.

The transition from fossil fuel-based energy systems to renewable sources such as solar, wind, and biogas is one of the primary domains where green entrepreneurship aligns with air quality improvement. The deployment of decentralized solar power systems by small-scale enterprises in off-grid areas not only reduces reliance on diesel generators but also expands access to clean energy in underserved regions (IRENA, 2023). As highlighted in the GIZ (2014) report, green businesses are particularly well-positioned to implement such localized energy solutions due to their flexibility and close engagement with community-level needs.

In the context of air quality, transportation and construction are among the highest contributors to urban pollution. Green entrepreneurs are responding with innovations such as battery-swapping stations for electric vehicles (EVs), micro-mobility platforms (e-scooters, bikes), sustainable logistics models, and eco-construction technologies (Mehta et al., 2024). These market solutions often emerge faster than state-led infrastructure developments and can rapidly scale with the support of green finance mechanisms, such as impact investment and climate bonds (Irfan et al., 2022).

Moreover, green entrepreneurship plays a transformative role in reconfiguring consumer markets and production models toward sustainability. The concept of the circular economy—emphasizing reuse, recycling, and waste minimization—is frequently embedded in green business models (Trapp & Kanbach, 2021). For example, ventures that promote biodegradable packaging, upcycled materials, or zero-waste retail systems reduce production-related emissions while also shifting consumer expectations toward environmentally friendly alternatives (Chandel, 2022). These behavioral and structural changes have indirect but lasting impacts on reducing environmental contaminants and particulate emissions (Ahmed et al., 2023; Zeng & Ren, 2022).

From a policy standpoint, international frameworks such as the United Nations Sustainable Development Goals (especially SDG 11: Sustainable Cities and Communities and SDG 13: Climate Action) endorse private-sector involvement in environmental improvement. Governments are beginning to recognize the potential of green entrepreneurship by offering fiscal incentives, easing bureaucratic barriers, and creating green innovation clusters (Ansari, 2025; OECD, 2022b). At the same time, digital platforms-such as those identified by Sobko et al. (2024), including Energy Web (www.energyweb.org), Climate-KIC (www.climate-kic.org), and Cleantech Group (www.cleantech.com) enable knowledge transfer, financing access, and mentorship networks for green entrepreneurs.

Quantitative trends also support the role of green entrepreneurship in air pollution mitigation. Investments in clean technologies and energy-efficient innovations have steadily increased over the past decade, with notable growth in sectors like clean transport, green buildings, and renewable energy infrastructure (Eurostat, 2024). Researches (Bye et al., 2023; Guo et al., 2025; Meirun et al., 2020) show that countries with higher rates of green innovation and entrepreneurship tend to report lower per capita emissions and better urban air quality indexes. This correlation points to the systemic value of entrepreneurship not only as a business function but also as a mechanism of ecological governance.

However, the pathway is not without challenges. Green entrepreneurs often encounter high initial capital costs, slow returns on investment, regulatory fragmentation, and limited consumer awareness in emerging markets (Chandel, 2022; GIZ, 2014). Moreover, the quantification of environmental impact from entrepreneurial activity remains a methodological hurdle. Few green startups have the capacity or tools to accurately measure their contribution to clean air outcomes, limiting both their strategic planning and appeal to green investors (Mukherjee et al., 2024a; Rizzitello et al., 2025).

To overcome these constraints, scholars advocate for hybrid business models that combine for-profit objectives with environmental mission statements, alongside partnerships with municipal governments, NGOs, and international donors (Ansari, 2025; Mehta et al., 2024). Additionally, entrepreneurial education and ecosystem building, through incubators, green accelerators, and sustainability curricula, are essential to cultivate a new generation of entrepreneurs who are not only economically savvy but also ecologically literate (Kesa et al., 2024; Santika et al., 2022).

In conclusion, green entrepreneurship offers a dynamic and impactful approach to addressing air pollution and advancing environmental sustainability. By fostering innovation, mobilizing resources, and challenging conventional industrial models, green entrepreneurs function as critical agents of change in the global movement toward clean air and low-carbon futures. As air quality continues to emerge as a public health and climate priority, the relevance of green entrepreneurship will only expand, demanding further research, targeted policy support, and integrated measurement systems to fully capture its environmental contributions.

GREEN ENTREPRENEURSHIP AND CLEAN AIR POLICIES: GLOBAL PRACTICE-BASED INSIGHTS

The 21st century has witnessed the convergence of two critical global priorities: fostering sustainable economic development through green entrepreneurship and mitigating the escalating impacts of air pollution through clean air governance. While both domains have independently gained attention in the academic and policy spheres, their intersection presents a powerful lens for understanding how entrepreneurial innovation and regulatory policy can jointly contribute to ecological resilience and public health (www.sdgs.un.org).

In this integrated landscape, green entrepreneurship—characterized by the creation of environmentally beneficial products, services, and business models often thrives in ecosystems supported by progressive environmental policies, while simultaneously influencing and being shaped by evolving clean air regulations. Conversely, clean air policies, traditionally shaped by public health concerns, are increasingly influenced by private sector innovations that offer scalable, data-driven, and cost-effective solutions. This section presents integrated global examples where these two trajectories converge, highlighting how governments, startups, NGOs, and communities co-create sustainable responses to air pollution and climate degradation (www.sdgs.un.org).

Sample Cases from Countries

Germany: Policy-Driven Green Innovation Ecosystem

Germany exemplifies how robust public policy frameworks can cultivate a dynamic green entrepreneurial ecosystem. Through the Energiewende initiative and GreenTech Innovation Strategy, the German government has incentivized innovation in renewable energy and energy efficiency. Institutions like the Fraunhofer Institutes partner with SMEs and startups to drive the commercialization of clean technologies. While Germany does not frame its environmental policy specifically under "clean air," the resulting reductions in particulate matter (PM) and CO_2 emissions from its renewable transition efforts highlight a synergistic relationship between entrepreneurial green solutions and improved air quality (MWK, 2022).

United States: Dual Pathways: Green Markets and Regulatory Reform

In the United States, the clean air and green entrepreneurship landscapes evolved along parallel tracks. On one side, the Clean Air Act (1970) provided a legislative backbone for emissions monitoring, air quality standards, and compliance enforcement. This policy, updated in 1990, is credited with reducing emissions of six key pollutants by 78% between 1970 and 2020, while the economy continued to grow (EPA, 2023).

Concurrently, the U.S. has nurtured market-based green entrepreneurship through venture capital, tax credits, and consumer-driven innovation. Firms like Tesla and Beyond Meat have leveraged sustainability concerns to disrupt traditional industries. These twin tracks—regulatory and market-led—have proven mutually reinforcing, demonstrating that environmental regulation can coexist with, and even stimulate, entrepreneurial dynamism.

China: State-Orchestrated Clean Air Reforms

China has taken a centralized, top-down approach to tackling air pollution. The National Air Quality Action Plan (2013) aimed to reduce PM2.5 levels by 25% in major cities, resulting in a 42% decline between 2013 and 2021. This success was largely driven by coal reduction, industrial relocation, and real-time emissions monitoring (AQLI, 2022).

Simultaneously, the Chinese government has fostered green entrepreneurship by supporting cleantech startups and renewable energy enterprises. Policies linking air quality targets to economic development planning have institutionalized sustainability as a national priority, aligning environmental and economic goals.

India: Dual Approach: NCAP and Grassroots Green Innovation

India's efforts span both macro-level clean air planning and grassroots green entrepreneurship. The National Clean Air Programme (NCAP), launched in 2019, aims to reduce urban PM2.5 and PM10 levels by 20–30% by 2024. This strategy empowers local authorities to develop city-specific action plans and expands the national air quality monitoring grid. Parallel to this, social enterprises such as SELCO India have pioneered affordable solar technologies for underserved populations. These innovations reduce reliance on kerosene and biomass, indirectly contributing to cleaner household air. India's model shows how decentralized entrepreneurship and centralized policy can collectively target both ambient and indoor air quality challenges (Moulin, 2025).

Poland: Urban Clean Air Innovation: Breathe Warsaw

Poland's capital city, Warsaw, launched the Breathe Warsaw initiative in 2019 to combat some of Europe's worst urban smog. The program introduced 165 air sensors, promoted the phasing out of coal heating, and established low-emission zones. These efforts are supported by EU environmental policy and funding (www.cleanairfund.org)

Though not explicitly entrepreneurial, this policy climate has encouraged startups to develop air quality monitoring technologies, electric mobility services, and green retrofitting solutions for buildings (UNEP, 2020).

South Korea: Green New Deal and AI-Driven Sustainability

South Korea's Green New Deal (MOEF, 2020) commits over $60 billion to digital and green transformation, emphasizing job creation in renewable energy, smart grids, and sustainable transport. Firms like Sungrow Korea develop AI-enhanced solar management systems.

The Korean model is notable for its precision: smart city infrastructure links real-time pollution data with traffic management and industrial output, creating feedback loops that enable both environmental regulation and business innovation (MOEF, 2020).

Nepal: EV Incentives for Urban Clean Air

Nepal utilizes its hydroelectric resources to power a rapid transition toward electric vehicles (EVs). The government offers tax incentives and imports subsidies for EVs while investing in charging infrastructure. This strategy has led to significant savings on fuel import and reductions in vehicular emissions in cities like Kathmandu (AP News, 2023). Nepal's experience illustrates how clean air policy can be directly aligned with energy transition and entrepreneurial opportunity, especially in the context of small economies.

To contextualize the country-specific practices discussed above, it is useful to examine how environmental performance varies across the globe. The following map presents a visual summary of the Environmental Performance Index (EPI) 2024 scores, offering comparative insights into the relative success of nations in achieving sustainability, air quality, and climate resilience goals.

Figure 1. Most Environmentally Friendly Countries-2025 (www.worldpopulationreview.com)

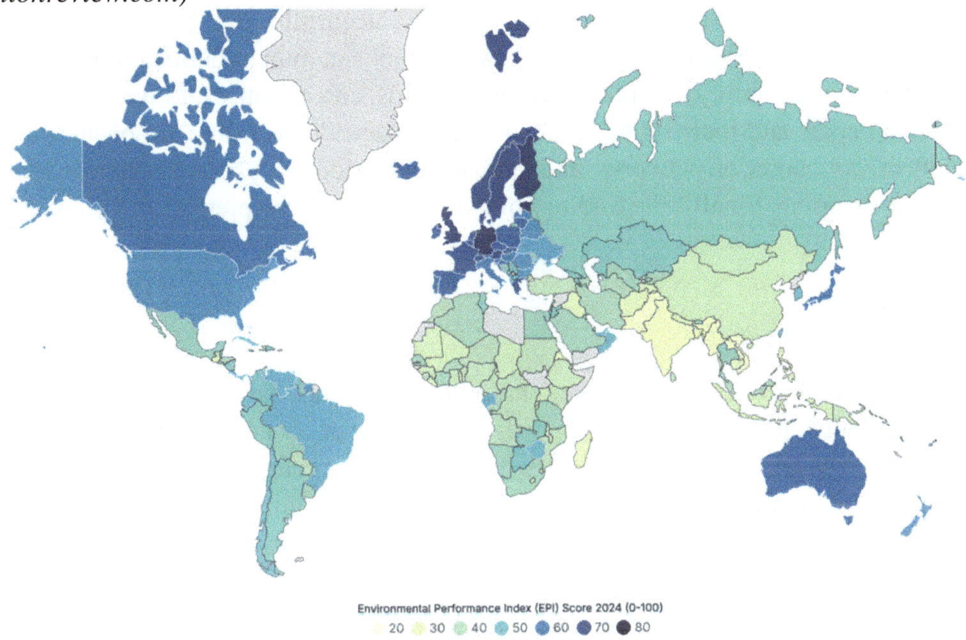

The map illustrates the Environmental Performance Index (EPI) scores of countries for 2024, reflecting their overall environmental health and sustainability efforts. According to the index, Nordic and Western European countries lead in environmental performance, while many developing nations face significant challenges due to air pollution, limited clean energy access, and weak regulatory frameworks.

METHODOLOGY

This research utilizes a bibliometric analysis to thoroughly assess the field of Green Entrepreneurship, drawing on data from the Web of Science database. The search is made in April 2025 and is centered on the term "Green Entrepreneurship" within the "Topic" field, which encompasses titles, abstracts, keywords, and author-defined keywords, to ensure a comprehensive review. A specific time frame (e.g., 1991–2025) is set to examine trends over the years. The search results are exported in either plain text (.txt) or CSV format, containing metadata such as titles, authors, affiliations, citations, keywords, abstracts, and references. A total of 1,190 documents have been identified, reflecting an annual growth rate of 12%. Among these, nearly 80% are journal articles. The average age of the documents is 4 years, indicating that publications on this topic have predominantly emerged in recent years. The analysis and visualization are conducted using the bibliometrix package in R, along with Biblioshiny, a web-based interface (Aria & Cuccurullo, 2017). Key analytical methods include performance analysis, which looks at annual publication trends, the most prolific authors, publications, institutions, and countries, as well as highly cited papers and journals to evaluate their impact. Furthermore, co-occurrence analysis, collaboration networks, and thematic evolution analysis are performed for science mapping. Finally, the findings of the analysis are interpreted and discussed.

FINDINGS

A total of 1,190 academic papers have been published on the subject of green entrepreneurship from 1991 to 2025, with 66% of these publications occurring within the last five years. This trend indicates that green entrepreneurship has emerged as a prominent area of interest among researchers globally in recent years. In 2024, the volume of published papers reached 250, and approximately 100 papers have been published in the first quarter of 2025.

Figure 2. Number of Publications Over Time

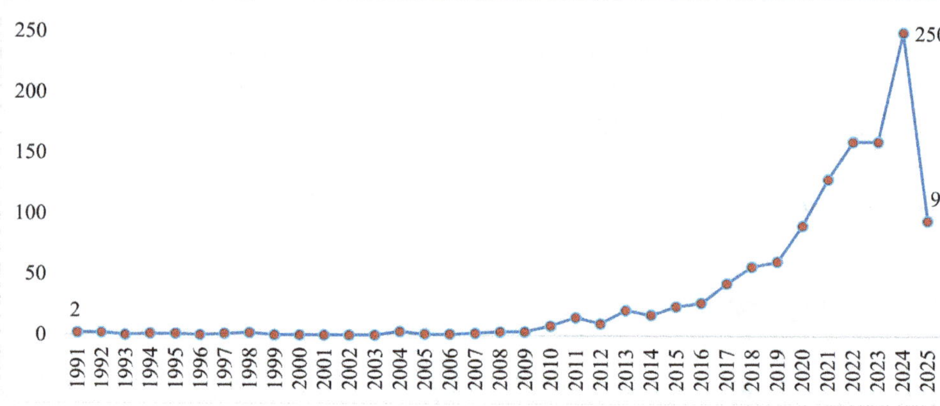

A total of green entrepreneurship has gained significant attention 3,273 authors have contributed to the field of green entrepreneurship. Among these, 65 authors (2%) have published more than three papers, while 257 authors (8%) have authored two papers, and a significant majority of 2,951 authors (90%) have published a single paper. The most prominent authors, defined as those with more than four publications, are illustrated in Figure 2. The data indicates that Singh is the leading author with seven published works on the subject, followed by Makhloufi and other authors with six publications, Zhang and other authors with five, and Wang and other authors with four publications in total.

Figure 3. The Most Relevant Authors

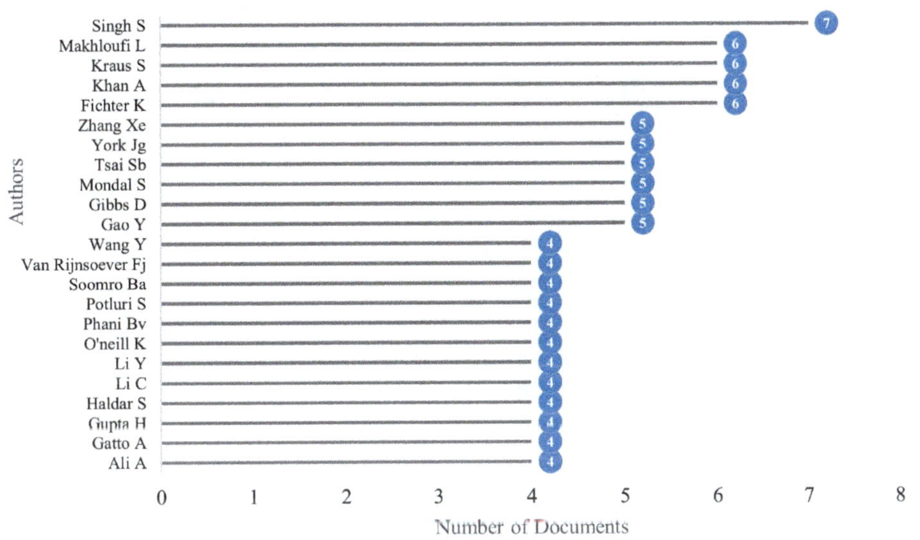

Figure 3 provides additional context for these contributions by utilizing a Sankey diagram to illustrate the connections among key cited references (CR), authors (AU), and predominant research themes (DE). Notably, Kraus and Singh are identified as pivotal nodes that connect foundational literature with essential concepts, including green entrepreneurship, sustainable development, and sustainable entrepreneurship. The intersection of frequently cited works, prominent authors, and persistent thematic areas highlights a dynamic and interdisciplinary research field, with a central focus on sustainability-oriented entrepreneurship and innovation.

Figure 4. Three-Field Plot

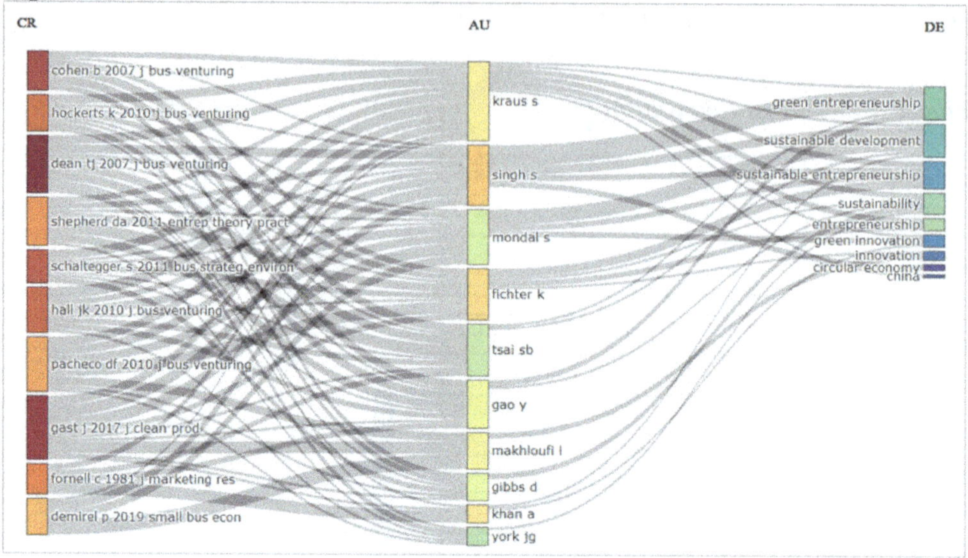

The examination of institutional affiliations reveals a heterogeneous array of contributors to the academic discourse surrounding green entrepreneurship. As depicted in Figure 4, the Ministry of Education and Science of Ukraine stands out as the most prolific institution, having published a total of 22 articles, thereby underscoring the critical role of governmental support in promoting sustainability-oriented research. Following this, Jiangsu University and the Indian Institute of Technology System (IIT System) have contributed 13 and 12 articles, respectively, highlighting the active involvement of prominent academic institutions in Asia.

A significant number of institutions, including the University of London, the University of California System, and King Faisal University, each produced 10 articles, reflecting a balanced and widespread interest in green entrepreneurship across various regions. The representation of institutions from Europe, Asia, the Middle East, and North America illustrates the global character of the research landscape in this domain. Furthermore, the inclusion of several post-Soviet institutions, such as the Ministry of Education of the Azerbaijan Republic and the Azerbaijan State University of Economics (UNEC), indicates an increasing academic focus on sustainable development within transitioning economies. The presence of both academic institutions and governmental entities emphasizes the interdisciplinary and policy-relevant aspects of research in green entrepreneurship. This institutional diversity not only signifies a broad scholarly interest but also suggests differing national priorities and capacities in addressing environmental challenges through entrepreneurial initiatives.

Figure 5. The Most Relevant Affiliations

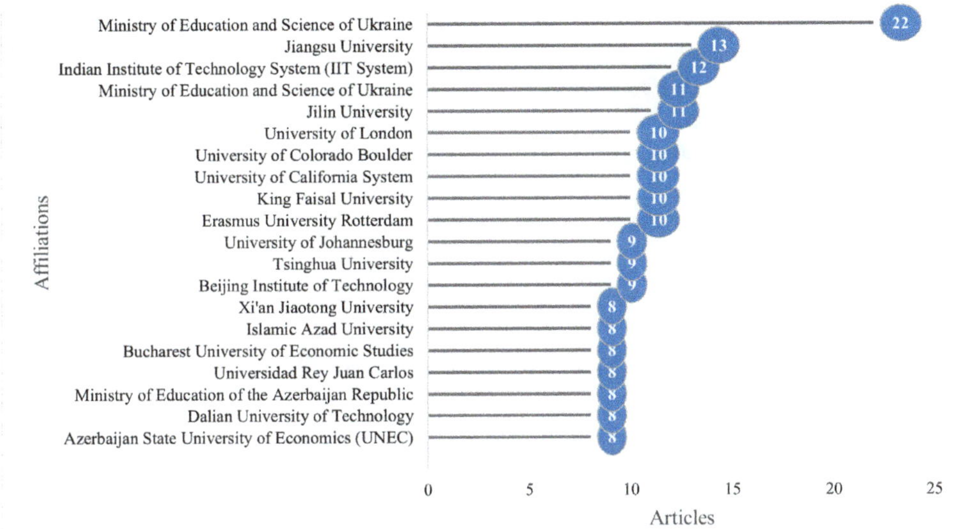

Figure 4 presents the distribution of scholarly output by country, distinguishing between Single Country Publications (SCP) and Multiple Country Publications (MCP). The analysis reveals that China is the dominant contributor to research in green entrepreneurship, with a total of 261 documents, of which 201 are SCPs and 60 are MCPs. This suggests that while China produces a high volume of research, much of it is conducted within national institutions, aligning with the earlier observation of the Ministry of Education and Science of Ukraine as the top individual institutional contributor. Together, these findings point to the strong role of domestic research capacity and government involvement in advancing sustainability studies.

Figure 6. The Most Relevant Countries

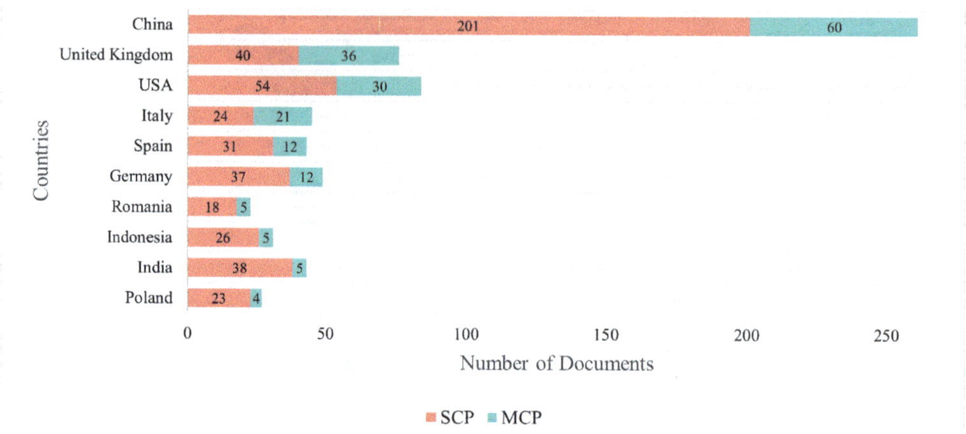

The United Kingdom and the United States follow with 76 and 84 documents, respectively, but with a significantly higher proportion of MCPs (36 and 30), indicating a greater degree of international collaboration. This pattern is also evident in countries such as Italy and Germany, which show more balanced ratios between SCP and MCP, reflecting their integration into global research networks. In contrast, countries such as India, Indonesia, and Romania display a predominance of SCPs and a limited number of MCPs, pointing to more locally concentrated research efforts and potentially fewer international partnerships. These trends are consistent with the institutional-level data, which also indicate a regional concentration of output within specific universities or ministries rather than collaborative research consortia.

The two figures collectively illustrate a bifurcated trend in the field of green entrepreneurship research. On one hand, certain nations emphasize the development of self-sufficient institutional outputs, while on the other hand, some are progressing through international collaboration. This differentiation is essential for comprehending both the quantity and the breadth of impact of green entrepreneurship research on a global scale. Enhancing international collaboration, especially among emerging and post-Soviet economies, has the potential to enrich the discipline by integrating varied perspectives and promoting collective innovation.

Figure 6 illustrates a co-occurrence network of keywords derived from document titles, which elucidates the thematic structure and interconnections present within the literature on green entrepreneurship. The network is categorized into distinct color-coded clusters, each signifying a specific thematic focus. The most prominent and central cluster, represented in blue, encompasses terms such as "green," "entrepreneurship," "sustainable," and "innovation," thereby underscoring the prevailing emphasis on sustainable entrepreneurial practices and technological advancements.

The red cluster, which includes keywords such as "students," "intention," and "education," indicates a focus on entrepreneurial intention and the influence of education in fostering pro-environmental behavior. The green cluster encompasses themes associated with digital transformation, circular economy, and economic growth, reflecting a growing interest in systemic changes and policy-driven innovation. Additionally, a smaller purple cluster, which includes terms like "literature" and "review," highlights methodological contributions through systematic reviews.

Figure 7. Co-occurrence Network of Keywords Extracted from Document Titles

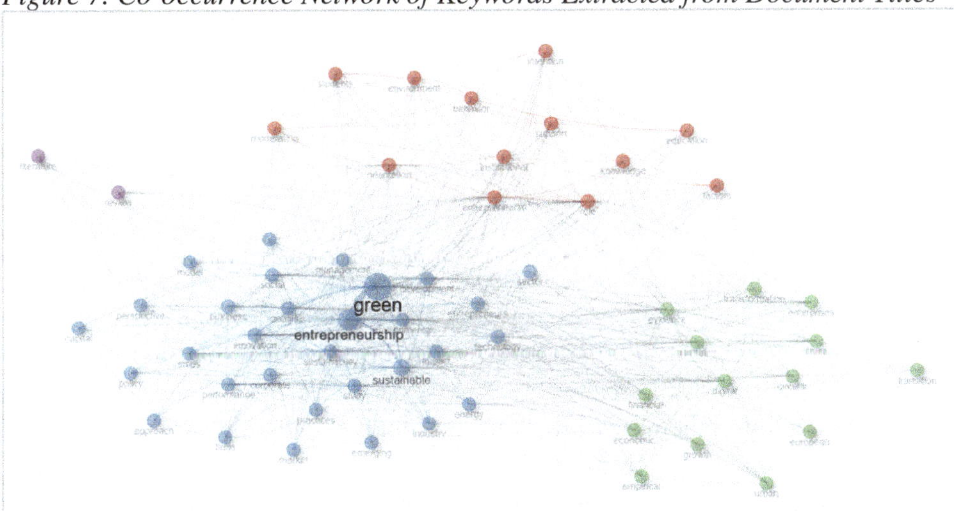

Collectively, the network demonstrates the multidisciplinary and interconnected nature of research in green entrepreneurship, linking sustainability, behavioral intention, and digital economic transformation.

Figure 7 presents a thematic map of keywords that categorizes research themes according to their relevance (centrality) and development (density) within the academic field. The map delineates four distinct quadrants: motor themes (characterized by high centrality and density), basic themes (high centrality, low density), niche themes (low centrality, high density), and emerging or declining themes (low in both dimensions). Notably, the themes of "firm performance," "dynamic capabilities," and "moderating role" are identified as well-developed and highly relevant motor themes, underscoring their pivotal role in enhancing both theoretical and practical understanding. The themes of "entrepreneurship," "growth," and "policy" are situated near the center of the map, indicating moderate development while maintaining strong connections to other themes.

Figure 8. Thematic Map of Keywords

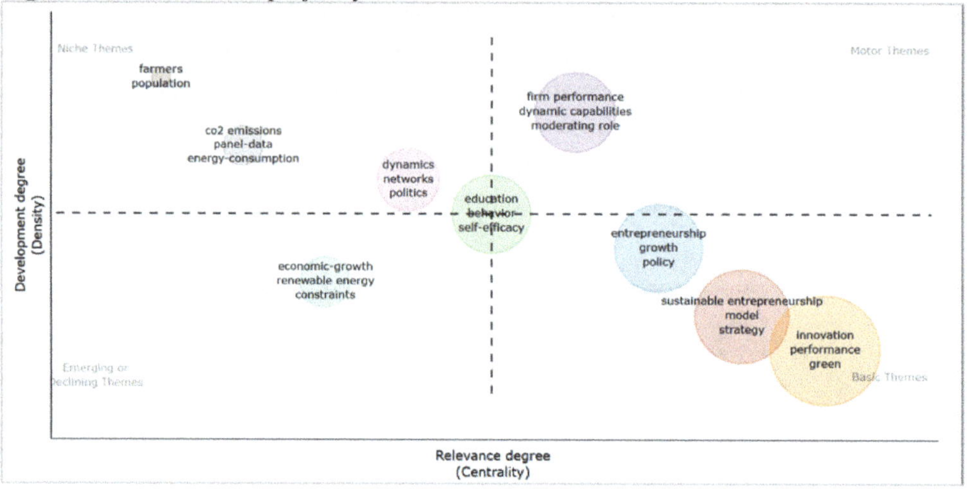

Conversely, "innovation," "performance," and "green" are categorized as basic themes, exhibiting high relevance but lower levels of development, thereby highlighting foundational areas that warrant further investigation. In contrast, topics such as "economic growth," "renewable energy," and "constraints" are classified as emerging or declining themes, while "farmers" and "population" are identified as niche themes, characterized by high internal coherence but limited broader impact. Overall, the map highlights key conceptual clusters and indicates potential directions for future research in green entrepreneurship and sustainability-related innovation.

DISCUSSION AND CONCLUSION

This chapter presents a comprehensive bibliometric analysis of green entrepreneurship, outlining its intellectual structure, regional distribution, principal contributors, and evolving thematic priorities related to environmental sustainability and clean air. The analysis of 1,190 publications indexed in the Web of Science (1991–2025) indicates a notable increase in scholarly output during the last five years, with 2024 marking the peak year (250 publications). This trend reflects the increasing relevance of green entrepreneurship in both scholarly and policy domains amidst global ecological challenges and the urgent need for low-carbon transitions.

The findings show that 3,273 authors have contributed to the green entrepreneurship domain, yet approximately 90% of them have published only once, indicating a broad but fragmented field. Although leading authors such as Singh, Makhloufi, and Wang demonstrate consistent engagement, the predominance of single contributions indicates a field that is still developing sustained scholarly continuity. This fragmentation may affect the accumulation of coherent theoretical models and impede the development of unified frameworks for empirical testing. Institutionally, the most active contributors are the Ministry of Education and Science of Ukraine, Jiangsu University, and the Indian Institute of Technology system, reflecting a combination of public and academic research initiatives with regionally concentrated outputs.

At the journal level, the most frequent outlets—Sustainability, the Journal of Cleaner Production, and the International Journal of Entrepreneurial Behavior and Research—signal a clear trend of multidisciplinary interest, where green entrepreneurship is situated at the intersection of environmental science, management studies, and entrepreneurial education. Citation analysis shows that the most frequently cited works emphasize concepts such as green entrepreneurial intention (Wang et al., 2021), sustainable innovation (Schaltegger & Wagner, 2011), and performance outcomes (Criado-Gomis, 2017), reflecting a growing scholarly interest in both the antecedents and outcomes of green entrepreneurship

Keyword co-occurrence analysis identified three main thematic clusters: (1) sustainable business models and innovation strategies, (2) behavioral constructs such as intention and motivation, and (3) systems-level enablers including digital transformation and circular economy. The blue cluster, dominated by terms such as 'green,' 'sustainable,' and 'entrepreneurship,' reflects a foundational concern with aligning business innovation to environmental goals. The red cluster, which includes 'education,' 'students,' and 'intention,' reinforces the role of human capital, suggesting that green entrepreneurial ecosystems are deeply shaped by educational infrastructures and cultural orientations. The green cluster, defined by 'digitalization,' 'circular economy,' and 'economic growth,' points to emerging systemic dimensions in which green entrepreneurship is increasingly nested.

The thematic map analysis confirms that motor themes such as ‘firm performance’ and ‘dynamic capabilities’ are positioned in the high-centrality and high-density quadrant, representing both theoretical maturity and practical significance. Conversely, 'clean air' and 'pollution reduction' did not appear as dominant keywords in any cluster, although their presence in niche or emerging theme zones suggests growing, albeit underdeveloped, interest. This confirms the study’s core hypothesis that green entrepreneurship’s role in air pollution mitigation is conceptually acknowledged but empirically underrepresented.

The country-level collaboration analysis provides further insights into the structure of knowledge production. China, while the most prolific country in terms of total publications (261), shows a predominance of single-country publications (SCP), suggesting a more insular approach to research. In contrast, countries like the United Kingdom (MCP = 36) and the United States (MCP = 30) exhibit higher proportions of multi-country collaborations, indicating their integration into transnational scientific networks. This imbalance underscores the need to strengthen both regional and international research collaborations to promote equitable knowledge exchange and innovation diffusion.

Institutional affiliation data underscores this regional disparity. While countries like Ukraine and India have government-backed institutional contributions, European and North American contributions often emerge from research-intensive universities. This divergence in institutional context affects the type of research questions being asked, the methodological approaches used, and ultimately, the applicability of findings to practice and policy. As suggested by Karadayı et al. (2025), lack of equitable institutional participation can skew the knowledge agenda toward high-income country contexts, potentially overlooking the unique sustainability challenges faced by emerging and transition economies.

Theoretical Contributions

The findings of this chapter enhance the theoretical framing of green entrepreneurship by building upon and further developing several conceptual models. One of the most prominent frameworks emphasized is the "Triple Bottom Line" (3P)—People, Planet, and Profit—which provides a multidimensional structure for evaluating the impact of green ventures (Elkington, 1997). Unlike conventional business models that prioritize financial outcomes, this framework recognizes that economic, social, and environmental values can and must coexist. The empirical trends and thematic patterns observed in the bibliometric analysis support this theoretical alignment, particularly through the growing use of performance metrics that capture sustainability indicators (Chandel, 2022; Schaltegger & Wagner, 2011).

In addition to reaffirming established models, this study enhances our understanding of the diversity of green entrepreneurial actors. The typology differentiating core and peripheral green entrepreneurs (GIZ, 2014) is expanded through recent studies that introduce the notion of 'adaptive green entrepreneurs (Karadayı et al., 2025). These actors operate in fluid regulatory and economic environments and adapt their practices in response to policy, market, and technological shifts. Recognizing this variation enables a more nuanced analysis of entrepreneurial behavior, particularly in relation to resilience, scalability, and context sensitivity. It also supports emerging work in entrepreneurship theory that stresses the importance of institutional embeddedness and entrepreneurial agency in shaping environmental outcomes.

Methodologically, this study advances sustainability entrepreneurship research by applying bibliometric science mapping to visualize intellectual networks and thematic evolution. Unlike traditional literature reviews, which may suffer from selection bias or thematic narrowness, bibliometric approaches offer a data-driven, scalable way to visualize intellectual networks, trace research evolution, and identify thematic gaps (Aria & Cuccurullo, 2017). By revealing underexplored intersections, such as green entrepreneurship's role in clean air solutions, this chapter positions bibliometric analysis as not only a descriptive tool but also a heuristic device for advancing theory.

In essence, these theoretical contributions help bridge the conceptual gap between sustainability goals and entrepreneurial practice, while providing new directions for scholars seeking to understand how values, capabilities, and institutions interact to shape environmentally meaningful innovation.

Practical and Policy Implications

The implications of this study extend beyond theory, offering important guidance for policy makers, practitioners, and ecosystem developers working to support green entrepreneurship. The findings point to the necessity of designing and maintaining multi-level entrepreneurial ecosystems that balance regulatory support, financial resources, and capacity building. One key area of intervention is the provision of green finance. Government subsidies, green bonds, and impact investment instruments have already demonstrated effectiveness in lowering entry barriers and de-risking innovation for environmentally oriented startups (Mukherjee et al., 2024b; OECD, 2022a).

However, access to finance alone is insufficient. Institutional capacity—particularly at the municipal and regional levels—plays a vital role in sustaining the momentum of green entrepreneurship. Ministries of environment, economic development agencies, universities, and local innovation hubs must work collaboratively to establish support structures such as incubators, mentorship programs, and accreditation systems that align entrepreneurial competencies with sustainability objectives (Santika et al., 2022).

Moreover, the findings highlight the growing importance of digital infrastructure in facilitating green innovation. Platforms like Energy Web, Climate-KIC, and Cleantech Group not only connect entrepreneurs with investors and experts but also serve as repositories of best practices and monitoring tools. These digital mechanisms are essential for improving transparency, accelerating knowledge diffusion, and ensuring accountability in environmental performance (Sobko et al., 2024).

Another crucial insight concerns the development of standardized metrics to assess the environmental impact of green ventures. While traditional business performance metrics focus on revenue and market share, green ventures must also report on indicators such as carbon footprint, emissions avoided, or air quality improvements. The absence of such metrics undermines both internal strategic planning and external legitimacy. Establishing common environmental performance standards would not only facilitate benchmarking but also increase the attractiveness of green ventures to investors and policy-makers seeking quantifiable results (Rizzitello et al., 2025).

Finally, the geographic asymmetries in green entrepreneurship—evidenced by the SCP/MCP publication imbalance—underline the necessity of fostering international collaboration, especially among emerging economies. South–South and South–North knowledge exchanges are crucial for ensuring that successful models are adapted and scaled in diverse institutional contexts, particularly where environmental challenges are most severe and resources are most constrained.

Future Research Agenda

Although this study advances our understanding of the green entrepreneurship landscape, several areas merit further exploration. First and foremost, future research should focus on quantifying the environmental outcomes of green ventures. While theoretical frameworks and case studies suggest a positive relationship between GE and environmental improvement, empirical evidence remains fragmented. Longitudinal, experimental, and comparative designs are needed to isolate the effects of entrepreneurial activity on environmental indicators, such as reductions in $PM_{2.5}$, CO_2, and VOC emissions (Cojoianu et al., 2024; Neumann, 2022).

Second, more work is needed to explore the role of education, culture, and behavioral intention in shaping green entrepreneurship. As highlighted by Wang et al. (2021) and Yi (2020), the translation of ecological values into entrepreneurial action is influenced by social norms, institutional support, and personal efficacy. Future studies should investigate how these factors vary across regions and demographic groups, and how targeted interventions—such as sustainability curricula or peer mentorship—can activate green entrepreneurial intention.

A third promising direction involves mapping inclusive entrepreneurship in the context of sustainability. Gender, race, and class often influence access to resources, legitimacy, and networks in entrepreneurial ecosystems. Abdelwahed et al. (2023) emphasize the need to foreground equity and social justice in the design of green innovation systems. Understanding how intersecting identities affect participation and success in GE can help design more equitable policy tools.

Finally, scholars should examine how green entrepreneurship contributes to a just transition—the process by which societies move toward sustainable economies without exacerbating social inequalities. This includes analyzing how green ventures support vulnerable populations, create decent jobs, and contribute to local resilience in the face of environmental shocks.

As research evolves, bibliometric monitoring should remain part of the methodological repertoire. Continued tracking of thematic clusters, author networks, and geographic shifts can provide an early-warning system for emerging gaps and signal new areas of innovation.

Conclusion

Green entrepreneurship stands out as an innovative approach that combines economic development and environmental responsibility in the face of environmental pollution, climate change and resource depletion, which are among the most fundamental problems of our age. This book chapter has comprehensively analyzed the relationship between green entrepreneurship and environmental sustainability, particularly clean air provision, at both theoretical and bibliometric levels. The findings show that green entrepreneurship offers not only environmental but also social and economic contributions; however, structural and methodological barriers must be overcome for these contributions to reach their full potential.

The results of the bibliometric analysis revealed that green entrepreneurship is a rapidly developing field in terms of scientific interest; and also detailed the transformation of collaboration models, thematic clusters and key concepts in the literature. The literature, especially developed around concepts such as "green innovation", "entrepreneurial intention", "firm performance", and "dynamic capabilities", demonstrates the multidisciplinary character and social impact of this field.

However, the link between measurable environmental outcomes such as clean air and green entrepreneurship is still not clear. While the positive effect demonstrated by Cojoianu et al. (2024) is observed at the local level, this relationship is seen to be weaker in studies conducted at the national level (Neumann, 2022). This can be explained by reasons such as the lack of measurement tools, the small scale of the initiatives, and the influence of other policy factors at the macro level.

However, green entrepreneurship is not only a practice focused on environmental outcomes; it has also become a systemic approach that affects many different areas, such as human development, regional development, entrepreneurship education, inclusiveness, and economic resilience. The literature indicates that green entrepreneurship is not only a business model but also a carrier of sustainable social transformation.

Going forward, it is critical for both researchers and policymakers to focus on three areas:

(1) Making the environmental impacts of green entrepreneurship directly measurable,
(2) Making entrepreneurship systems inclusive, regional, and fair,
(3) Designing ecosystems in a structure that is open to international cooperation and digitalization.

In conclusion, green entrepreneurship is a powerful tool that can reverse environmental destruction and achieve sustainable development. However, the realization of this potential will be possible with institutional support, educational policies, financial incentives, and scientific contributions. This study aims to provide a solid foundation for future strategies of green entrepreneurship by deeply analyzing both the theoretical framework and the structure of academic literature. In this context, green entrepreneurship will be one of the fundamental dynamics that will determine the development vision not only of today but also of the future.

REFERENCES

Abdelwahed, N. A. A., Al Doghan, M. A., Saraih, U. N., & Soomro, B. A. (2023). Green entrepreneurship in Saudi Arabia: Shaping the landscape of the greener economy. *Journal of Small Business and Enterprise Development*, *30*(7), 1352–1376. DOI: 10.1108/JSBED-05-2023-0239

Ahmed, I., Irshad, A., Zafar, S., Khan, B. A., Raza, M., & Ali, P. R. (2023). The role of environmental initiatives and green value co-creation as mediators: Promoting corporate entrepreneurship and green innovation. *SN Business & Economics*, *3*(4), 85. DOI: 10.1007/s43546-023-00465-w

Albort-Morant, G., Henseler, J., Leal-Millán, A., & Cepeda-Carrión, G. (2017). Mapping the field: A bibliometric analysis of green innovation. *Sustainability (Basel)*, *9*(6), 1011. DOI: 10.3390/su9061011

Ansari, M. S. A. (2025). Innovation and green entrepreneurship as catalysts for sustainability in Oman. *Cleaner Waste Systems*, *11*, 100281. DOI: 10.1016/j.clwas.2025.100281

AQLI. (2022). *China's War on Air Pollution: The Results Are In.* Air Quality Life Index. https://aqli.epic.uchicago.edu

Aria, M., & Cuccurullo, C. (2017). Bibliometrix: An R-tool for comprehensive science mapping analysis. *Journal of Informetrics*, *11*(4), 959–975. DOI: 10.1016/j.joi.2017.08.007

Bansal, P., & Clelland, I. (2004). Talking trash: Legitimacy, impression management, and unsystematic risk in the context of the natural environment. *Academy of Management Journal*, *47*(1), 93–103. DOI: 10.2307/20159562

Belz, F. M., & Binder, J. K. (2017). Sustainable entrepreneurship: A convergent process model. *Business Strategy and the Environment*, *26*(1), 1–17. DOI: 10.1002/bse.1887

Buttel, F. H. (2000). Ecological modernization as social theory. *Geoforum*, *31*(1), 57–65. DOI: 10.1016/S0016-7185(99)00044-5

Bye, B., Kaushal, K. R., Rosnes, O., Turner, K., & Yonezawa, H. (2023). The road to a low emission society: Costs of interacting climate regulations. *Environmental and Resource Economics*, *86*(3), 565–603. DOI: 10.1007/s10640-023-00804-3

Chandel, A. F. (2022). *Green entrepreneurship and sustainable development.* ResearchGate. https://www.researchgate.net/publication/362856766_Green_Entrepreneurship_and_Sustainable_Development

Cho, M. H. (2024). Key Factors Affecting Startups' Contribution to SDGs for a Sustainable Future: Integrating a Triple Bottom Line (TBL) Theory. *Journal of Ecohumanism, 3*(7), 615–631. DOI: 10.62754/joe.v3i7.4232

Chu, F., Zhang, W., & Jiang, Y. (2021). How does policy perception affect green entrepreneurship behavior? An empirical analysis from China. *Discrete Dynamics in Nature and Society, 2021*, 1–9. DOI: 10.1155/2021/7973046

Cojoianu, T. F., Hoepner, A. G., Hu, X., Ramadan, M., Veneri, P., & Wójcik, D. (2024). Are cities venturing green? A global analysis of the impact of green entrepreneurship on city air pollution. *Small Business Economics, 62*(2), 523–540. DOI: 10.1007/s11187-023-00764-4

Criado-Gomis, A., Cervera-Taulet, A., & Iniesta-Bonillo, M.-A. (2017). Sustainable Entrepreneurial Orientation: A Business Strategic Approach for Sustainable Development. *Sustainability (Basel), 9*(9), 1667. DOI: 10.3390/su9091667

Dean, T. J., & McMullen, J. S. (2007). Toward a theory of sustainable entrepreneurship: Reducing environmental degradation through entrepreneurial action. *Journal of Business Venturing, 22*(1), 50–76. DOI: 10.1016/j.jbusvent.2005.09.003

Diale, C. D., Kanakana-Katumba, G., & Maladzhi, R. W. (2019, December). *IEEE International Conference on Industrial Engineering and Engineering Management (IEEM)* (pp. 384-389). IEEE. Doi: DOI: 10.1109/IEEM44572.2019.8978804

DiMaggio, P. J., & Powell, W. W. (1983). The iron cage revisited: Institutional isomorphism and collective rationality. *American Sociological Review, 48*(2), 147–160. DOI: 10.2307/2095101

Duran, C. (2024). Bibliometric Analysis of Green Entrepreneurship. *Beykoz Akademi Dergisi, 12*(2), 250–282. DOI: 10.14514/beykozad.1519660

Elkington, J. (1997). *Cannibals with forks: The triple bottom line of 21st-century business.* Capstone.

EPA. (2023). *Progress Cleaning the Air and Improving People's Health.* United States Environmental Protection Agency. https://www.epa.gov/clean-air-act-overview/progress-cleaning-air-and-improving-peoples-health

Esqueda, H. (2019). A profile of the popular entrepreneur: A value-based approach. *Academia (Caracas), 32*(2), 267–281. DOI: 10.1108/ARLA-03-2018-0055

European Environment Agency. (2024). Green bonds in Europe - European Environment Agency, https://www.eea.europa.eu/en/analysis/indicators/green-bonds-8th-eap

Eurostat. (2024). *Green bond issuance and investments in clean technologies (EU-27)*. https://ec.europa.eu/eurostat

Fernando, S., & Lawrence, S. (2014). A theoretical framework for CSR practices: Integrating legitimacy theory, stakeholder theory and institutional theory. *Journal of theoretical accounting research, 10*(1), 149-178.

Gao, Y., Tsai, S. B. J., Zhang, S., & Li, G. (2018). Green entrepreneurship in transitional economies: Breaking through the constraints of legitimacy. In *Green Production Strategies for Sustainability* (pp. 184–209). IGI Global., DOI: 10.4018/978-1-5225-3537-9.ch011

GIZ (Deutsche Gesellschaft für Internationale Zusammenarbeit). (2014). *Green entrepreneurship: A promising path towards a sustainable future in Tunisia and elsewhere*. Retrieved from https://www.greenpolicyplatform.org/sites/default/files/downloads/resource/ GIZ_%20Green%20Entrepreneurship%20in%20Tunisia.pdf

Gu, W., Pan, H., Hu, Z., & Liu, Z. (2022). The triple bottom line of sustainable entrepreneurship and economic policy uncertainty: An empirical evidence from 22 countries. *International Journal of Environmental Research and Public Health, 19*(13), 7758. DOI: 10.3390/ijerph19137758

Guo, L., Cheng, Z., Tani, M., Cook, S., Zhao, J., & Chen, X. (2025). Air pollution and entrepreneurship. *China Economic Review, 89*, 102327. DOI: 10.1016/j.chieco.2024.102327

Hall, J. K., Daneke, G. A., & Lenox, M. J. (2010). Sustainable development and entrepreneurship: Past contributions and future directions. *Journal of Business Venturing, 25*(5), 439–448. DOI: 10.1016/j.jbusvent.2010.01.002

Hameed, I., Zaman, U., Waris, I., & Shafique, O. (2021). A serial-mediation model to link entrepreneurship education and green entrepreneurial behavior: Application of resource-based view and flow theory. *International Journal of Environmental Research and Public Health, 18*(2), 571. DOI: 10.3390/ijerph18020550

Huang, J., Liu, R., Wang, W., Wang, Z. A., Wang, C., & Jin, Y. J. (2024). Unleashing Fintech's potential: A catalyst for green bonds issuance. *Journal of International Financial Markets, Institutions and Money, 93*, 102009. DOI: 10.1016/j.intfin.2024.102009

Iqbal, N., Khan, A., Gill, A. S., & Abbas, Q. (2020). Nexus between sustainable entrepreneurship and environmental pollution: Evidence from developing economy. *Environmental Science and Pollution Research International, 27*(29), 36242–36253. DOI: 10.1007/s11356-020-09642-y

IRENA – International Renewable Energy Agency. (2023). *Decentralised solar PV: A gender perspective*. https://www.irena.org/publications

Irfan, M., Razzaq, A., Sharif, A., & Yang, X. (2022). Influence mechanism between green finance and green innovation: Exploring regional policy intervention effects in China. *Technological Forecasting and Social Change, 182*, 121882. DOI: 10.1016/j.techfore.2022.121882

Jennings, P. D., & Zandbergen, P. A. (1995). Ecologically sustainable organizations: An institutional approach. *Academy of Management Review, 20*(4), 1015–1052. DOI: 10.2307/258964

Kachynska, N., Prakhovnik, N., Zemlyanska, O., Ilchuk, O., & Kovtun, A. (2022). Ecological modernization of enterprises: environmental risk management. *Health Education and Health Promotion, 10*(1), 175-182. https://hehp.modares.ac.ir/article-5-59219-en.html

Karadayı, T., Yazıcı, S., & Akdemir Ömür, G. (2025). Climate Change and Entrepreneurship: A Review and Research Agenda. *Global Business and Organizational Excellence, 44*(4), 16–43. DOI: 10.1002/joe.22281

Kesa, D. D., Wu, M., Abdillah, F., Harjadi, D., & Ningsih, A. (2024). The Impact of Digital Entrepreneurship Courses and Motivation on Fostering a Green Entrepreneurial Spirit in Students' Perceptions. *International Journal of Educational Qualitative Quantitative Research, 3*(2), 46–55. DOI: 10.58418/ijeqqr.v3i2.120

Meadows, D. H. (2008). *Thinking in systems: A primer*. Chelsea Green Publishing.

Mehta, A. K., Chandel, P. S., Shukla, R., & Selvakumar, P. (2024). Opportunities for green entrepreneurship. In Mehta, A. (Ed.), *Green innovation and sustainability practices* (pp. 223–238). IGI Global., DOI: 10.4018/979-8-3693-7046-9.ch009

Meirun, T., Makhloufi, L., & Ghozali Hassan, M. (2020). Environmental outcomes of green entrepreneurship harmonization. *Sustainability (Basel), 12*(24), 10615. DOI: 10.3390/su122410615

MOEF. (2020). *Korean New Deal: National Strategy for a Great Transformation*. Republic of Korea Ministry of Economy and Finance. https://stip.oecd.org/stip/interactive-dashboards/policy-initiatives/2023%2Fdata%2FpolicyInitiatives%2F99991604

Mol, A. P., & Sonnenfeld, D. A. (2000). *Ecological modernization around the world: Perspectives and critical debates*. Routledge.

Mondal, S., & Singh, S. (2025). A bibliometric and thematic analysis of green entrepreneurship in business research: Current status and future research directions. *Environment, Development and Sustainability*, *27*(1), 67–139. DOI: 10.1007/s10668-023-03905-7

Most Environmentally Friendly Countries. (2025). https://worldpopulationreview.com/country-rankings/most-environmentally-friendly-countries

Moulin, O. (2025). India's grassroot innovations are powering global climate resilience – Dipa Bagai, NRDC India, https://etedge-insights.com/sdgs-and-esg/climate-change/indias-grassroot-innovations-are-powering-global-climate-resilience-dipa-bagai-nrdc-india/

Mukherjee, A., Owen, R., Scott, J. M., & Lyon, F. (2024a). Financing green innovation startups: A systematic literature review on early-stage SME funding. *Venture Capital*, •••, 1–27. DOI: 10.1080/13691066.2024.2410730

Muo, I., & Azeez, A. A. (2019). Green entrepreneurship: Literature review and agenda for future research. *International Journal of Entrepreneurial Knowledge*, *7*(2). Advance online publication. DOI: 10.37335/ijek.v7i2.90

MWK. (2022). *GreenTech Innovation Strategy*. Federal Ministry for Economic Affairs and Climate Action. https://www.bmwk.de

Nadhifa, S. R., Prabarini, A. M., & Wati, I. R. (2024). Green entrepreneurial behavior: A review and research agenda. *Research Horizon*, *4*(3), 295–304. https://journal.lifescifi.com/index.php/RH/article/view/278

Neumann, T. (2022). Impact of green entrepreneurship on sustainable development: An ex-post empirical analysis. *Journal of Cleaner Production*, *377*, 134317. DOI: 10.1016/j.jclepro.2022.134317

News, A. P. (2023). *EV sales boom in Nepal, helping to save on oil imports and alleviate smog*. https://apnews.com/article/nepal-ev-energy-hydropower241d486f228051c73ff2793e043b0cca

Odeyemi, O., Usman, F. O., Mhlongo, N. Z., Elufioye, O. A., & Ike, C. U. (2024). Sustainable entrepreneurship: A review of green business practices and environmental impact. *World Journal of Advanced Research and Reviews*, *21*(2), 346–358. DOI: 10.30574/wjarr.2024.21.2.0461

OECD. (2022a). *Climate-resilient finance and investment.* OECD Environment Working Papers. https://www.oecd.org/en/publications/climate-resilient-finance-and-investment_223ad3b9-en.html

OECD. (2022b). *Policies to Support Green Entrepreneurship.* OECD Report. https://www.oecd.org/en/publications/policies-to-support-green-entrepreneurship_e92b1946-en.html

Onnoghen, N. U., Onwuzurike, U., & Anabaraonye, B. (2024). Enhancing climate resilience through green entrepreneurship education in Nigeria. *J Civ Eng Appl, 5*(1), 33–36.

Rizzitello, E., Piazza, M., & Perrone, G. (2025). Unlocking green startup investments: How environmental policy pressures drive Venture Capital funding decisions. *Technological Forecasting and Social Change, 217*, 1–17. DOI: 10.1016/j.techfore.2025.124158

Santika, I. W., Wardana, I. M., Setiawan, P. Y., & Widagda, I. G. (2022). Entrepreneurship education and green entrepreneurial intention: A conceptual framework. *Linguistics and Culture Review*, 797-810. https://doi.org/DOI: 10.21744/lingcure.v6nS1.2159

Savitz, A. (2013). *The triple bottom line: How today's best-run companies are achieving economic, social and environmental success.* John Wiley & Sons.

Schaltegger, S., & Wagner, M. (2011). Sustainable entrepreneurship and sustainability innovation: Categories and interactions. *Business Strategy and the Environment, 20*(4), 222–237. DOI: 10.1002/bse.682

SDGS-UN. (n.d.) https://sdgs.un.org/partnerships

Shepherd, D. A., & Patzelt, H. (2011). The new field of sustainable entrepreneurship: Studying entrepreneurial action linking "what is to be sustained" with "what is to be developed". *Entrepreneurship Theory and Practice, 35*(1), 137–163. DOI: 10.1111/j.1540-6520.2010.00426.x

Sobko, O., Gavkalova, N., Kurbatska, L., Boichyk, V., & Krysovatyy, I. (2024). Green entrepreneurship strategies: Integrating eco-friendly energy technologies with the digital economy. *Grassroots Journal of Natural Resources, 7*(3), s92–s114. DOI: 10.33002/nr2581.6853.0703ukr05

Tien, N. H., Tien, N. V., Mai, N. P., & Duc, L. D. M. (2023). Green entrepreneurship: A game changer in Vietnam business landscape. *International Journal of Entrepreneurship and Small Business, 48*(4), 408–431. DOI: 10.1504/IJESB.2023.130829

Trapp, C. T., & Kanbach, D. K. (2021). Green entrepreneurship and business models: Deriving green technology business model archetypes. *Journal of Cleaner Production, 297*, 126694. DOI: 10.1016/j.jclepro.2021.126694

UNEP. (2020). *Five Cities Tackling Air Pollution*. https://www.unep.org

United Nations. (2015). *Transforming our world: The 2030 Agenda for Sustainable Development*. A/70/L.1 https://www.refworld.org/legal/resolution/unga/2015/en/111816

Wang, W., Cao, Q., Zhuo, C., Mou, Y., Pu, Z., & Zhou, Y. (2021). COVID-19 to green entrepreneurial intention: Role of green entrepreneurial self-efficacy, optimism, ecological values, social responsibility, and green entrepreneurial motivation. *Frontiers in Psychology, 12*, 732904. DOI: 10.3389/fpsyg.2021.732904

WCED (World Commission on Environment and Development). (1987). *Our Common Future*. Oxford University Press.

Wei, X., Jiang, Y., & Gan, T. (2024). Air pollution and entrepreneurship: Evidence from China. *Applied Economics, 56*(24), 2860–2874. DOI: 10.1080/00036846.2023.2203454

WHO-World Health Organization. (2023). *Air pollution*. https://www.who.int/news-room/fact-sheets/detail/ambient-(outdoor)-air-quality-and-health

Yasir, N., Babar, M., Mehmood, H. S., Xie, R., & Guo, G. (2023). The environmental values play a role in the development of green entrepreneurship to achieve sustainable entrepreneurial intention. *Sustainability (Basel), 15*(8), 6451. DOI: 10.3390/su15086451

Yi, G. (2020). From green entrepreneurial intentions to green entrepreneurial behaviors: The role of university entrepreneurial support and external institutional support. *The International Entrepreneurship and Management Journal, 17*(2), 963–979. DOI: 10.1007/s11365-020-00649-y

Zeng, J., & Ren, J. (2022). How does green entrepreneurship affect environmental improvement? Empirical findings from 293 enterprises. *The International Entrepreneurship and Management Journal, 18*(1), 409–434. DOI: 10.1007/s11365-021-00780-4

Chapter 10
Global Trends in Smog and Air Pollution Controls

Kashvi Chaturvedi
https://orcid.org/0009-0005-3192-3400
Ajeenkya D.Y. Patil University, India

Susanta Das
https://orcid.org/0000-0002-9314-3988
Ajeenkya D.Y. Patil University, India

Manavi Gilotra
Ajeenkya D.Y. Patil University, India

Samina Firoz Waglawala
https://orcid.org/0000-0002-7521-7605
Ajeenkya D.Y. Patil University, India

ABSTRACT

This chapter examines the causes, historical development, health effects, environmental repercussions, and mitigation techniques associated with air pollution and smog. Based on international research, it emphasizes how transportation, urbanization, and industry have made air quality issues worse, particularly in Asia's megacities. Particular focus is placed on how pollutants like sulfur dioxide, nitrogen oxides, ozone, and particulate matter affect respiratory and cardiovascular health. The chapter also looks at legislative frameworks for lowering pollution exposure, biological remediation methods, and modern modeling and forecasting tools. It highlights the pressing need for coordinated, scientifically based solutions using case studies from China, India, Pakistan, and Poland. The study focus areas

DOI: 10.4018/979-8-3373-2469-2.ch010

and major contributions in this field are further revealed by bibliometric analysis, which highlights the connections between environmental systems, human health, and regulatory action in the fight against global air pollution.

1. INTRODUCTION

Since ancient times, human development and air pollution, one of the most enduring and urgent environmental issues have been linked. Its effects, which range from personal health to global climate systems, have gotten worse as a result of growing urbanization, industrialization, and dependency on fossil fuels. Although people have long been aware of the negative impacts of air pollution, such as respiratory distress, offensive smells, and decreased vision, scientific knowledge of its causes, constituent chemicals, and health consequences has only greatly increased in the last several decades (Fowler et al., 2020; Bai et al., 2018).

There is ample historical evidence of the detrimental impacts of air pollution. Although they lacked the modern ability to monitor or analyse specific pollutants, people in previous ages were aware of the unpleasant and detrimental effects of smoky or contaminated air. By the 17th to 19th century, European scientists, especially in the UK, had started to distinguish particles and gaseous chemicals in the atmosphere, and early written documents indicate complaints about smoke in medieval cities. The foundation for comprehending the causes of pollution and its possible effects on health was established by these investigations. Early calls for regulation were sparked by the industrial revolution's increased awareness, particularly in urban areas that relied heavily on coal. The issue of acid rain in the 1960s led to concerted efforts in North America and Europe to monitor the air. These networks helped governments create focused mitigation plans by supplying vital information on precipitation and air quality (Fowler et al., 2020).

Both natural and man-made factors contribute to air pollution. Particulate matter and gases in the air are produced by natural processes such as dust storms, forest fires, and volcanic eruptions. However, human activity is the primary contributor in the majority of locations, particularly in urban and industrial areas. Burning fossil fuels in power plants, factories, and transportation networks, heating homes with coal and biomass, and agricultural activities like open-field crop burning are important anthropogenic sources. Particulate matter (PM2.5 and PM10), sulphur dioxide (SO_2), nitrogen dioxide (NO_2), carbon monoxide (CO), carbon dioxide (CO_2), and ground-level ozone (O_3) are among the dangerous pollutants released by these sources (Bai et al., 2018).

Air pollution is still a major problem in Poland, a nation that has historically relied on coal-based electricity. It's interesting to note that huge, professionally run power plants, which frequently use emission controls, do not produce the worst pollution. Instead, the most dangerous emissions originate from the transportation and residential sectors, where pollutants are discharged at ground level and have an immediate impact on the local population. This scenario shows how small-scale, dispersed pollutants can collectively affect urban air quality more significantly than large-scale, regulated sources. Poland has acknowledged the socio-political aspect of air quality management by implementing Low-Emission Economy Plans (LEEPs) and Low-Altitude Emission Reduction Programs (LAERPs), which are attempts to combine local initiatives with more comprehensive national strategies (Burchard-Dziubińska, 2019).

Air pollution has numerous and well-documented negative effects on health. Exposure to ambient pollutants, particularly ozone and fine particulate matter, has been linked to a variety of health issues, according to numerous studies. These include cardiovascular disorders like heart attacks, strokes, and hypertension, as well as respiratory disorders including asthma, bronchitis, and chronic obstructive pulmonary disease (COPD). Negative pregnancy outcomes, several types of cancer, and neurological diseases are also linked to long-term exposure (Chen & Kan, 2008; Sierra-vargas & Teran, 2012; Mishra, 2017). Due to their weakened immune systems or increased air intake per body weight, children, the elderly, and people with pre-existing medical illnesses are especially at risk (Chen & Kan, 2008).

The extensive effects of extended exposure to air pollution are still being demonstrated by scientific data. For instance, research indicates that cutting emissions from diesel engines might dramatically reduce the number of heart attacks and hospitalizations for respiratory conditions, ultimately saving thousands of lives each year (Sierra-vargas & Teran, 2012). The cardiovascular effects of air pollution are also highlighted by recent studies, which show a strong correlation between increased mortality from heart-related disorders and both short-term exposure to coarse particulate matter (PM10) and long-term exposure to fine particles (PM2.5). In order to minimize exposure and lower health hazards, these findings highlight the necessity of both systemic controls and individual preventive measures (Mishra, 2017).

Air pollution has equally dire effects on the ecology. Global warming is mostly caused by pollutants like CO_2 and methane, whereas acid rain and ecosystem damage are caused by nitrogen oxides and sulphur dioxide. According to Bai et al. (2018), ozone layer depletion, which is associated with specific air pollutants, increases the amount of UV light that reaches the Earth's surface, increasing the risk of cataracts, skin cancer, and harm to aquatic ecosystems.

Smog, a mix of smoke and fog created when pollutants react with atmospheric components, is one of the most obvious types of air pollution. There are two main types of smog that are commonly known: photochemical or Los Angeles smog, which is produced by sunlight-driven reactions between nitrogen oxides and hydrocarbons, and classical or London smog, which is produced by sulphur compounds and is common in cold, humid conditions (Isha et al., 2024; Rani et al., 2011). In particular, photochemical smog is linked to sunny climates, heavy traffic, and industrial emissions. Numerous cities worldwide, including those in China, India, and Australia, have it (Rani et al., 2011).

High ground-level ozone levels, which are particularly harmful to respiratory health, are a result of this kind of smog. Smog emergencies have frequently occurred in Pakistani cities like Lahore and Faisalabad, especially in the winter when emissions and weather patterns combine to trap pollutants close to the ground. These cities' PM2.5 concentrations are more than ten times higher than what the World Health Organization (WHO) recommends. The government has responded by outlawing the burning of crops, controlling industrial activities, and raising public awareness through educational programs, particularly for medical students involved in community health outreach (Butt et al., 2018).

Another example is China, where serious air quality degradation, especially in its northern cities, has resulted from rapid economic growth. China has seen multiple haze outbreaks since 2011, marked by exceptionally high levels of PM2.5 and related pollutants. The Atmospheric Pollution Prevention and Control Action Plan was introduced in 2013 as a result of public concern about the negative health impacts. This program promoted greener energy use and focused on lowering emissions from transportation and industries. An epidemiological study conducted in Jinan, a significant city in eastern China, found a correlation between daily mortality and air pollution levels. Given that variations in population susceptibility and pollutant composition can result in region-specific health consequences, the study underlined the need of localized data (Zhang et al., 2017).

Air pollution issues have been made worse by urbanization, particularly in megacities, urban areas with a population of 10 million or more. With a predicted population of nearly 730 million, there will be 41 megacities by 2030, up from 10 in 1990. Because of the high concentration of automobile traffic, industrial activity, and building, many densely populated cities have disproportionately high levels of air pollution. In these cities, almost 96% of the population is exposed to PM2.5 concentrations that are higher than WHO safety standards, and air pollution is directly responsible for 7% of the illness burden. In South and East Asian cities, where exposure per capita and related health effects are among the highest worldwide, this burden is even more noticeable (Marlier et al., 2016).

Accurate modelling and forecasting technologies are crucial for managing and reducing air pollution. The two categories of forecasting methods are deterministic models, which model atmospheric chemical and physical processes, and parametric models, which employ patterns in past data. In-depth simulations of pollutant interactions under various environmental conditions are possible with sophisticated tools such as the Community Multiscale Air Quality (CMAQ) model (Bai et al., 2018). Though their effectiveness varies by region, neural networks and machine learning models, which have been introduced by recent breakthroughs in artificial intelligence, improve forecast accuracy. According to Daly and Zannetti (2007), these technologies are essential for creating timely notifications, educating the public, and assisting with policy decisions.

Additionally, modelling clarifies the fundamental dynamics of pollution dispersal. Diffusion, the way pollutants travel through turbulent air, chemical transformation, the way they interact with other substances, deposition, the way they land on surfaces, and atmospheric transport, the way they follow wind patterns, are important processes. Our knowledge of diffusion, particularly with regard to tiny particles in turbulent air layers, is still a relatively recent scientific endeavour, despite the fact that atmospheric transport was researched decades ago (Daly & Zannetti, 2007).

In order to convert complicated pollutant data into useful insights, accurate air quality indicators are required. For this, one of the most popular tools is the Air Quality Index (AQI). However, inconsistent measuring methods and poor cross-regional comparability can harm AQIs. Researchers have suggested a new paradigm that focuses on six fundamental pollutants: PM2.5, ozone, sulphur dioxide, nitrogen oxides, carbon monoxide, and volatile organic compounds, in order to get around these restrictions. To facilitate more precise cross-country comparisons and encourage more informed policy development, this method makes use of population-weighted global statistics and standardized units. Moreover, this paradigm can be combined with atmospheric models to predict the potential effects of different regulation approaches on pollution levels and health consequences. It gives decision-makers strong instruments to evaluate environmental interventions and improve public health tactics in a way that is grounded in science (Wolf et al., 2022).

Lastly, environmentally friendly methods for enhancing air quality, such as bioremediation, are becoming more and more popular. These techniques eliminate, neutralize, or break down air pollutants by using living things like fungi, bacteria, algae, and plants (Isha et al., 2024). Two important methods in this field that provide environmentally friendly substitutes for traditional mechanical or chemical filtering systems are phytoremediation and biosorption. Because of its affordability and versatility, bioremediation holds particular promise for environments with limited resources. Individual measures can also help lower health risks from air pollution, such as wearing well-made face masks, remaining indoors on days with heavy pollution, and include foods high in antioxidants in one's diet (Mishra, 2017).

This introduction presents a comprehensive overview of air pollution's historical evolution, its primary sources and pollutants, associated health and environmental impacts, modelling and forecasting techniques, and regional examples of mitigation efforts. It incorporates a diverse array of referenced studies to illustrate how modern societies are addressing this global concern through scientific innovation and policy reform. Figure 1 illustrates the chapter flow visually.

Figure 1. Flowchart Illustrating Flow of Chapter

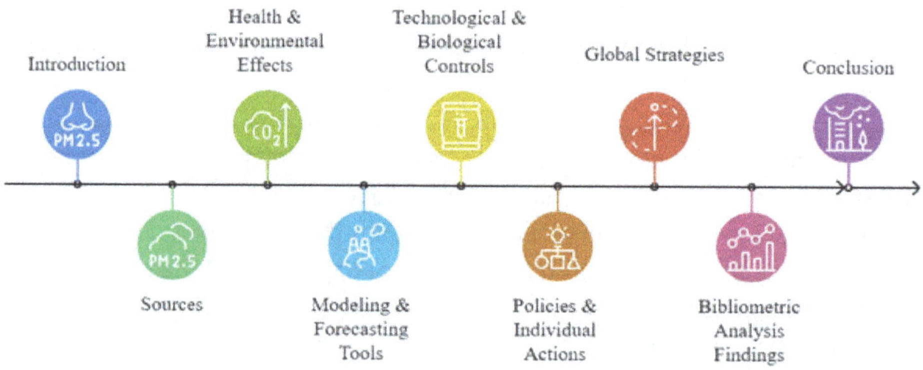

2. LITERATURE REVIEW

The negative health effects of indoor smoke and poor urban air quality were noted in early documents, demonstrating the ancient human awareness of air pollution. Hippocrates and other ancient Greek writers acknowledged the connection between air pollution and illness, and attempts to enhance interior conditions were reflected in developments like the chimney in the 12th century. Although scientific knowledge was still in its infancy, the medicinal effects of air pollution were recognized in a number of cultures, including Roman, Chinese, and Arab civilizations. In the Middle Ages, popular thought linked illness to "miasma," or unpleasant smells, instead of particular chemical substances. The earliest regulations appeared in medieval England, and the regulatory responses were initially limited and driven by economic considerations. The extent of pollution did not require more methodical answers until the industrial revolution, which was marked by metropolitan areas that relied heavily on coal (Fowler et al., 2020).

The Great London Smog of 1952, one of the most important incidents in the history of air pollution, killed thousands of people and sparked extensive legislative changes, including the Clean Air Act of 1956. This was a significant event that resulted in lower smoke and sulphur dioxide emissions and set the stage for further advancements in air quality. With the increasing effectiveness of urban pollution control, regional phenomena such as acid rain came into focus. Acid deposition from long-range air transport harmed crops, lakes, and forests, especially in North America and Northern Europe. International agreements to reduce sulphur and nitrogen emissions were prompted by scientific findings that connected industrial emissions to transboundary contamination. These activities developed into international air quality programs that target pollutants that endanger ecosystems and the stability of the climate, including methane, ammonia, and ground-level ozone (Fowler et al., 2020).

Even while many wealthy countries have made headway in lowering their emissions, growing pollution levels in some Asian countries still pose major health hazards. The intricate relationships between air quality and climate change are exacerbated by rising emissions of greenhouse gases and volatile organic compounds. However, there is cautious hope for long-term improvements in air quality worldwide due to rising environmental consciousness, technology advancements, and governmental changes (Fowler et al., 2020).

Air pollution has become a major public health and economic concern in addition to an environmental one. According to the WHO, air pollution is responsible for one in eight fatalities worldwide, primarily from respiratory and cardiovascular diseases. Nitrogen oxides (NOx), ozone (O_3), sulphur dioxide (SO_2), and fine particulate matter (PM2.5 and PM10) are the main pollutants of concern. Because PM can enter the circulation through the lungs and cause oxidative stress and systemic inflammation, it poses serious health hazards. Major sources of PM, including as home heating, industrial operations, and automobile traffic, have been designated by the European Union as the main priorities for mitigation. Air pollution has a significant and growing economic impact, including medical costs, lost productivity, and environmental damage. Poland has some of the worst air pollution in Europe. The extensive use of subpar fuels, antiquated heating systems, excessive emissions from transportation and industry, and poor waste management are all contributing issues. Lax standards and poor enforcement persist despite the Clean Air Programme and EU environmental directives being put into effect. Poland's air quality continues to fall short of WHO and EU standards. Peak smog occurrences have been linked to higher rates of cancer, strokes, cardiovascular and respiratory disorders, and other public health issues. The hazards are exacerbated by toxic emissions, such as carcinogens and heavy metals, necessitating more thorough interventions and stronger controls (Burchard-Dziubińska, 2019).

around the past, a lot of cities had very high pollution levels around the middle of the 20th century, but in areas with strong pollution management, these levels have considerably dropped. However, a lot of developing and transitional economies still have high levels of PM, SO_2, NO_2, and O_3 emissions, which are mostly caused by burning fossil fuels. Strong correlations between exposure to pollution and an increased risk of cardiovascular and respiratory disorders have been confirmed by research. Interestingly, there is no safe level at which contaminants stop being harmful to human health. The health effects of even little increases in pollutant concentrations are quantifiable, especially for vulnerable groups such as children, the elderly, and people with underlying medical disorders (Chen & Kan, 2008).

While chronic exposure leads to long-term diseases and noticeably higher mortality rates from heart and lung disorders, short-term exposure to air pollution increases daily mortality and hospital admissions. Reduced pollution has been shown to improve public health. Interventions in Hong Kong and Dublin, for instance, showed that quick drops in air pollution were associated with instantaneous gains in population health. In order to promote incremental improvements in developing nations, the WHO created air quality recommendations with intermediate targets after realizing these hazards (Chen & Kan, 2008).

Research on air pollution covers a wide range of topics, including regulatory evaluations, health effects, emission inventories, and pollution control. Emission inventories (EIs), which provide thorough records of pollutant emissions over particular geographies and time periods, are essential tools in this discipline. Models of air quality, policy formulation, and compliance strategies are all influenced by these inventories. For instance, the U.S. Environmental Protection Agency uses activity rates and emission factors to measure pollution levels in its standardized emissions data reporting systems. Pollution has a wide range of negative health effects, from higher mortality rates from cardiovascular disease to decreased sperm quality and increased childhood eczema. It is crucial to assess the socioeconomic costs of air pollution, quantifying the effects on industry, public health, and agriculture using techniques including engineering cost estimates and market value methods. Another research topic is pollution control efficiency, where methods such as data envelopment analysis are employed to evaluate how well different industries lower emissions. In order to notify the public of pollution spikes and start emergency responses, early warning systems that combine weather forecasts with air quality data are essential. In order to simulate the dispersion of pollutants in complicated terrains more accurately, modern air pollution modelling has moved from two-dimensional to three-dimensional systems. To improve forecasting accuracy, hybrid models integrate multiple forecasting techniques, such as geographic information systems and artificial intelligence. In order to provide a more comprehensive understanding of the dynamics of air quality, models such as BlueSky and WRF-Chem are made to predict emissions from natural occurrences like volcanic activity and wildfires (Bai et al., 2018).

When coal, gasoline, fuel, and wood are burned, particulate matter a significant contributor to urban air pollution is released into the atmosphere. PM may also emerge from secondary chemical reactions in the atmosphere involving nitrogen oxides and Volatile Organic Compounds (VOC). Size, makeup, length of exposure, and personal vulnerability all affect its health effects. According to Sierra-vargas and Teran (2012), symptoms can include coughing, eye irritation, worsening of chronic respiratory conditions, and early death. The interaction of sunlight with organic molecules and nitrogen oxides produces ozone (O_3), which has a negative impact on lung function and makes disorders like asthma worse. Particularly for those who are sensitive, physical activity during times of high ozone can exacerbate these effects (Sierra-vargas & Teran, 2012).

Because their immune systems are still developing and their breathing rates are faster, children are more susceptible to air pollution. Pollutants can exacerbate respiratory conditions and allergic reactions. For example, it has been demonstrated that diesel exhaust particles interact with allergens and exacerbate inflammation of the airways. Oxidative stress, in which reactive oxygen and nitrogen species damage cellular proteins and tissues, is one of the main mechanisms of pollution-induced damage. According to Sierra-vargas and Teran (2012), this procedure raises the risk of cancer, chronic inflammation, and other degenerative disorders. Reductions in PM exposure have been linked to better respiratory outcomes and longer life expectancies, especially in susceptible populations, according to encouraging research (Sierra-vargas & Teran, 2012).

Coordination between individual and governmental efforts is necessary to reduce air pollution. Personal tactics include enhancing indoor air circulation, particularly in homes that use biomass fuels, and minimizing outdoor activities during pollution peaks. Government initiatives like the implementation of renewable energy, green infrastructure, and more stringent automobile pollution regulations are essential. For instance, brief increases in air quality during the 2008 Beijing Olympics were associated with fewer hospitalizations for asthma. By escalating pollution concentrations through rising temperatures and changed weather patterns, climate change makes air quality even more problematic. Therefore, comprehensive climate and pollution regulations are required, but they are still difficult to police and execute (Sierra-vargas & Teran, 2012).

The study of pollution dispersion from point sources, particularly industrial stacks, has benefited greatly from models such as the Gaussian Plume Model. The model, which was created using empirical observations, helps establish stack heights for regulatory compliance and simulates how pollutants spread under different atmospheric conditions. Lagrangian and Eulerian models, which follow pollutants along wind paths or inside predetermined spatial grids, respectively, were later developments. Global simulations, long-distance transport studies, and the creation of hybrid methodologies have all been made possible by these models (Daly & Zannetti, 2007).

These days, models like AERMOD and CALPUFF are frequently utilized. CALPUFF mimics the behaviour of pollutants in complicated terrains and under non-steady situations, whereas AERMOD is a steady-state Gaussian model that incorporates meteorological data. In order to mimic pollutant transformations over various geographical and temporal scales, advanced photochemical models like CMAQ and CAMx employ mathematical representations of atmospheric chemistry. These simulations are improved by meteorological inputs from models such as MM5 and CALMET. Furthermore, statistical and particle-based models can explain dispersion, particularly in situations where it is challenging to establish clear cause-and-effect links (Daly & Zannetti, 2007).

Reducing air pollution has been repeatedly linked to major benefits in public health, particularly for vulnerable groups like children, the elderly, and people with pre-existing medical illnesses. Research indicates that improved respiratory function and longer life expectancy are linked to reduced particulate matter concentrations. Therefore, governments have a crucial role to play in managing air quality. Expanding green urban design, switching to cleaner energy sources, enforcing strict emission regulations, and assisting with public awareness campaigns are examples of effective policy actions. Outpatient visits for respiratory illnesses significantly decreased during the Beijing Olympics, demonstrating the practical benefits of short-term air quality measures (Sierra-vargas & Teran, 2012).

To lessen the negative health effects of air pollution, comprehensive methods that incorporate community involvement and state regulation are required. Individuals can experience modest but significant respite from smog by taking simple steps like wearing protective masks during smog events, reducing exposure to heavy traffic corridors, and using indoor air purifiers. In order to support these initiatives, policymakers should enforce cleaner fuel standards, encourage public transit, and provide subsidies for eco-friendly home appliances. As pollution and climate change continue to interact, these initiatives become increasingly more critical because, particularly in warmer weather circumstances, more frequent photochemical smog occurrences could degrade air quality. Countries like Mexico have enacted climate laws to cut carbon emissions and encourage renewable energy, but there are still significant obstacles due to enforcement issues and inadequate urban planning (Sierra-vargas & Teran, 2012).

Models such as the Gaussian Plume Model, which was initially developed to estimate the spread of contaminants from industrial stack emissions, have significantly advanced scientific understanding of air pollution dispersion. In order to help regulators choose the right stack heights to reduce ground-level concentrations, this model computes the horizontal and vertical spread of contaminants. The model simulates real-world dispersion conditions by using weather and atmospheric stability data. It developed over time to support line and area sources and incorporate multiple reflections. The scientific community had moved away from regional pollution control and toward acknowledging air pollution as a worldwide issue by the 1970s.

The significance of long-range air transport was highlighted when it was shown that tall emission stacks, which were previously believed to lessen local effects, actually contributed to regional acidification (Daly & Zannetti, 2007).

The creation of increasingly complex modelling systems was spurred by this realization. Lagrangian approaches track pollutant packages as they travel, whereas Eulerian models split a region into fixed grid cells to examine pollution within each segment. Accuracy has improved in both urban and regional settings as a result of the combination of these methods. In regulatory and planning applications, contemporary models like CALPUFF, a non-steady-state model that works well in changeable weather circumstances, and AERMOD, a steady-state Gaussian dispersion tool, have become indispensable. The intricate chemistry of atmospheric processes is taken into consideration by photochemical models such as CAMx and CMAQ (Community Multiscale Air Quality), which model pollutant interactions and transformations throughout time (Daly & Zannetti, 2007).

The atmospheric characteristics required for these dispersion models are provided by meteorological models like MM5 and CALMET. To provide more accurate models of pollution plumes across different terrains, many additionally incorporate modules for particle behaviour and plume ascent. Particularly for localized air quality assessments, statistical models that use ambient concentration data without presuming causation are frequently used (Daly & Zannetti, 2007).

Research conducted in China in recent years has brought attention to the substantial impact that rising air pollution places on public health. Since 2011, haze outbreaks in cities like Jinan have alarmingly increased, mostly due to rising levels of PM2.5, PM10, SO_2, and NO_2. In the past, the accuracy of health effect assessments was constrained by a lack of high-resolution PM2.5 data. Recent research, however, has shown definite relationships between pollution levels and mortality rates utilizing better monitoring and longitudinal data. Rising PM10 and NO_2 concentrations were linked to a considerable increase in daily death rates, according to the Jinan study, which was carried out between 2011 and 2015. In winter, when smog seasons were marked by stagnant air and high emissions, the health effects were particularly pronounced. The most vulnerable groups were determined to be women and the elderly, with the greatest impact on cardiovascular and respiratory mortality. Crucially, this study demonstrated the severe impact that smog outbreaks have on public health, as mortality increases happen only a few days after pollution peaks. The thorough five-year dataset confirmed the validity of these conclusions and emphasized the significance of local data in developing successful air quality regulations (Zhang et al., 2017).

A common tool for informing the public about pollution levels is the air quality index (AQI). These indices combine information from different pollutants, including CO, NO_2, PM2.5, and PM10, to give a simplified picture of air quality. To direct public health responses, regulatory agencies such as the U.S. Environmental Protection Agency establish pollution thresholds under laws like the Clean Air Act. For instance, in Pakistan, places like Abbottabad have attained improved air standards, but Lahore continues to rank among the most polluted cities in the world because of industrial and vehicle emissions (Isha et al., 2024).

Bangladesh has the worst levels of smog in the world, and nations like China, India, Pakistan, Indonesia, and Bangladesh continue to have disproportionately high rates of air pollution-related deaths. The United States, New Zealand, and Canada, on the other hand, record much lower levels of pollutants than countries with stricter regulations. Industrial intensity, transportation regulations, and waste management effectiveness are some of the factors influencing these disparities (Isha et al., 2024). Smog is a mixture of dangerous gasses and fine particulates that can seriously harm people's health. It increases cardiovascular events, worsens respiratory conditions, and may even have an impact on reproductive health. Smog particles can increase the risk of antibiotic-resistant bacteria by entering the lungs, according to recent research. New information about the connections between air quality and health was made available by the COVID-19 pandemic. High levels of air pollution have been linked in studies to higher incidence of postpartum depression, respiratory discomfort, and gestational diabetes in expectant mothers. Furthermore, a large number of industrial chemicals that the EPA has determined are carcinogenic, and exposure to these airborne pollutants has been linked to COVID-19 issues on multiple occasions (Isha et al., 2024).

Biological remediation techniques are emerging remedies for air pollution. Pseudomonas putida and other microorganisms are used in bacterial bioremediation to convert nitrogen oxides and hydrocarbons into less toxic byproducts. Although these bacteria are susceptible to environmental restrictions and may be blocked by high pollutant concentrations, they have shown efficacy in degrading pollutants under particular conditions. Additionally, fungi have intriguing remedial properties. Fungal enzymes are used in microembolisation processes to break down complex contaminants, such as heavy metals and hydrocarbons. White-rot fungi have demonstrated remarkable efficacy in breaking down a wide range of organic substances. Although these tactics can be used both in situ and out of situ, their effectiveness frequently depends on certain environmental factors (Isha et al., 2024).

An other frontier in pollution control is nanotechnology. By capturing or neutralizing a wide range of contaminants, engineered nanoparticles can improve on conventional bioremediation methods. That being said, some nanomaterials are harmful and persistent in the environment, which calls for more investigation and regulation. A plant-based method for air purification is called phytoremediation, which uses species that have the ability to absorb pollutants. Cost-effective and adaptable to urban landscaping, this approach's effectiveness is contingent upon biological and climatic conditions. Despite the potential of these novel techniques, obstacles like the complexity of pollutants, environmental unpredictability, and possible secondary consequences prevent their general adoption. Integration with regulatory frameworks, urban planning, and continuing genetic engineering and nanoscience research are necessary to overcome these obstacles (Isha et al., 2024).

Although bioremediation techniques show great promise, a number of obstacles still prevent their widespread use. The effectiveness of biological decomposition is limited by the trace amounts of many contaminants. Moreover, specific environmental parameters like temperature, humidity, and pollution load are frequently necessary for bioremediation to be successful. During the degrading process, there is also a chance of producing hazardous secondary byproducts. These drawbacks highlight the necessity of interdisciplinary integration to improve the practical efficacy of bioremediation by fusing biotechnology, nanoscience, urban planning, and environmental policy. Innovations in genetic engineering and nanotechnology are probably going to be very important in enhancing these biological systems' efficiency, specificity, and robustness in both urban and rural settings (Isha et al., 2024).

Global ecosystems and human well-being are under risk due to the interconnected environmental problems of air pollution and climate change. The burning of fossil fuels is one of the main anthropogenic activities that contributes to greenhouse gas emissions, which raise the levels of carbon dioxide, methane, and nitrous oxide in the atmosphere. Sea level rise, biodiversity loss, weather pattern disturbances, and challenges to food and water security are all consequences of these gases' acceleration of climatic volatility and driving of global warming. By lowering air quality and increasing the overall environmental burden, pollution, especially in the form of airborne gases and particles, makes these problems worse. Although the main causes are automobile exhaust and industrial pollution, natural occurrences like dust storms and volcano eruptions also contribute (Kandpal, 2024).

One of the most dangerous forms of air pollution is smog, which is usually a dense yellow or brown haze made up of particulate matter and ground-level ozone. It is caused by sunlight interacting with nitrogen oxides and volatile organic chemicals, and it is particularly common in crowded urban environments. Exposure to smog can cause everything from minor irritations like eye, nose, and throat pain to serious respiratory and cardiovascular conditions, reduced lung function, and even early mortality. Smog has a direct impact on ecological and economic sustainability since it impairs photosynthesis, lowers crop output, and deteriorates vision. The WHO has released thorough air quality standards that include safe concentration limits for main pollutants like PM2.5, PM10, CO, NO_2, SO_2, and O_3 in order to combat these effects. Governments use these recommendations as a baseline for creating national policies and public health initiatives. The recent tightening of PM2.5 regulations to lower exposure and associated mortality is only one example of how updated limits frequently take into account new scientific knowledge about the dangers of pollution. Since ground-level ozone is a strong oxidant created mostly by industrial and vehicle emissions in direct sunlight, it continues to be a crucial area of concern (Kandpal, 2024).

Air quality issues have gotten worse due to rapid urbanization and the growth of megacities, which are urban agglomerations with more than 10 million residents. Megacities increased from 10 to 28 between 1990 and 2014, and by 2030, there are expected to be 41 of them, with a combined population of nearly 730 million. In Asia, where cities like Delhi, Beijing, Dhaka, and Karachi routinely report PM2.5 concentrations far beyond global health limits, the environmental pressure from this growth is particularly noticeable. Millions of people die each year from pollution-related causes in these areas, where the majority of urban dwellers are exposed to dangerously high levels of air pollution on a daily basis (Marlier et al., 2016).

Globally, pollution trends differ. Other places, particularly in South and East Asia, have seen rising pollution levels as a result of industrial expansion and inadequate enforcement, while other regions, such as the eastern United States, have successfully decreased PM2.5 levels by regulatory initiatives between 1998 and 2012. Vehicle emissions, industrial operations, burning in agriculture, and the use of fossil fuels are some of the main causes. To monitor pollutant concentrations, assess the efficacy of policies, and predict pollution extremes, satellite data, ground-based monitoring stations, and atmospheric modelling methods are combined (Marlier et al., 2016).

The effects of urban air pollution on health are widespread. Chronic exposure is associated with developmental abnormalities in children, reduced labour productivity, cognitive impairment, and cardiovascular and respiratory ailments. Exposure during pregnancy or early childhood, for example, may have long-term consequences on intellectual development and lung capacity. Despite efforts to reduce emissions, urban centres like Beijing and Shanghai often surpass international pollution regulations. On the other hand, stricter enforcement of regulations, cleaner transportation, and effective urban planning have allowed places like Tokyo to preserve comparatively cleaner air (Marlier et al., 2016).

Countries' efforts to reduce urban pollution differ greatly from one another. To cut emissions, some governments have put in place legislative frameworks, environmental fees, and infrastructural improvements like wider metro systems and automobile limitations. For instance, Delhi's odd-even car rule and Beijing's alternate-day driving program have had varying degrees of effectiveness. However, such actions are not necessarily guaranteed to be equitable. Richer people may frequently avoid regulations or protect themselves from pollution, which lessens public pressure for systemic change. Therefore, to guarantee broad support for air quality measures, inclusive and socially equitable strategies are required (Marlier et al., 2016).

Interestingly, air pollution mitigation also contributes to climate change adaptation. Numerous contaminants are also climate forcers, including ozone precursors and black carbon. By lowering their concentrations, global warming is moderated and public health is enhanced. These interactions, though, can be intricate. Certain actions taken to lower one pollutant may unintentionally increase another or change the dynamics of the atmosphere in unforeseen ways. For example, non-linear chemical processes may cause aggressive NOx reductions to occasionally raise ozone levels. These interdependencies must therefore be carefully taken into account when designing policies (Marlier et al., 2016).

One of the main causes of air pollution worldwide is transportation. Since World War II, the number of vehicles produced annually has increased dramatically from 5 million to about 70 million, with nations such as China currently dominating the manufacture of both cars and motorcycles. Personal automobiles have been more and more in demand as the world's population has grown, especially in urban areas. Over 1 billion vehicles were in use worldwide by 2002, and this number has been rising at a rate of 27 million new cars per year (Global Trends in Motor Vehicle Pollution Control: A 2011 Update, 2011).

PM, NOx, CO, SO_2, and VOCs are among the dangerous pollutants that are released by motor vehicles. Smog, respiratory ailments, heart disease, and climate change are all exacerbated by these. Because fine particulate matter (PM2.5 and PM10) can enter the bloodstream and lungs deeply, it presents unique hazards. There is no threshold below which these particles are deemed harmless, according to the WHO. Similarly, respiratory discomfort and asthma episodes are linked to ground-level ozone produced by NOx and VOCs. Disproportionately impacted are children, the elderly, and people with pre-existing ailments (Global Trends in Motor Vehicle Pollution Control: A 2011 Update, 2011).

In addition to their negative health impacts, automobile emissions lower crop yields, decrease visibility, and hasten structure degradation. Acid rain is caused by pollutants like SO_2 and NOx, which harm infrastructure and ecosystems. In the meantime, global warming is intensified by greenhouse gases like CO_2 and black carbon. Rapid motorization is causing emissions to rise in poorer countries, despite the fact that wealthy countries have implemented cleaner technologies and more stringent laws. International collaboration, technological advancement, and strict enforcement are necessary to meet these difficulties (Global Trends in Motor Vehicle Pollution Control: A 2011 Update, 2011).

The health risks of uncontrolled emissions are highlighted by past pollution incidents. Extremely high levels of SO_2 and suspended particles during the 1985 smog crisis in North Rhine-Westphalia, West Germany, prompted emergency responses. During the haze period, there was a significant increase in hospital admissions, ambulance calls, and cardiovascular fatalities, particularly among people who already had heart issues (Wichmann et al., 1989). Comparably, the Great London Smog of 1952 sparked the creation of contemporary air quality laws and caused over 4,000 immediate deaths. Current research supports the effects of air pollution on the cardiovascular system. Even brief exposure to PM2.5 increases the risk of arrhythmias, heart attacks, and strokes. Smokers, diabetics, and the elderly are the groups most impacted. Atherosclerosis and endothelial dysfunction are caused by fine particles that enter the bloodstream and cause systemic inflammation. Risk can be decreased by taking precautions like wearing masks, utilizing air purifiers, and remaining indoors during smog episodes, but more extensive mitigation calls for stricter laws and better technology (Mishra, 2017).

A global public health emergency is the unequal burden of air pollution in low- and middle-income nations. Even while many affluent countries have cut their emissions, more than 90% of people on Earth still live in places with pollution levels higher than those recommended by the WHO. Accurately assessing exposure is a major obstacle in resolving this issue. Population density is not taken into consideration by conventional measures that rely on national averages. A fuller picture of the true health hazards is provided by a more equitable approach called population-weighted exposure (Wolf et al., 2022).

This method was used in a recent study that assessed exposure to six pollutants in more than 200 countries: PM2.5, O_3, NOx, SO_2, CO, and VOCs. The findings showed glaring differences. While legislative measures have led to advancements in certain countries, such as China and India, others continue to face increased exposure. The study also emphasized intricate relationships between pollutants, including how atmospheric feedback loops may cause decreases in NOx and VOCs to inadvertently raise ozone levels. These results assist governments anticipate trade-offs and improve their environmental plans by informing policy design. Importantly, countries can monitor progress and enact more fair, efficient air quality regulations when routinely updated, population-relevant exposure data is available (Wolf et al., 2022).

3. BIBLIOMETRIC ANALYSIS

The co-occurrence network map in Figure 2, which was created from the Lens bibliographic dataset and displayed using VOSViewer, depicts the intellectual framework of international research on SMOG and air pollution control from 1990 to 2025. Depending on the data layer under analysis, the map displays a constellation of connected nodes, each of which represents a term, author, or referenced source. The modularity-based clustering technique of VOSViewer determines cluster membership, whereas node size represents frequency or citation strength. Four to five prominent clusters that match the field's main thematic categories are highlighted in the graphic. The largest cluster, which is usually colored red or yellow, focuses on "air pollution," "PM2.5," "particulate matter," and "health effects," suggesting that health-related studies are now a significant source of research interest. Keywords like "emission sources," "combustion," and "atmospheric chemistry" make up another cluster that represents the basic physical and chemical study on SMOG creation. The technological aspect of managing air quality is reflected in a third cluster that frequently links "sensor technology," "remote sensing," and "monitoring systems." Further clusters that capture governance and intervention studies are connected to "policy," "urban planning," and "mitigation." As environmental sciences combine with public health, engineering, and data analytics, the inter-cluster links demonstrate increasing interdisciplinarity. All things considered, the network visualization shows how SMOG and air pollution research have developed into a highly interwoven field where environmental policy, technology, and health converge.

Figure 2. Bibliometric Analysis on Global Trends in SMOG and Air Pollution Control made using VOSViewer and data from Lens: Network Clustering (Lens, 2025; Van Eck & Waltman, 2022).

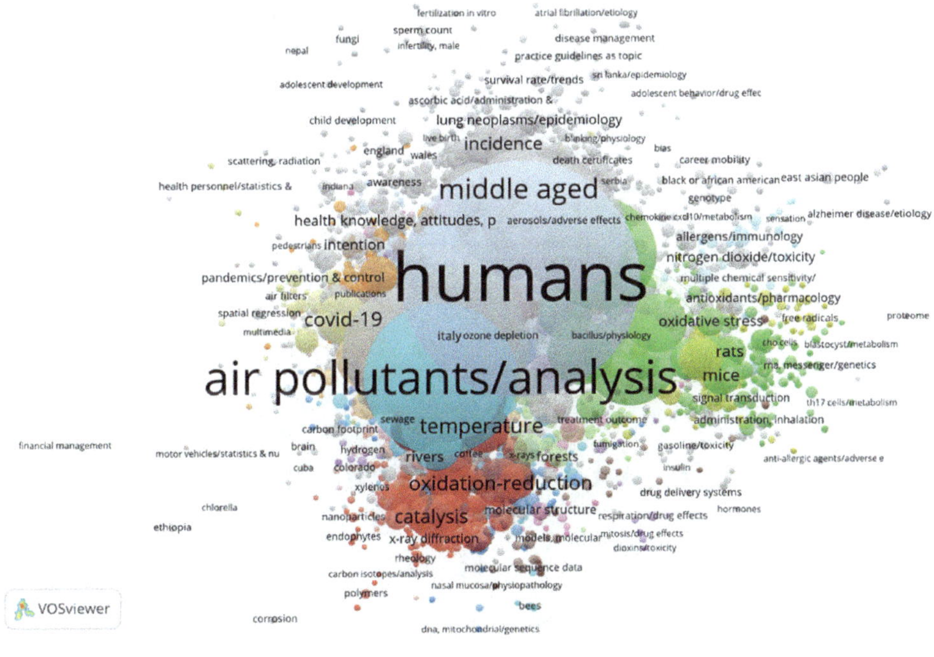

A temporal-overlay visualization from VOSViewer that shows the evolution of research themes from 1990 to 2025 is shown in Figure 3. Each keyword's average publishing year is shown by the overlay color gradient, which goes from dark blue for earlier years to yellow for more recent ones. The older portion of the map in the early 1990s is occupied by dominating phrases like "air quality," "ozone," and "photochemical smog," which show the time when focus was on the chemistry of pollutants and the measurement of criterion gases. New words including "PM2.5," "particulate matter," "exposure assessment," and "respiratory disease" were introduced in the 2000s in response to the growth of extensive epidemiological studies that demonstrated connections between fine particles and outcomes related to human health. The overlay shows diversification toward phrases like "climate change," "satellite monitoring," "remote sensing," and "urban air pollution" during the 2010–2020 decade, indicating the integration of atmospheric research with data analytics and sustainability studies. Emerging terminology like "machine learning," "low-cost sensors," and "policy assessment" appear in the most recent years (2020–2025), indicating that policy and technology advancements are now shaping the study frontier. This temporal layering highlights the field's evolution from traditional air chemistry to multidisciplinary frameworks that include environmental governance, smart monitoring, and health risk.

Figure 3. Bibliometric Analysis on Global Trends in SMOG and Air Pollution Control made using VOSViewer and data from Lens: Timeline Chart (Lens, 2025; Van Eck & Waltman, 2022).

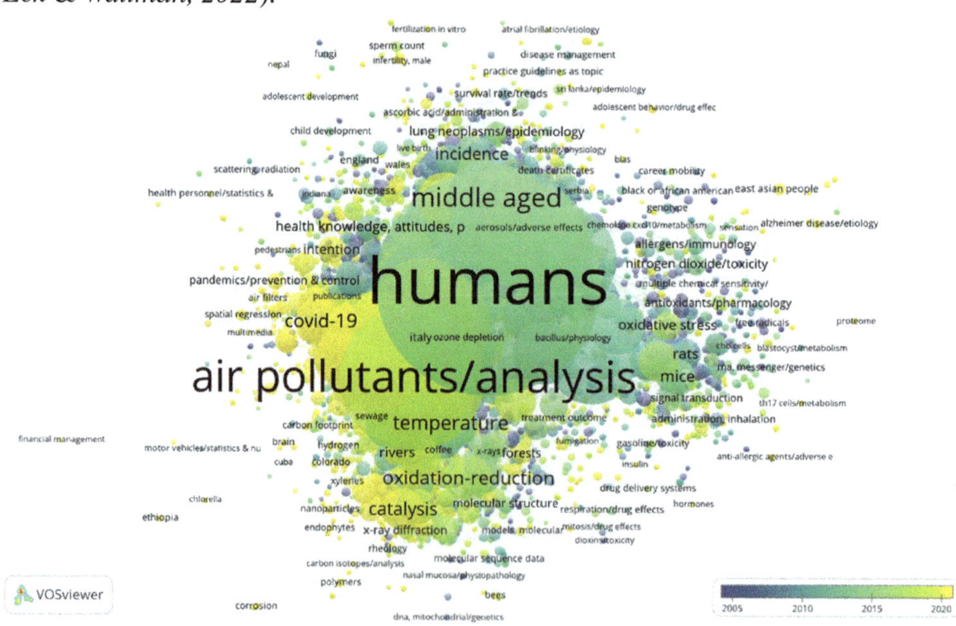

A VOSViewer density map, a supplementary visualization highlighting the concentration and intensity of research activity across keywords, is shown in Figure 4. High co-occurrence frequency areas in this plot are bright or warm-colored, whilst less active themes are displayed in cooler tones. The densest area on the image, which is frequently shown in bright yellow, is centered on the terms "PM2.5," "air pollution," "health impact," and "emission control." These phrases, which highlight the importance of particulate matter and its effects on health, form the basis of academic discourse in SMOG-related studies. Around subjects like "aerosol composition," "modeling," "climate interaction," and "sensor technology," secondary zones of intermediate density emerge, reflecting supporting fields of study focused on the causes of pollution production and its wider environmental connections. Emerging subjects including "electric vehicles," "renewable energy transition," and "urban policy" are included on the map's periphery, indicating a shift in emphasis toward integrated sustainability solutions. The field's intellectual development is demonstrated by the gradual dispersion of density toward new subjects over time, from systemic approaches addressing mitigation, technology deployment, and public health to descriptive air-quality assessment. As a guide for both established and new research goals, this density visualization successfully pinpoints "hotspots" where scholarly attention and citation impact are focused.

Figure 4. Bibliometric Analysis on Global Trends in SMOG and Air Pollution Control made using VOSViewer and data from Lens: Density Chart (Lens, 2025; Van Eck & Waltman, 2022).

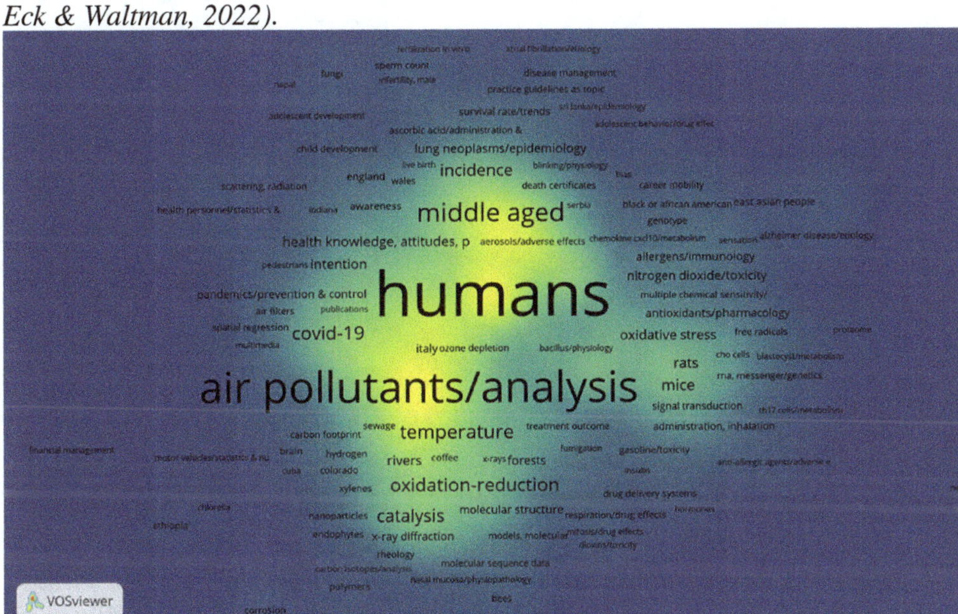

The annual publication trend of SMOG and air pollution control research is shown in Figure 5, which is a Python visualization that was taken from Lens. From 1990 to 2025, the time series shows a distinct and consistent rising trend in scientific production. The low baseline of activity in the early 1990s reflects the early stage of research, when only a small number of papers addressed photochemical pollution and SMOG production. As the health impacts of particulate matter are acknowledged in international scientific and policy conversations, publication counts start to rise somewhat in the 2000s. After 2010, there is a noticeable acceleration that corresponds with increased air quality monitoring, fast urbanization, and increased global concern over PM2.5 exposure in nations like China and India. Between 2015 and 2022, there was a sharp increase in multidisciplinary publications due to links between environmental health and the Sustainable Development Goals (SDGs). Although interest is still high, a little variation in recent years may be due to delays in database indexing or brief disruptions in research during the COVID-19 pandemic. The general trend demonstrates that SMOG and air pollution have emerged as enduring worldwide concerns, with study growing in fields such as environmental engineering, epidemiology, data science, and policy.

Figure 5. Line Chart on Publications on SMOG and Air Pollution between 1990-2025 with data from Lens (Lens, 2025).

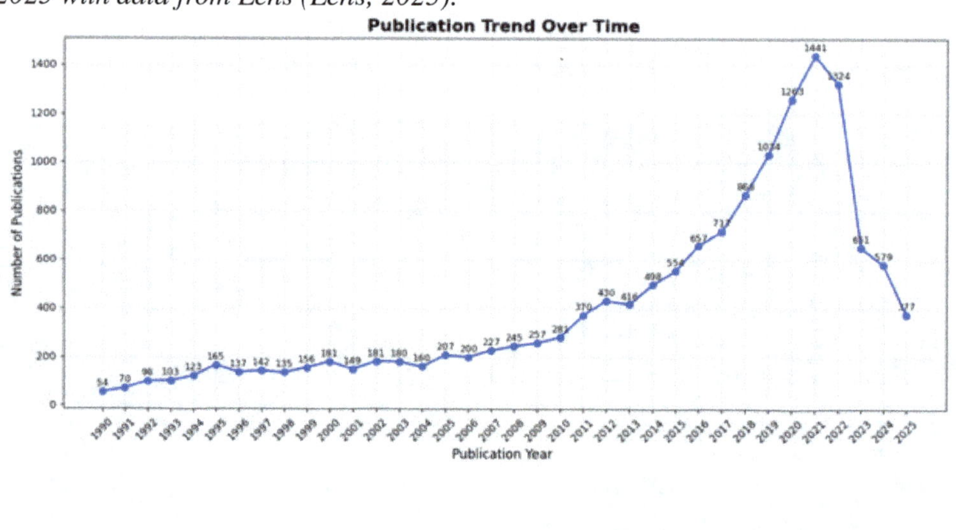

A Python-generated word cloud that graphically depicts the frequency of terms appearing in the dataset between 1990 and 2025 is shown in Figure 6. The worldwide research corpus's thematic focus is communicated by larger, bolder words, which have greater occurrence frequencies. The prevalent keywords—air pollution, PM2.5, health, emissions, exposure, and particulate matter—show that the main focus of the literature is on identifying the sources of pollution, measuring exposure, and evaluating the effects on human health. National case studies and regulatory research are crucial to the discipline, as seen by the use of geographic and policy-related terminology like China, urban, industrial emissions, and air quality standards.

Figure 6. Word Cloud on Keywords on SMOG and Air Pollution (Lens, 2025).

The growing use of cutting-edge technologies and data-driven methods is highlighted by emerging words like climate change, remote sensing, machine learning, and mitigation techniques. Thus, the term "cloud" encapsulates a harmony between conventional atmospheric-science study and the more recent, multidisciplinary approaches that incorporate technology, health, and politics. The visual predominance of phrases connected to health and the environment highlights how SMOG research has moved from localized air-quality control to global public health and sustainability views.

Based on author affiliation data from Lens, Figure 7 shows a globe map created with Python that shows the geographical distribution of publications on SMOG and air pollution control from 1990 to 2025. A clear concentration of research activity in a few key places is shown by the graphic. Global output is dominated by the United States, China, and European nations (especially the United Kingdom, Germany, and Italy), which together account for a sizable portion of publications. Due to severe

SMOG occurrences and national clean-air programs that increased financing for research, China's growth trajectory has been especially rapid since the mid-2000s. Through comprehensive epidemiological investigations and environmental policy research conducted within the framework of the Clean Air Act, the United States consistently maintains its leadership position.

Figure 7. World Graph Depicting Countries' Contribution on Research (Lens, 2025).

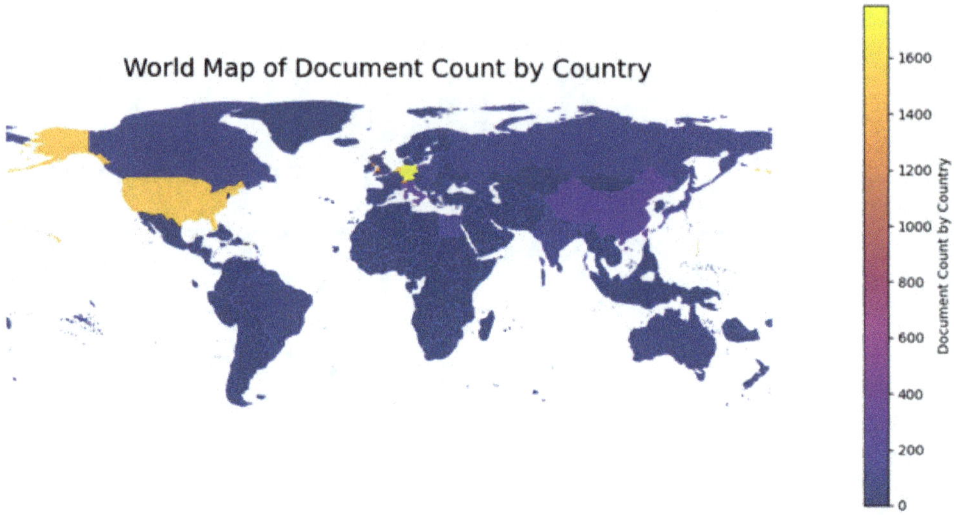

Transboundary pollution, emission modeling, and the integration of climate and air quality policies are the main areas of focus for European research. Growing awareness and local policy responses to deteriorating urban air pollution are reflected in the emergence of participation from South Korea, Brazil, India, and Southeast Asian countries. Strong international ties are indicated by collaboration lines, which are not displayed in this static graphic but can be inferred from co-authorship data, particularly between North American and Asian institutions. Nonetheless, under-represented regions including parts of Latin America and Africa indicate possible opportunities for capacity growth. All things considered, the geographic pattern shows how SMOG research has evolved from a few developed countries to a more widely dispersed endeavor tackling the universal problem of air pollution.

The distribution of publications in the field of Global Trends in SMOG and Air Pollution Control across major academic publishers for the years 2000–2025 is shown in Figure 8, a tree map visualization made with Origin software using data taken from the Lens database. With 3,618 papers, Springer Science and Business Media LLC holds the top spot in the visualization, demonstrating its critical role in spreading atmospheric and environmental research. With 1,498 articles, MDPI

AG comes in second, demonstrating the robust existence of open-access platforms that encourage quick publication and interdisciplinary cooperation in environmental sciences. Additional noteworthy donors are Wiley (273), Elsevier BV (957), and Copernicus GmbH (984), which constitute the well-established core of scientific publishers actively funding research on pollution monitoring, air quality, and climate interactions.

Figure 8. Publishers' Contribution in Publishing Topic Related Papers (Lens, 2025; OriginLabs, 2025).

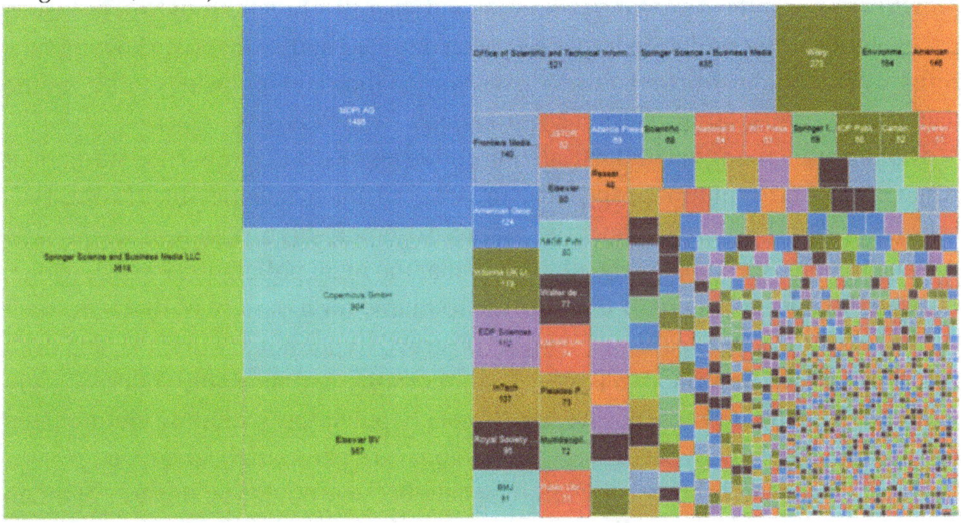

The existence of publishers like Frontiers Media, SAGE Publications, and Informa UK highlights the variety of channels that promote research on sustainability solutions, environmental policy, and air pollution. Smaller contributors that publish specialized but significant works on atmospheric modeling, chemical processes of pollutants, and technical advancements in air-quality control include Atlantis Press, IOP Publishing, and the Royal Society of Chemistry. The tree map successfully illustrates the hierarchical concentration of publications, where many smaller organizations contribute to the field's subject depth and specialization but a small number of powerful publishers account for the majority of global research.

Major academic publishers prioritize environmental research, public health, and pollution reduction, which is reflected in this concentration pattern. In order to solve transboundary pollution and climate concerns, there is a global shift toward more data transparency, collaborative research, and accessibility, which is highlighted by the substantial representation of open-access publishers. Overall, the graphic shows how SMOG and air pollution control have developed into important fields of study, backed by a variety of publishing houses dedicated to expanding knowledge, encouraging creativity, and influencing legislation for a more sustainable and cleaner environment.

4. CONCLUSION

One of the most complicated and urgent environmental issues facing the modern world is still air pollution. It still has a significant impact on the environment and human health in spite of growing awareness and scientific breakthroughs. The historical evolution of air pollution issues has been outlined in this chapter, showing how initial worries about smoke and soot gave way to thorough knowledge of airborne chemical elements, atmospheric behaviour, and their extensive effects.

Rapidly urbanizing and industrializing countries bear a disproportionate amount of the cost of air pollution, especially in Asia, where megacities like Delhi, Beijing, Dhaka, and Lahore frequently endure extremely high pollution levels. The information provided unequivocally shows that the main pollutants, PM2.5, ozone, nitrogen dioxide, and sulphur dioxide, are to blame for a variety of diseases, such as heart attacks, strokes, asthma, chronic obstructive pulmonary disease, and even neurological and reproductive issues. With mortality and morbidity rates strongly correlated with pollution exposure, vulnerable populations, including children, the elderly, and people with pre-existing conditions, are particularly at danger.

Both classical and photochemical smog have become obvious indicators of urban environmental deterioration. Because photochemical smog is linked to ground-level ozone, a highly reactive and harmful gas, it is especially deadly. Photochemical smog is caused by industrial activities and automobile emissions. Repeated smog crises in nations like China and Pakistan have prompted both grassroots calls for cleaner air and official involvement.

Technologies for modelling and forecasting, such CMAQ, CALPUFF, and AERMOD, provide useful resources for comprehending the dispersion of pollutants and anticipating pollution occurrences. The accuracy and utility of air quality prediction have significantly improved with the addition of AI and machine learning to forecasting algorithms. Policymakers are now able to make more informed decisions on pollution control because to these models, enhanced emission inventories, and real-time monitoring.

On the policy front, nations have adopted a range of tactics, from fuel switching and automobile limits to public transit spending and the promotion of renewable energy. Enforcement is still uneven, though, and in many areas, environmental preservation is still subordinated to economic expansion. Addressing the environmental and socioeconomic aspects of air pollution calls for a more equitable and coordinated strategy.

Promising avenues for the future are presented by novel biological techniques including bioremediation, phytoremediation, and pollution control based on nanotechnology. Even though there are still obstacles to overcome, like environmental constraints and the intricacy of pollutant combinations, ongoing research and interdisciplinary cooperation can be helpful. Short-term protection can also be obtained through individual behavioural modifications, including as wearing face masks, enhancing indoor air quality, and eating a diet high in antioxidants, particularly in areas with high exposure levels.

Bibliometric analysis is also presented to show the necessity of researching for solutions to combat air pollution and SMOG by combining various sources, i.e. VOSViewer, Python and OriginLabs, to produce figures for visual representation. This analysis helps in understanding how the research trend has focused on the topic in last few years. It shows some very important keywords that should be used in any paper or chapter as can be inferred by other chapters which are taken as references for the analysis.

Combating air pollution necessitates a multifaceted strategy that incorporates public involvement, science, policy, and technology. Collaborative international action is necessary due to the global character of air pollution, including its transboundary movement, climate effects, and economic impacts. Significant improvement is achievable, as this chapter has demonstrated, but in order to guarantee cleaner air and healthier communities in the next decades, persistent dedication, creativity, and equity-driven initiatives will be essential.

REFERENCES

Bai, L., Wang, J., Ma, X., & Lu, H. (2018). Air Pollution Forecasts: An Overview. *International Journal of Environmental Research and Public Health, 15*(4), 780. DOI: 10.3390/ijerph15040780

Burchard-Dziubińska, M. (2019). Air pollution and health in Poland: Anti-smog movement in the most polluted polish cities. *Ecology & Environment, 69*(2), 15.

Butt, M. U., Waseef, R. F., & Ahmed, H. (2018). Perception about the factors associated with smog among medical students. *Biomedica, 34*(4).

Chen, B., & Kan, H. (2008). Air pollution and population health: A global challenge. *Environmental Health and Preventive Medicine, 13*(2), 94–101. DOI: 10.1007/s12199-007-0018-5

Daly, A., & Zannetti, P. (2007). Air Pollution Modeling – An Overview. In P. Zannetti, D. Al-Ajmi, & S. Al-Rashied (Eds.), *AMBIENT AIR POLLUTION*. The Arab School for Science and Technology (ASST). https://home.iitk.ac.in/~anubha/Modeling.pdf

Fowler, D., Brimblecombe, P., Burrows, J., Heal, M. R., Grennfelt, P., Stevenson, D. S., Jowett, A., Nemitz, E., Coyle, M., Liu, X., Chang, Y., Fuller, G. W., Sutton, M. A., Klimont, Z., Unsworth, M. H., & Vieno, M. (2020). A chronology of global air quality. *Philosophical Transactions. Series A, Mathematical, Physical, and Engineering Sciences, 378*(2183), 20190314. DOI: 10.1098/rsta.2019.0314

Global trends in motor vehicle pollution control: a 2011 update. (2011). In *COMBUSTION ENGINES*, 106–107.

Isha, N., Ali, S., Khalid, A., Naseer, I. A., Raza, H., & Chang, Y. (2024). Bioremediation of smog: Current trends and future perspectives. *Processes (Basel, Switzerland), 12*(10), 2266. DOI: 10.3390/pr12102266

Kandpal, P. C. (2024). *Combating air pollution: Comparisons between Delhi and Mexico City*. Springer Cham., DOI: 10.1007/978-3-031-68027-4

Lens. *Lens.* Retrieved October, 2025. https://www.lens.org/lens/

Marlier, M. E., Jina, A. S., Kinney, P. L., & DeFries, R. S. (2016). Extreme air pollution in global megacities. *Current Climate Change Reports, 2*(1), 15–27. DOI: 10.1007/s40641-016-0032-z

Mishra, S. (2017). Is smog innocuous? Air pollution and cardiovascular disease. *Indian Heart Journal, 69*(4), 425–429. DOI: 10.1016/j.ihj.2017.07.016

OriginLab. (2025). *Origin and OriginPro - data analysis and graphing software*. Retrieved October, 2025, from https://www.originlab.com/

Rani, B., Singh, U., Chuhan, A. K., Sharma, D., & Maheshwari, R. (2011). Photochemical smog pollution and its mitigation measures. In *Journal of Advanced Scientific Research*. http://www.sciensage.info/jasr

Sierra-vargas, M. P., & Teran, L. M. (2012). Air pollution: Impact and prevention. *Respirology (Carlton, Vic.)*, *17*(7), 1031–1038. DOI: 10.1111/j.1440-1843.2012.02213.x

Van Eck, N. J., & Waltman, L. (2022). *VOSviewer* manual (Version 1.6.18). Leiden University, CWTS. Retrieved from https://www.vosviewer.com/documentation/Manual_VOSviewer_1.6.18.pdf

Wichmann, H. E., Mueller, W., Allhoff, P., Beckmann, M., Bocter, N., Csicsaky, M. J., Jung, M., Molik, B., & Schoeneberg, G. (1989). Health effects during a smog episode in West Germany in 1985. *Environmental Health Perspectives*, *79*, 89–99. DOI: 10.1289/ehp.897989

Wolf, M. J., Esty, D. C., Kim, H., Bell, M. L., Brigham, S., Nortonsmith, Q., Zaharieva, S., Wendling, Z. A., De Sherbinin, A., & Emerson, J. W. (2022). New insights for tracking global and local trends in exposure to air pollutants. *Environmental Science & Technology*, *56*(7), 3984–3996. DOI: 10.1021/acs.est.1c08080

Zhang, J., Liu, Y., Cui, L., Liu, S., Yin, X., & Li, H. (2017). Ambient air pollution, smog episodes and mortality in Jinan, China. *Scientific Reports*, *7*(1), 11209. Advance online publication. DOI: 10.1038/s41598-017-11338-2

ADDITIONAL READING

Haagen-Smit, A. J. (1964). The Control of Air Pollution. *Scientific American*, *210*(1), 24–31. https://www.jstor.org/stable/24935982. DOI: 10.1038/scientificamerican0164-24

Haagen-Smit, A. J. (1970). A Lesson from the Smog Capital of the World. *Proceedings of the National Academy of Sciences of the United States of America*, *67*(2), 887–897. DOI: 10.1073/pnas.67.2.887

Hu, D., & Jiang, J. (2013). A Study of Smog Issues and PM_{2.5} Pollutant Control Strategies in China. *Journal of Environmental Protection*, *04*(07), 746–752. DOI: 10.4236/jep.2013.47086

Wang, Y., Li, M., Wang, L., Wang, H., Zeng, M., Zeng, B., Qiu, F., & Sun, C. (2019). Can remotely delivered electricity really alleviate smog? An assessment of China's use of ultra-high voltage transmission for air pollution prevention and control. *Journal of Cleaner Production*, *242*, 118430. DOI: 10.1016/j.jclepro.2019.118430

Watkins, A., Danilewitz, M., Kusha, M., Massé, S., Urch, B., Quadros, K., Spears, D., Farid, T., & Nanthakumar, K. (2012). Air pollution and arrhythmic risk: The smog is yet to clear. *The Canadian Journal of Cardiology*, *29*(6), 734–741. DOI: 10.1016/j.cjca.2012.09.005

KEY TERMS AND DEFINITIONS

Air Pollution: The existence of dangerous compounds in the air that endanger both the environment and human health.

Air Quality Modelling: Using computer models to forecast the movement and changes of pollutants in the atmosphere

Emission Sources: The places where pollutants are emitted into the atmosphere, including industry, power plants, and automobiles.

Environmental Policy: Laws and rules aimed at lowering pollution and safeguarding public health and air quality.

Health Effects: The detrimental effects that air pollution has on people's health, such as heart and respiratory conditions.

Particulate Matter: Dust and soot are examples of particulate matter, which are tiny solid or liquid airborne particles that can enter the lungs and cause major health issues.

Photochemical SMOG: It is a form of air pollution created when sunlight combines with pollutants such as car exhaust to create ozone and other compounds.

Smog: It is a thick, frequently observable type of air pollution that is produced by a combination of chemical pollutants, smoke, and fog.

Chapter 11

Global Pathways to Smog Mitigation: Technological Innovations, Policy Frameworks, and Community Action for Cleaner Air

Veena Shree
https://orcid.org/0009-0002-2131-9080
Vellore Technical University, India

ABSTRACT

This chapter, per the authors, explores the complex challenges posed by air pollution and smog, highlighting their impacts on public health, urban environments, and global sustainability. It examines how technological innovation, renewable energy adoption, and sustainable transport systems can mitigate environmental risks. The chapter also discusses the role of policy frameworks, governance, industrial emission control, and community engagement in promoting behavioral change and advancing the green economy. By linking these strategies to the Sustainable Development Goals and climate action, the chapter emphasizes practical approaches for achieving environmental sustainability in rapidly urbanizing societies.

CHAPTER-1 INTRODUCTION

Smog has become one of the defining environmental challenges of the twenty-first century. Initially perceived as a localized outcome of industrialization, it has now evolved into a transnational public health and sustainability crisis. Historical events such as the Great Smog of London in 1952—which led to over 4,000 excess

DOI: 10.4018/979-8-3373-2469-2.ch011

deaths in just a few days—highlighted the lethal effects of air pollution and prompted stronger regulation in Western Europe (Bell & Davis, 2001). Although cleaner technologies and stricter environmental policies have reduced the severity of smog in North America and parts of Europe, rapidly industrializing regions such as South Asia, East Asia, and Africa continue to experience recurring hazardous episodes that affect millions (Guttikunda & Gurjar, 2012; Li et al., 2017).

Smog is not merely an atmospheric problem but a complex phenomenon rooted in economic activity, energy use, and urbanization. Its primary drivers include fossil fuel combustion, industrial emissions, vehicular traffic, biomass burning, and poor waste management practices (Seinfeld & Pandis, 2016). These processes release pollutants such as nitrogen oxides, sulfur dioxide, and volatile organic compounds, which interact under sunlight to produce secondary pollutants like ozone and fine particulate matter (PM2.5 and PM10) (Jacob, 1999). The health consequences are grave. The World Health Organization (2023) attributes more than seven million premature deaths annually to air pollution exposure, with respiratory illness, cardiovascular disease, and lung cancer being the major outcomes. Beyond health, smog reduces agricultural yields, damages ecosystems, corrodes infrastructure, and disrupts transportation systems (Mannucci & Franchini, 2017).

Urbanization patterns have intensified the smog problem. Cities such as Delhi, Lahore, Beijing, and Jakarta exemplify how industrial growth, motorization, and coal-based energy lock societies into high-pollution pathways (Gurjar et al., 2016). Even developed economies continue to face challenges: Los Angeles has struggled with photochemical smog since the mid-twentieth century, despite decades of stringent regulation (Burke et al., 2020). More recently, climate-linked wildfires in California and Australia have produced transboundary haze, demonstrating how smog transcends geographic boundaries through atmospheric circulation (Johnston et al., 2019).

Although the atmospheric science of smog is well understood, responses have often been fragmented. Technological solutions such as industrial scrubbers, catalytic converters, and renewable energy transitions have advanced but remain unevenly distributed between high-income and low-income regions (Sharma & Dikshit, 2016). Policy frameworks such as the U.S. Clean Air Act, China's Air Pollution Prevention and Control Action Plan, and India's National Clean Air Programme have shown progress, yet weak enforcement, inadequate resources, and competing economic interests limit their impact (Andrews, 2020; Zhang et al., 2019). Community initiatives, including Delhi's "Odd-Even" traffic rationing scheme and campaigns against agricultural stubble burning, demonstrate the importance of grassroots engagement, but these measures are difficult to sustain without institutional backing (Narain & Krupnick, 2016).

The fragmented nature of current approaches underscores the need for integrated strategies that combine scientific innovation, robust governance, and civic participation. Smog mitigation requires cooperation across borders, sectors, and institutions. International frameworks such as the Paris Agreement and the United Nations Sustainable Development Goals (SDGs) explicitly link air quality to climate action (SDG 13), sustainable cities (SDG 11), and reduced inequalities (SDG 10) (United Nations, 2015). Moreover, the economic costs are staggering: the World Bank (2020) estimates that air pollution causes over US$5 trillion in annual welfare losses, disproportionately burdening low- and middle-income countries.

This chapter argues that effective smog mitigation demands a holistic, multi-stakeholder approach. By comparing lessons from Beijing's state-led interventions, California's market-driven mechanisms, and Europe's community-oriented strategies, it seeks to identify effective practices and their limitations. The chapter pursues three goals: (1) to trace the historical evolution of smog as an environmental and socio-economic challenge; (2) to critically assess existing mitigation strategies, including technological, policy, and community initiatives; and (3) to propose an integrated framework that aligns innovation, regulation, and citizen engagement with the global sustainability agenda.

Ultimately, smog mitigation is not only about cleaner air but also about protecting human health, strengthening economic resilience, and advancing sustainable development. By weaving together technology, governance, and community action, this chapter aims to contribute both a scholarly analysis and a practical roadmap for building healthier and more equitable societies.

CHAPTER-2. HISTORICAL PERSPECTIVES – FROM THE INDUSTRIAL REVOLUTION TO THE 21ST CENTURY

The trajectory of smog reflects how industrialization, urban growth, and technological change have shaped environmental health across centuries. Its history demonstrates that while societies have repeatedly encountered smog crises, responses have often lagged behind the pace of industrial and economic expansion.

Early Industrial Era

The Industrial Revolution in eighteenth- and nineteenth-century Britain marked the first major escalation of smog. Widespread coal combustion powered factories, railways, and households, producing dense smoke that blanketed urban landscapes. London, Manchester, and Birmingham became synonymous with polluted skies, where respiratory diseases and reduced life expectancy were common (Mosley, 2014).

During this period, smoke was paradoxically considered both a sign of prosperity and an unavoidable nuisance, with little state intervention to address its health impacts.

Global Spread of Industrial Smog

As industrialization spread to continental Europe and North America, similar conditions emerged. Pittsburgh and Chicago in the United States, as well as the Ruhr Valley in Germany, became heavily polluted centers where skies darkened with soot and sulfur emissions. In many of these regions, industry was prioritized over health, and air quality regulation remained minimal until the twentieth century (Brimblecombe, 2011). This stage illustrates how smog was historically perceived not as an environmental crisis but as a byproduct of economic progress.

Mid-20th Century Turning Points

The mid-twentieth century brought several landmark crises that redefined public and political perceptions of air pollution. The Great Smog of London in 1952 caused thousands of premature deaths within a single week, forcing governments to recognize smog as a severe health hazard rather than a minor inconvenience. This tragedy directly influenced the UK Clean Air Act of 1956, one of the earliest comprehensive legislative responses to urban air pollution (Royal College of Physicians, 2016).

Japan's post-war economic boom also came at a heavy environmental cost. The infamous Yokkaichi asthma outbreak of the 1960s, caused by petrochemical emissions, illustrated the dangers of unchecked industrial expansion and catalyzed citizen movements demanding stricter regulation (Fujikura, 2001). Similarly, in the United States, Los Angeles experienced persistent photochemical smog linked to rising automobile usage, leading to the establishment of the Environmental Protection Agency (EPA) in 1970 and the passage of the Clean Air Act the same year.

Late 20th to 21st Century Shifts

By the late twentieth century, many industrialized countries had reduced visible smog levels through cleaner technologies and stronger enforcement. However, this progress was uneven. As environmental regulations tightened in developed nations, pollution-intensive industries relocated to emerging economies. Cities such as Beijing, Delhi, and Jakarta became new epicentres of smog, reflecting the uneven global geography of environmental risk (Aunan & Pan, 2004). These regions experienced air quality crises reminiscent of nineteenth-century Europe but on a far larger scale due to higher population densities and rapid motorization.

At the same time, new transboundary dimensions emerged. Wildfires in Australia and California, dust storms in the Middle East, and haze episodes from Southeast Asian peatland burning illustrated that smog was no longer confined to urban industrial centres. Instead, it became a regional and even global concern, carried across borders by atmospheric circulation (Kim Oanh et al., 2006).

Historical Lessons

The historical evolution of smog highlights recurring patterns: societies often tolerate rising pollution during periods of rapid economic growth, only addressing it after catastrophic events trigger public outcry. This lag between cause and response has led to repeated cycles of health crises, regulation, and gradual technological adaptation. Learning from this history is crucial for designing proactive strategies in the twenty-first century, particularly in rapidly urbanizing regions where the risks remain acute.

CHAPTER-3. DRIVERS OF SMOG FORMATION – INDUSTRY, TRANSPORT, URBANIZATION, WASTE MANAGEMENT

Smog is not the result of a single pollutant or isolated source, but rather the convergence of multiple human-driven activities that release harmful emissions into the atmosphere. Understanding these drivers is crucial to developing policies and technologies that address the root causes rather than just the visible symptoms. Four interlinked sectors—industry, transport, urbanization, and waste management—play central roles in the persistence of smog worldwide.

Industry

Industrial processes remain among the most significant contributors to smog formation. Manufacturing sectors, especially those dependent on coal, oil, and biomass, release nitrogen oxides, sulphur dioxide, particulate matter, and volatile organic compounds. In many developing economies, outdated equipment and limited emission controls amplify the problem. Steel plants, thermal power stations, cement factories, and petrochemical complexes often operate close to urban centres, exposing densely populated areas to high concentrations of pollutants. Even where regulations exist, enforcement gaps and cost pressures frequently delay the adoption of cleaner technologies.

Transport

The rapid expansion of motorized transport has intensified smog in cities across the globe. Automobiles, buses, and trucks powered by diesel and petrol emit large volumes of nitrogen oxides, carbon monoxide, and hydrocarbons that combine under sunlight to form photochemical smog. In megacities such as Delhi, Jakarta, and Mexico City, vehicular emissions are now the dominant source of air pollution. Traffic congestion worsens this impact, as idling vehicles emit more pollutants per kilometer than those moving at steady speeds. Although cleaner fuel standards and electric mobility provide alternatives, affordability and infrastructure constraints have slowed their widespread adoption.

Urbanization

Unplanned urban growth exacerbates smog by concentrating both people and polluting activities in confined spaces. Expanding city boundaries bring industries, highways, and residential zones into close proximity, while high population densities increase energy demand for heating, cooling, and cooking. Informal settlements often rely on biomass or kerosene for daily needs, contributing additional pollutants. The vertical and horizontal spread of cities also influences local meteorology, creating "urban heat islands" that trap pollutants close to the ground and intensify smog episodes.

Waste Management

Inadequate waste management practices add another dimension to smog formation. Open burning of municipal waste, a common practice in many low- and middle-income countries, releases particulate matter, carbon monoxide, and toxic chemicals such as dioxins and furans. Agricultural residue burning, particularly in South Asia, has become a recurring seasonal driver of smog that blankets entire regions. Landfills, when poorly managed, emit methane and other gases that worsen local air quality while contributing to broader climate challenges.

Interconnected Dynamics

These drivers do not operate in isolation but interact in ways that amplify smog. For example, rapid urbanization increases demand for both industrial production and motorized transport, while insufficient waste management practices add seasonal spikes in emissions. Together, these dynamics create persistent air pollution that resists piecemeal interventions. A systemic understanding of these interconnections is therefore essential for designing effective strategies that target multiple drivers simultaneously rather than treating them as separate problems.

CHAPTER-4. SMOG AND THE SUSTAINABLE DEVELOPMENT GOALS (SDGS)

The global response to smog cannot be separated from the broader international agenda of sustainable development. The United Nations Sustainable Development Goals (SDGs), adopted in 2015, provide a comprehensive framework for linking air quality improvements to global priorities such as health, climate action, sustainable cities, and economic growth. Smog, as a multidimensional issue, intersects with several of these goals, making it both a barrier to progress and an opportunity for integrated solutions.

Health and Well-Being (SDG 3)

Smog directly undermines SDG 3, which aims to ensure healthy lives and promote well-being for all ages. Exposure to fine particulate matter (PM2.5) and ground-level ozone contributes to respiratory infections, asthma, cardiovascular diseases, and premature deaths. In countries with recurring smog episodes, children and the elderly are particularly vulnerable, creating a significant burden on public health systems. Reducing air pollution is therefore indispensable for achieving universal health coverage and lowering preventable mortality rates.

Sustainable Cities and Communities (SDG 11)

Urban areas are the epicenters of smog formation, making SDG 11—focused on sustainable cities and communities—central to mitigation efforts. Poorly planned urban growth, congested transport systems, and reliance on fossil fuels exacerbate smog episodes. Policies that prioritize green infrastructure, efficient public transport, and renewable energy can simultaneously reduce emissions and improve urban livability. Clean air initiatives also support social equity, as marginalized groups often live in the most polluted neighborhoods.

Climate Action (SDG 13)

Smog and climate change are closely linked. Many smog-forming pollutants, such as black carbon and methane, are also short-lived climate pollutants that contribute to global warming. Efforts to control smog can therefore yield co-benefits for

SDG 13 by slowing climate change while improving local air quality. Conversely, climate change can worsen smog by altering weather patterns and increasing the frequency of wildfires, underscoring the need for integrated policies that address both challenges simultaneously.

Economic Growth and Decent Work (SDG 8)

Air pollution is often perceived as an unavoidable byproduct of economic growth. However, smog imposes heavy economic costs through lost productivity, increased healthcare expenses, and reduced agricultural yields. By investing in clean energy and sustainable technologies, countries can align economic growth with environmental protection. Smog mitigation can therefore advance SDG 8 by fostering green jobs, reducing productivity losses, and ensuring that growth is inclusive and sustainable.

Partnerships for the Goals (SDG 17)

Because smog frequently crosses borders, effective solutions require international cooperation, knowledge-sharing, and financial support. SDG 17 emphasizes partnerships across governments, private sector actors, and civil society. Collaborative platforms such as the Climate and Clean Air Coalition (CCAC) illustrate how collective action can accelerate progress by pooling resources and expertise. Global partnerships also play a key role in making advanced technologies accessible to low- and middle-income countries.

An Integrated Perspective

Smog's intersection with multiple SDGs highlights its systemic nature. Tackling it effectively demands integrated strategies that combine public health, climate, economic, and social objectives. Instead of treating clean air as a standalone issue, embedding smog mitigation into the broader SDG agenda ensures that solutions are sustainable, equitable, and globally coordinated.

CHAPTER-5. TECHNOLOGICAL INNOVATIONS – AI MONITORING, RENEWABLE ENERGY, SUSTAINABLE TRANSPORT, INDUSTRIAL CONTROLS

Technological innovations are at the heart of global strategies to mitigate smog. While the drivers of smog are well documented, it is through advancements in science and engineering that practical solutions emerge. Over the past two decades, countries have increasingly turned to new technologies not only to monitor and predict air quality but also to reduce emissions at their source. These innovations—ranging from digital surveillance of pollutants to clean energy systems—are redefining how societies can balance development with environmental sustainability.

Artificial Intelligence and Air Quality Monitoring

Artificial Intelligence (AI) has transformed how cities track and respond to air pollution. AI-based models can integrate data from ground sensors, satellites, and meteorological systems to forecast smog episodes with remarkable accuracy. By detecting pollution hotspots in real time, these tools enable governments to issue health advisories, implement emergency traffic restrictions, and target enforcement measures. AI also supports long-term planning by identifying trends that may not be visible through conventional monitoring methods.

Renewable Energy Systems

One of the most effective ways to address smog is by reducing reliance on fossil fuels. Renewable energy technologies such as solar, wind, and hydroelectric power play a crucial role in displacing coal and oil-based energy generation. Countries like China and India, which struggle with recurring smog episodes, are investing heavily in solar and wind capacity to reduce emissions from power plants. Beyond emission reduction, renewables also provide energy security and long-term cost benefits, making them central to sustainable smog mitigation strategies.

Sustainable Transport Solutions

The transportation sector is a leading contributor to smog, particularly in densely populated cities. Technological innovations in this sector are reshaping mobility patterns. Electric vehicles (EVs), hybrid systems, and clean public transit networks reduce dependence on gasoline and diesel fuels. Additionally, smart traffic management systems—supported by AI and the Internet of Things (IoT)—help minimize congestion, thereby cutting emissions. Investments in cycling infrastructure and pedestrian-friendly designs further support the transition to low-emission transport models.

Industrial Emission Controls

Industries remain one of the largest sources of smog-forming pollutants. Technological innovations in emission controls—such as electrostatic precipitators, flue-gas desulfurization, and catalytic converters—help capture harmful particles before they enter the atmosphere. In parallel, cleaner production technologies and circular economy practices are being adopted to minimize waste and resource use. By combining end-of-pipe controls with systemic changes in industrial processes, industries can significantly lower their contribution to urban smog levels.

An Integrated Technological Approach

While each innovation is impactful on its own, the real potential lies in integrating these technologies into broader systems. For example, renewable energy expansion can be paired with electrified public transport, while AI-based monitoring can guide both industrial regulation and urban planning. By embedding innovation into national development strategies, countries can achieve cleaner air without sacrificing economic progress.

CHAPTER-6. POLICY FRAMEWORKS AND GOVERNANCE – INTERNATIONAL, REGIONAL, AND NATIONAL STRATEGIES

Policy frameworks and governance structures play a decisive role in shaping the success of smog mitigation efforts. While technological solutions provide the tools to address emissions, policies determine how effectively these tools are implemented and scaled. Over the years, governance has evolved from fragmented regulations to more comprehensive frameworks that combine international agreements, regional collaborations, and national legislation. These approaches highlight the importance of collective responsibility and coordinated action in tackling air pollution.

International Agreements and Cooperation

At the international level, smog is recognized as both an environmental and public health issue that transcends borders. Global treaties and conventions have addressed transboundary air pollution by encouraging cooperation on emission reductions, scientific research, and technology transfer. Agreements under organizations such as the United Nations Environment Programme (UNEP) and World Health Organization (WHO) have set global standards and monitoring frameworks. These mechanisms foster accountability and create opportunities for knowledge-sharing among countries at different stages of development.

Regional Frameworks

Regional collaboration often bridges the gap between global commitments and national action. Initiatives in Asia, Europe, and North America have shown that shared strategies can help address cross-border smog challenges. For instance, regional air quality networks enable data harmonization, joint monitoring, and coordinated policy responses. Such frameworks are particularly relevant in areas where industrial and transport emissions affect multiple countries simultaneously. By aligning regulations across borders, regional efforts reduce inconsistencies and promote collective resilience against smog formation.

National Policies and Regulations

At the national level, governments are adopting a range of legal and institutional measures to combat smog. These include emission standards for industries and vehicles, restrictions on polluting fuels, and the promotion of renewable energy. National air quality acts and environmental protection laws establish enforcement mechanisms that compel industries and individuals to comply with emission limits. In addition, urban governance strategies—such as congestion pricing, clean energy incentives, and city-level monitoring systems—help localize national policy frameworks.

Governance Challenges and Opportunities

Despite these advances, gaps remain in policy enforcement and institutional capacity. Corruption, limited resources, and weak compliance systems often hinder effective governance. However, opportunities for improvement lie in integrating air quality objectives into broader development agendas, leveraging digital technologies for monitoring, and fostering public participation. Transparent governance, combined with strong legal accountability, can transform policies from paper commitments into tangible improvements in air quality.

Towards a Multilevel Governance Model

The complexity of smog requires governance models that operate simultaneously at global, regional, and national levels. Integration across these layers ensures coherence, avoids policy duplication, and enhances long-term sustainability. Multilevel governance, grounded in legal accountability and supported by technological innovation, represents the most viable pathway to cleaner air and healthier communities.

7. COMMUNITY AND BEHAVIORAL DIMENSIONS – AWARENESS, LIFESTYLE SHIFTS, GRASSROOTS MOVEMENTS

Technological innovations and policy reforms form critical pillars of smog mitigation, but they cannot achieve long-term results without community engagement and behavioral transformation. Since smog is a product of collective activities—ranging from industrial operations to daily commuting—the solutions must equally involve collective participation. This chapter highlights how awareness campaigns, lifestyle modifications, and grassroots movements contribute to air quality improvement by fostering societal ownership of environmental responsibility.

Awareness and Education

Public awareness serves as the foundation for Behavioral change. Community-based education programs raise understanding of the health risks associated with smog, including respiratory illnesses, cardiovascular diseases, and reduced life expectancy. Campaigns led by schools, NGOs, and media outlets often translate complex scientific information into accessible knowledge, empowering citizens to make informed decisions. When awareness is integrated into public education systems, it creates a long-term cultural shift toward prioritizing clean air as a basic right.

Lifestyle Shifts

Individual choices collectively influence urban air quality. Reductions in private vehicle use, adoption of public transportation, preference for clean cooking fuels, and energy-efficient practices all contribute to lowering smog levels. Government incentives for sustainable consumption—such as subsidies for electric vehicles or penalties on high-emission activities—help reinforce these lifestyle shifts. Over time, such changes normalize sustainable habits, turning them into everyday practices rather than temporary adjustments.

Grassroots and Community Movements

Grassroots movements play a vital role in holding governments and industries accountable while mobilizing communities around air quality issues. Localized protests, citizen-led monitoring networks, and advocacy groups have pressured policymakers to strengthen emission standards and implement sustainable urban plans. Community participation also enables innovative local solutions, such as neighborhood clean energy cooperatives and urban greening projects. These movements are particularly powerful in regions where institutional responses have been slow or inadequate.

Behavioral Economics and Social Norms

Shaping behaviour also requires understanding economic incentives and social psychology. Studies show that individuals are more likely to adopt eco-friendly behaviors when such actions are reinforced by community norms, peer influence, or financial rewards. Integrating Behavioral economics into environmental governance allows policymakers to design interventions that resonate with everyday decision-making. By reframing sustainable actions as socially desirable and financially beneficial, communities are more likely to sustain long-term Behavioral change.

Building a Culture of Collective Responsibility

Community action and behaviour change are not substitutes for policy or technology but rather complementary forces. When citizens feel empowered and socially responsible, they amplify the effectiveness of institutional strategies. Building a culture where clean air is viewed as a shared responsibility—supported by awareness, lifestyle adjustments, and grassroots initiatives—ensures that smog mitigation efforts are inclusive, equitable, and sustainable

8. ECONOMIC DIMENSIONS – COSTS, PRODUCTIVITY, AND GREEN ECONOMY OPPORTUNITIES

Air pollution and smog are not only public health and environmental challenges but also carry significant economic implications. Their effects extend beyond individual well-being to influence labor productivity, healthcare systems, infrastructure, and overall economic growth. This chapter examines the financial costs of smog, its impact on economic activities, and the emerging opportunities within the transition toward a green economy.

Economic Costs of Smog

The direct economic burden of smog arises from healthcare expenditures linked to pollution-related illnesses such as asthma, bronchitis, and cardiovascular conditions. Governments often allocate substantial resources to medical care, reducing funds available for education, infrastructure, and other development priorities. Indirect costs include lost productivity due to absenteeism, premature deaths, and diminished labor capacity, all of which contribute to reduced economic output.

Productivity and Workforce Implications

Smog significantly affects workforce performance. Prolonged exposure to polluted air reduces workers' cognitive and physical abilities, leading to inefficiencies across sectors ranging from manufacturing to services. Studies indicate that urban centers with persistently high smog levels experience measurable reductions in GDP growth due to lower productivity and increased labor market disruptions. Moreover, small-scale workers, daily wage earners, and informal sector employees—who lack healthcare coverage—bear disproportionate economic risks.

Infrastructure and Urban Development

Smog also imposes financial strain on infrastructure and urban systems. Corrosion of buildings, damage to cultural heritage, and reduced visibility affecting transportation networks create additional economic challenges. Delays in flights, road congestion, and shipping disruptions due to low visibility increase operational costs and hinder trade efficiency. These losses illustrate the interconnected nature of smog, where environmental degradation translates into economic vulnerabilities.

Transition to a Green Economy

Despite these challenges, the push for smog mitigation presents economic opportunities through the development of a green economy. Investment in renewable energy, electric mobility, energy-efficient buildings, and circular economy practices not only reduces emissions but also generates employment and stimulates innovation. Transitioning to cleaner technologies allows governments and businesses to decouple economic growth from environmental harm, positioning sustainability as a driver of competitiveness.

Policy and Market Incentives

Effective market-based mechanisms—such as carbon pricing, emissions trading systems, and green subsidies—play a vital role in steering economies toward cleaner pathways. These measures create incentives for industries to adopt environmentally responsible practices while promoting innovation in green technologies. International cooperation and financing mechanisms further strengthen these opportunities, especially in developing economies where the costs of transitioning are often high.

Economic Resilience through Sustainable Models

Ultimately, integrating smog mitigation into economic planning fosters resilience by reducing healthcare costs, protecting labor productivity, and ensuring long-term growth. Green economy opportunities offer a pathway where environmental protection aligns with prosperity, proving that clean air is not merely a health goal but also an economic imperative.

9. CASE STUDIES – SOUTH ASIA, CHINA, EU, AND NORTH AMERICA

Understanding smog as a global phenomenon requires analysing regional experiences. Each region demonstrates distinct socio-economic drivers, policy approaches, and public health impacts. Case studies provide valuable lessons for shaping adaptable and context-specific strategies.

South Asia

South Asia faces some of the highest smog levels worldwide, with cities like Delhi, Lahore, and Dhaka consistently recording severe air quality. Contributing factors include rapid urbanization, coal dependency, stubble burning, and weak enforcement of environmental regulations. Regional initiatives, such as the *National Clean Air Programme (NCAP)* in India, seek to reduce particulate matter concentrations, but challenges remain due to limited monitoring networks and economic constraints.

China

China represents a transformation from being one of the most polluted nations to a global leader in aggressive mitigation. Policies such as the *Air Pollution Prevention and Control Action Plan (2013)* introduced strict emission standards, capped coal consumption, and promoted renewable energy. Heavy investments in electric vehicles and urban air monitoring systems have significantly improved air quality in cities like Beijing, although industrial regions continue to face periodic smog events.

European Union (EU)

The EU has implemented some of the most stringent air quality policies, guided by the *Air Quality Directive*. Cross-border cooperation ensures coordinated reduction of pollutants like NOx and SO_2. Member states integrate air quality targets with climate commitments under the European Green Deal, creating a holistic framework. Public participation, clean transport, and strict penalties for violators highlight the EU's model of balancing enforcement with citizen awareness.

North America

North America demonstrates how long-term regulation can deliver results. The *Clean Air Act* in the United States has led to drastic reductions in pollutants since the 1970s, supported by strong enforcement from the Environmental Protection Agency (EPA). Canada's policies focus on both domestic emission reductions and managing transboundary smog with the U.S. Collaborative mechanisms, such as the Air Quality Agreement (1991), show the importance of international cooperation in addressing shared challenges.

Comparative Insights

While South Asia struggles with capacity constraints, China showcases how political commitment can drive rapid improvements. The EU emphasizes integration of air quality and climate goals, whereas North America demonstrates the value of legal enforcement sustained over decades. Collectively, these case studies highlight that smog mitigation requires not a single template but adaptive approaches tailored to economic, social, and political contexts.

10. INTEGRATED PATHWAYS FOR SMOG MITIGATION

Smog mitigation cannot rely on isolated interventions. Sustainable solutions require integrated approaches that combine technological, policy, economic, and social dimensions. By merging science with governance and public participation, societies can move toward cleaner and healthier environments.

Technological Integration

Technological innovations such as renewable energy adoption, electric mobility, and AI-driven monitoring systems provide the backbone for reducing emissions. However, their effectiveness increases when they are supported by complementary policy frameworks and financial incentives. Cleaner industrial production and waste-to-energy technologies further reduce localized smog sources.

Policy and Governance Alignment

A strong policy environment ensures that technological progress translates into real-world impact. Integrated air quality management frameworks encourage collaboration across ministries, regions, and even national borders. Policies that link air quality goals with climate action, such as carbon pricing and emission trading systems, create a unified regulatory approach.

Economic and Market-Based Instruments

Economic tools such as green subsidies, pollution taxes, and carbon markets provide industries with incentives to adopt cleaner technologies. At the same time, investment in green infrastructure, such as public transport and renewable power, creates long-term economic benefits by reducing productivity losses from pollution-related health issues.

Community Engagement and Behavioral Change

Technological and policy interventions are incomplete without societal participation. Awareness campaigns, citizen monitoring networks, and lifestyle shifts—such as reducing car dependency—make mitigation efforts more sustainable. Public-private partnerships strengthen capacity building and foster innovation by linking communities, businesses, and governments.

Multi-Stakeholder Collaboration

True integration emerges when governments, industries, civil society, and international organizations coordinate their actions. Partnerships between developed and developing nations, knowledge-sharing platforms, and funding mechanisms for clean technology transfer are essential for creating a global movement against smog.

Pathway to Sustainable Futures

An integrated model ensures that smog mitigation is not treated as a short-term emergency response but as part of a broader sustainability agenda. When air quality improvement aligns with climate change commitments and health policies, societies can achieve long-lasting benefits.

11. FUTURE DIRECTIONS – CLIMATE CHANGE, INTERNATIONAL COOPERATION, SUSTAINABLE FUTURES

The future of smog mitigation will be shaped by how societies respond to the intertwined challenges of climate change, rapid urbanization, and global inequality. Smog cannot be addressed in isolation; instead, it must be integrated into broader sustainability frameworks that connect environmental protection, human health, and economic resilience.

Climate Change Linkages

Smog and climate change are driven by many of the same pollutants, such as black carbon, methane, and ground-level ozone. Addressing these pollutants simultaneously not only improves air quality but also reduces global warming. Future strategies will increasingly focus on climate–air quality co-benefits, ensuring that mitigation actions contribute to both cleaner air and a stable climate.

International Cooperation

Air pollution knows no borders, and atmospheric circulation often carries smog across regions and even continents. The future of mitigation therefore depends on strengthening international agreements, data-sharing platforms, and cross-border policies. Enhanced cooperation under frameworks like the Paris Agreement and regional clean air alliances will be crucial for developing joint monitoring systems and coordinated responses.

Technological and Scientific Advances

Emerging technologies will continue to transform mitigation efforts. Real-time AI-based monitoring, satellite surveillance, and predictive modeling will make it possible to anticipate smog episodes before they become crises. The growth of renewable energy, electric transport systems, and carbon-neutral industrial processes will further reduce emission sources. However, equitable access to these technologies will remain an important challenge for low- and middle-income countries.

Pathways to Sustainable Futures

The long-term future of smog mitigation requires aligning policy, technology, and community action with the Sustainable Development Goals (SDGs). Investments in green growth models, urban resilience planning, and public health infrastructure will help societies transition toward sustainable futures. At the same time, education and public awareness will remain central to building a culture of responsibility and behavioral change.

Conclusion

The next phase of smog mitigation must be globally coordinated, science-driven, and socially inclusive. By linking smog reduction with climate change mitigation, fostering international partnerships, and ensuring equitable access to technology, the world can move closer to sustainable futures where clean air is a universal right rather than a privilege.

12. CONCLUSION – LESSONS LEARNED AND RECOMMENDATIONS

Smog has evolved from a localized environmental hazard to a global health, economic, and climate challenge. The chapters preceding this one have shown that while the causes of smog are rooted in industrialization, transport dependence, rapid urbanization, and waste mismanagement, its consequences are far-reaching, undermining human health, productivity, and sustainability.

Key Lessons Learned

1. **Historical Evolution Matters** – The shift from the Great Smog of London (1952) to recurring crises in Asian megacities reveals that while technology and governance can reduce pollution, progress is uneven and context-specific.
2. **Drivers Are Interconnected** – Industrial growth, transportation, and energy dependence amplify each other, making smog mitigation a multidimensional challenge.
3. **SDG Linkages Are Clear** – Smog directly undermines progress toward health (SDG 3), sustainable cities (SDG 11), and climate action (SDG 13), reinforcing the need for integrated approaches.
4. **Technology Alone Is Insufficient** – While innovations like AI-based monitoring and renewable energy hold promise, their impact depends on affordability, accessibility, and effective implementation.
5. **Policy and Governance Are Central** – Regulatory frameworks succeed only when supported by strong enforcement, adequate funding, and international cooperation.

6. **Communities Play a Critical Role** – Behavioral change, awareness, and grassroots initiatives complement top-down strategies, ensuring long-term sustainability.
7. **Economic Realities Cannot Be Ignored** – The trillions lost annually to air pollution highlight that clean air is not a cost but an investment in health, resilience, and green growth.

Recommendations

- **Adopt Integrated Frameworks**: Policies should link climate action, energy transition, and public health with air quality management.
- **Strengthen International Cooperation**: Cross-border pollution requires shared monitoring systems, regional agreements, and global accountability.
- **Promote Equity in Technology Access**: Ensure that innovations such as electric mobility and advanced monitoring tools reach low- and middle-income countries.
- **Invest in Public Awareness**: Governments and NGOs must scale up campaigns that foster behavioral change and encourage citizen participation.
- **Incentivize Green Growth Models**: Economic policies should reward cleaner industries, sustainable transport, and renewable energy adoption.
- **Future-Oriented Planning**: Align air quality policies with long-term resilience strategies, ensuring that cities prepare for the combined pressures of smog, climate change, and urbanization.

Final Reflection

The struggle against smog is ultimately a struggle for human well-being, equity, and sustainability. Clean air must be treated as a fundamental right, central to public health, economic prosperity, and climate stability. By combining science, policy, technology, and community action, societies can move beyond fragmented efforts and towards sustainable futures where urban growth and clean air coexist.

REFERENCES

Andrews, R. N. L. (2020). *Managing the environment, managing ourselves: A history of American environmental policy* (3rd ed.). Yale University Press.

Aunan, K., & Pan, X. (2004). Exposure–response functions for health effects of ambient air pollution applicable for China—A meta-analysis. *The Science of the Total Environment*, *329*(1–3), 3–16. DOI: 10.1016/j.scitotenv.2004.03.008 PMID: 15262154

Bell, M. L., & Davis, D. L. (2001). Reassessment of the lethal London fog of 1952: Novel indicators of acute and chronic consequences of acute exposure to air pollution. *Environmental Health Perspectives*, *109*(S3), 389–394. DOI: 10.1289/ehp.01109s3389

Brimblecombe, P. (2011). *The Big Smoke: A History of Air Pollution in London since Medieval Times*. Routledge.

Burke, M., Driscoll, A., Heft-Neal, S., & Erhart, J. (2020). The changing risk and burden of wildfire in the United States. *Proceedings of the National Academy of Sciences of the United States of America*, *117*(46), 24609–24617. DOI: 10.1073/pnas.2011048117 PMID: 33431571

Fujikura, R. (2001). Public participation, stakeholder involvement and consensus building in Japan's environmental policy-making. *Environmental Impact Assessment Review*, *21*(5), 435–453. DOI: 10.1016/S0195-9255(01)00081-1

Gurjar, B. R., Jain, A., Sharma, A., Agarwal, A., Gupta, P., Nagpure, A. S., & Lelieveld, J. (2016). Human health risks in megacities due to air pollution. *Atmospheric Environment*, *142*, 475–495. DOI: 10.1016/j.atmosenv.2016.06.030

Gurjar, B. R., Ravindra, K., & Nagpure, A. S. (2016). Air pollution trends over Indian megacities and their local-to-global implications. *Atmospheric Environment*, *142*, 475–495. DOI: 10.1016/j.atmosenv.2016.06.030

Guttikunda, S. K., & Gurjar, B. R. (2012). Role of meteorology in seasonality of air pollution in megacity Delhi, India. *Environmental Monitoring and Assessment*, *184*(5), 3199–3211. DOI: 10.1007/s10661-011-2182-8 PMID: 21713474

Jacob, D. J. (1999). *Introduction to atmospheric chemistry*. Princeton University Press. DOI: 10.2307/j.ctt7t8hg

Johnston, F. H., Henderson, S. B., Chen, Y., Randerson, J. T., Marlier, M., DeFries, R. S., Kinney, P., Bowman, D. M. J. S., & Brauer, M. (2019). Estimated global mortality attributable to smoke from landscape fires. *Environmental Health Perspectives*, *120*(5), 695–701. DOI: 10.1289/ehp.1104422 PMID: 22456494

Kim Oanh, N. T., Upadhyay, N., Zhuang, Y.-H., Hao, Z.-P., Murthy, D. V. S., Lestari, P., Villarin, J. T., Chengchua, K., Co, H. X., & Dung, N. T. (2006). Particulate air pollution in six Asian cities: Spatial and temporal distributions, and associated sources. *Atmospheric Environment*, *40*(18), 3367–3380. DOI: 10.1016/j.atmosenv.2006.01.050

Li, T., Zhang, Y., Wang, J., Xu, D., & Yin, Z. (2017). All-cause mortality risk associated with long-term exposure to ambient PM2.5 in China: A cohort study. *The Lancet. Public Health*, *2*(12), e547–e555. DOI: 10.1016/S2468-2667(17)30241-4 PMID: 30314593

Mannucci, P. M., & Franchini, M. (2017). Health effects of ambient air pollution in developing countries. *International Journal of Environmental Research and Public Health*, *14*(9), 1048. DOI: 10.3390/ijerph14091048 PMID: 28895888

Mosley, S. (2014). Environmental history of air pollution and protection. In *The Oxford Handbook of Environmental History* (pp. 1–27). Oxford University Press. DOI: 10.1007/978-3-319-09180-8_5

Narain, U., & Krupnick, A. (2016). The impact of Delhi's Odd-Even driving policy on air pollution. *Resources for the Future*. https://www.rff.org/publications/working-papers/the-impact-of-delhis-odd-even-driving-policy-on-air-pollution

Royal College of Physicians. (2016). *Every breath we take: The lifelong impact of air pollution*. RCP.

Seinfeld, J. H., & Pandis, S. N. (2016). *Atmospheric chemistry and physics: From air pollution to climate change* (3rd ed.). Wiley.

Sharma, M., & Dikshit, O. (2016). Comprehensive study on air pollution and greenhouse gases in India. *Indian Journal of Environmental Protection*, *36*(5), 402–411.

United Nations. (2015). *Transforming our world: The 2030 Agenda for Sustainable Development*. UN Publishing. https://sdgs.un.org/2030agenda

United Nations Environment Programme (UNEP). (2019). *Air pollution in Asia and the Pacific: Science-based solutions*. Nairobi: UNEP. https://www.unep.org/resources/report/air-pollution-asia-and-pacific-science-based-solutions

World Bank. (2020). *The cost of air pollution: Strengthening the economic case for action*. World Bank Publications. https://openknowledge.worldbank.org/handle/10986/25013

World Health Organization. (2023). *Air pollution*. https://www.who.int/health-topics/air-pollution

Zhang, Q., Zheng, Y., Tong, D., Shao, M., Wang, S., Zhang, Y., Xu, X., Wang, J., He, H., Liu, W., Ding, Y., Lei, Y., Li, J., Wang, Z., Zhang, X., Wang, Y., Cheng, J., Liu, Y., Shi, Q., & Hao, J. (2019). Drivers of improved PM2.5 air quality in China from 2013 to 2017. *Proceedings of the National Academy of Sciences of the United States of America, 116*(49), 24463–24469. DOI: 10.1073/pnas.1907956116 PMID: 31740599

Zhang, X., Zhang, X., & Chen, X. (2018). The impact of exposure to air pollution on labor supply: Evidence from China. *Journal of Environmental Economics and Management, 90*, 293–309. DOI: 10.1016/j.jeem.2018.05.001

Zheng, S., Pozzer, A., Cao, C. X., & Lelieveld, J. (2018). Long-term (2001–2012) PM2.5 exposures and associated mortality burdens in China. *Proceedings of the National Academy of Sciences of the United States of America, 112*(11), 3527–3532. DOI: 10.1073/pnas.1422939112

Chapter 12

Global Trends in Combating Air Pollution and the Universal Nature of the Right to Environment

Talha Turhan
https://orcid.org/0000-0002-6638-0929
Erciyes University, Turkey

ABSTRACT

This study investigates the conceptual and normative foundations of environmental law with a specific focus on the human right to a healthy environment. Methodologically, the article adopts an interdisciplinary legal analysis that integrates environmental science, human rights theory, and public international law. The inquiry begins with a definitional framework of the environment as a composite of biotic and abiotic elements, followed by an exploration of its socio-legal dimensions. Ultimately, the article advances a prescriptive methodology, arguing for the recognition of an enforceable universal right to a healthy environment. This methodological design combines descriptive legal mapping with evaluative critique, proposing that sustainable governance requires a holistic integration of environmental law into the human rights system. The study thus positions itself at the intersection of environmental protection and fundamental rights, offering a framework for both academic discourse and policy-making.

DOI: 10.4018/979-8-3373-2469-2.ch012

INTRODUCTION

Since ancient times, human development and air pollution, one of the most enduring and urgent environmental issues have been linked. Its effects, which range from personal health to global climate systems, have gotten worse as a result of growing urbanization, industrialization, and dependency on fossil fuels. Although people have long been aware of the negative impacts of air pollution, such as respiratory distress, offensive smells, and decreased vision, scientific knowledge of its causes, constituent chemicals, and health consequences has only greatly increased in the last several decades.

There is ample historical evidence of the detrimental impacts of air pollution. Although they lacked the modern ability to monitor or analyze specific pollutants, people in previous ages were aware of the unpleasant and detrimental effects of smoky or contaminated air. By the 17th to 19th century, European scientists, especially in the UK, had started to distinguish particles and gaseous chemicals in the atmosphere, and early written documents indicate complaints about smoke in medieval cities. The foundation for comprehending the causes of pollution and its possible effects on health was established by these investigations. Early calls for regulation were sparked by the industrial revolution's increased awareness, particularly in urban areas that relied heavily on coal. The issue of acid rain in the 1960s led to concerted efforts in North America and Europe to monitor the air. These networks helped governments create focused mitigation plans by supplying vital information on precipitation and air quality (Fowler et al., 2020).

Both natural and man-made factors contribute to air pollution. Particulate matter and gases in the air are produced by natural processes such as dust storms, forest fires, and volcanic eruptions. However, human activity is the primary contributor in the majority of locations, particularly in urban and industrial areas. Burning fossil fuels in power plants, factories, and transportation networks, heating homes with coal and biomass, and agricultural activities like open-field crop burning are important anthropogenic sources. Particulate matter (PM2.5 and PM10), sulfur dioxide (SO_2), nitrogen dioxide (NO_2), carbon monoxide (CO), carbon dioxide (CO_2), and ground-level ozone (O_3) are among the dangerous pollutants released by these sources (Bai et al., 2018).

Air pollution is still a major problem in Poland, a nation that has historically relied on coal-based electricity. It's interesting to note that huge, professionally run power plants, which frequently use emission controls, do not produce the worst pollution. Instead, the most dangerous emissions originate from the transportation and residential sectors, where pollutants are discharged at ground level and have an immediate impact on the local population. This scenario shows how small-scale, dispersed pollutants can collectively affect urban air quality more significantly than large-scale, regulated sources. Poland has acknowledged the socio-political aspect of air quality management by implementing Low-Emission Economy Plans (LEEPs) and Low-Altitude Emission Reduction Programs (LAERPs), which are attempts to combine local initiatives with more comprehensive national strategies (Burchard-Dziubińska, 2019).

China's Atmospheric Pollution Prevention and Control Action Plan and Poland's Low-Emission Economy Plans are both represent a pure political will, abstract capacity building structure, enforcement rigor, and public compliance. While several developing contexts, weak regulatory oversight, informal economic sectors, and competing development priorities undermine implementation. Citing example, Pakistn has banned crop burning in Punjab, yet their enforcement remains inconsistent due to limited monitoring and agricultural dependence on low-cost disposal methods. On the other hand, the EU's binding air quality directives succeed partly due to supranational accountability mechanisms and financial support for municipal upgrades—highlighting how governance structures critically shape policy outcomes.

Air pollution has numerous and well-documented negative effects on health. Exposure to ambient pollutants, particularly ozone and fine particulate matter, has been linked to a variety of health issues, according to numerous studies. These include cardiovascular disorders like heart attacks, strokes, and hypertension, as well as respiratory disorders including asthma, bronchitis, and chronic obstructive pulmonary disease (COPD). Negative pregnancy outcomes, several types of cancer, and neurological diseases are also linked to long-term exposure (Chen & Kan, 2008; Sierra-vargas & Teran, 2012; Mishra, 2017). Due to their weakened immune systems or increased air intake per body weight, children, the elderly, and people with pre-existing medical illnesses are especially at risk (Chen & Kan, 2008).

The extensive effects of extended exposure to air pollution are still being demonstrated by scientific data. For instance, research indicates that cutting emissions from diesel engines might dramatically reduce the number of heart attacks and hospitalizations for respiratory conditions, ultimately saving thousands of lives each year (Sierra-vargas & Teran, 2012). The cardiovascular effects of air pollution are also highlighted by recent studies, which show a strong correlation between increased mortality from heart-related disorders and both short-term exposure to coarse particulate matter (PM10) and long-term exposure to fine particles (PM2.5). In order to minimize exposure and lower health hazards, these findings highlight the necessity of both systemic controls and individual preventive measures (Mishra, 2017).

Air pollution has equally dire effects on the ecology. Global warming is mostly caused by pollutants like CO_2 and methane, whereas acid rain and ecosystem damage are caused by nitrogen oxides and sulfur dioxide. According to Bai et al. (2018), ozone layer depletion, which is associated with specific air pollutants, increases the amount of UV light that reaches the Earth's surface, increasing the risk of cataracts, skin cancer, and harm to aquatic ecosystems.

Smog, a mix of smoke and fog created when pollutants react with atmospheric components, is one of the most obvious types of air pollution. There are two main types of smog that are commonly known: photochemical or Los Angeles smog, which is produced by sunlight-driven reactions between nitrogen oxides and hydrocarbons, and classical or London smog, which is produced by sulfur compounds and is common in cold, humid conditions (Isha et al., 2024; Rani et al., 2011). In particular, photochemical smog is linked to sunny climates, heavy traffic, and industrial emissions. Numerous cities worldwide, including those in China, India, and Australia, have it (Rani et al., 2011).

High ground-level ozone levels, which are particularly harmful to respiratory health, are a result of this kind of smog. Smog emergencies have frequently occurred in Pakistani cities like Lahore and Faisalabad, especially in the winter when emissions and weather patterns combine to trap pollutants close to the ground. These cities' PM2.5 concentrations are more than ten times higher than what the World Health Organization (WHO) recommends. The government has responded by outlawing the burning of crops, controlling industrial activities, and raising public awareness through educational programs, particularly for medical students involved in community health outreach (Butt et al., 2018).

Another example is China, where serious air quality degradation, especially in its northern cities, has resulted from rapid economic growth. China has seen multiple haze outbreaks since 2011, marked by exceptionally high levels of PM2.5 and related pollutants. The Atmospheric Pollution Prevention and Control Action Plan was introduced in 2013 as a result of public concern about the negative health impacts. This program promoted greener energy use and focused on lowering emissions from transportation and industries. An epidemiological study conducted in Jinan, a significant city in eastern China, found a correlation between daily mortality and air pollution levels. Given that variations in population susceptibility and pollutant composition can result in region-specific health consequences, the study underlined the need of localized data (Zhang et al., 2017).

Air pollution issues have been made worse by urbanization, particularly in megacities, urban areas with a population of 10 million or more. With a predicted population of nearly 730 million, there will be 41 megacities by 2030, up from 10 in 1990. Because of the high concentration of automobile traffic, industrial activity, and building, many densely populated cities have disproportionately high levels of air pollution. In these cities, almost 96% of the population is exposed to PM2.5 concentrations that are higher than WHO safety standards, and air pollution is directly responsible for 7% of the illness burden. In South and East Asian cities, where exposure per capita and related health effects are among the highest worldwide, this burden is even more noticeable (Marlier et al., 2016).

Accurate modeling and forecasting technologies are crucial for managing and reducing air pollution. The two categories of forecasting methods are deterministic models, which model atmospheric chemical and physical processes, and parametric models, which employ patterns in past data. In-depth simulations of pollutant interactions under various environmental conditions are possible with sophisticated tools such as the Community Multiscale Air Quality (CMAQ) model (Bai et al., 2018). Though their effectiveness varies by region, neural networks and machine learning models, which have been introduced by recent breakthroughs in artificial intelligence, improve forecast accuracy. According to Daly and Zannetti (2007), these technologies are essential for creating timely notifications, educating the public, and assisting with policy decisions.

Additionally, modeling clarifies the fundamental dynamics of pollution dispersal. Diffusion, the way pollutants travel through turbulent air, chemical transformation, the way they interact with other substances, deposition, the way they land on surfaces, and atmospheric transport, the way they follow wind patterns, are important processes. Our knowledge of diffusion, particularly with regard to tiny particles in turbulent air layers, is still a relatively recent scientific endeavor, despite the fact that atmospheric transport was researched decades ago (Daly & Zannetti, 2007).

In order to convert complicated pollutant data into useful insights, accurate air quality indicators are required. For this, one of the most popular tools is the Air Quality Index (AQI). However, inconsistent measuring methods and poor cross-regional comparability can harm AQIs. Researchers have suggested a new paradigm that focuses on six fundamental pollutants: PM2.5, ozone, sulfur dioxide, nitrogen oxides, carbon monoxide, and volatile organic compounds, in order to get around these restrictions. To facilitate more precise cross-country comparisons and encourage more informed policy development, this method makes use of population-weighted global statistics and standardized units. Moreover, this paradigm can be combined with atmospheric models to predict the potential effects of different regulation approaches on pollution levels and health consequences. It gives decision-makers strong instruments to evaluate environmental interventions and improve public health tactics in a way that is grounded in science (Wolf et al., 2022).

Lastly, environmentally friendly methods for enhancing air quality, such as bioremediation, are becoming more and more popular. These techniques eliminate, neutralize, or break down air pollutants by using living things like fungi, bacteria, algae, and plants (Isha et al., 2024). Two important methods in this field that provide environmentally friendly substitutes for traditional mechanical or chemical filtering systems are phytoremediation and biosorption. Because of its affordability and versatility, bioremediation holds particular promise for environments with limited resources. Individual measures can also help lower health risks from air pollution, such as wearing well-made face masks, remaining indoors on days with heavy pollution, and include foods high in antioxidants in one's diet (Mishra, 2017).

This introduction presents a comprehensive overview of air pollution's historical evolution, its primary sources and pollutants, associated health and environmental impacts, modeling and forecasting techniques, and regional examples of mitigation efforts. It incorporates a diverse array of referenced studies to illustrate how modern societies are addressing this global concern through scientific innovation and policy reform.

LITERATURE REVIEW

The negative health effects of indoor smoke and poor urban air quality were noted in early documents, demonstrating the ancient human awareness of air pollution. Hippocrates and other ancient Greek writers acknowledged the connection between air pollution and illness, and attempts to enhance interior conditions were reflected in developments like the chimney in the 12th century. Although scientific knowledge was still in its infancy, the medicinal effects of air pollution were recognized in a number of cultures, including Roman, Chinese, and Arab civilizations. In the Middle Ages, popular thought linked illness to "miasma," or unpleasant smells, instead of particular chemical substances. The earliest regulations appeared in medieval England, and the regulatory responses were initially limited and driven by economic considerations. The extent of pollution did not require more methodical answers until the industrial revolution, which was marked by metropolitan areas that relied heavily on coal (Fowler et al., 2020).

The Great London Smog of 1952, one of the most important incidents in the history of air pollution, killed thousands of people and sparked extensive legislative changes, including the Clean Air Act of 1956. This was a significant event that resulted in lower smoke and sulfur dioxide emissions and set the stage for further advancements in air quality. With the increasing effectiveness of urban pollution control, regional phenomena such as acid rain came into focus. Acid deposition from long-range air transport harmed crops, lakes, and forests, especially in North America and Northern Europe. International agreements to reduce sulfur and nitrogen emissions were prompted by scientific findings that connected industrial emissions to transboundary contamination. These activities developed into international air quality programs that target pollutants that endanger ecosystems and the stability of the climate, including methane, ammonia, and ground-level ozone (Fowler et al., 2020).

Air pollution and regional cooperation's are two inseparable issue that needs to be handle simultaneously. Given the fact that there is transboundary nature of air pollution, the UN Economic Commission for Europe Convention on Long-Range Transboundary Air Pollution laid the groundwork for protocols in 1979. This protocal aimed to decrease nitrogen emissions across Europe. In the same logic, the ASEAN Agreement on Transboundary Haze Pollution (2002) aimed to address recurrent smoke haze from agricultural burning in Southeast Asia. However, non-binding nature if agreement has limited enforcement. Similarly, UNEP's Clean Air Initiative also projected to increase knowledge-sharing and capacity-building among cities in Africa, Asia, and Latin America. All of these mechanisms clearly examine both the promise and challenges of multilateral environmental governance in a fragmented geopolitical landscape, not to mention fats nexus between air pollution and global cooperation.

Even while many wealthy countries have made headway in lowering their emissions, growing pollution levels in some Asian countries still pose major health hazards. The intricate relationships between air quality and climate change are exacerbated by rising emissions of greenhouse gases and volatile organic compounds. However, there is cautious hope for long-term improvements in air quality worldwide due to rising environmental consciousness, technology advancements, and governmental changes (Fowler et al., 2020).

Air pollution has become a major public health and economic concern in addition to an environmental one. According to the WHO, air pollution is responsible for one in eight fatalities worldwide, primarily from respiratory and cardiovascular diseases. Nitrogen oxides (NOx), ozone (O_3), sulfur dioxide (SO_2), and fine particulate matter (PM2.5 and PM10) are the main pollutants of concern. Because PM can enter the circulation through the lungs and cause oxidative stress and systemic inflammation, it poses serious health hazards. Major sources of PM, including as home heating, industrial operations, and automobile traffic, have been designated by the European Union as the main priorities for mitigation. Air pollution has a significant and growing economic impact, including medical costs, lost productivity, and environmental damage. Poland has some of the worst air pollution in Europe. The extensive use of subpar fuels, antiquated heating systems, excessive emissions from transportation and industry, and poor waste management are all contributing issues. Lax standards and poor enforcement persist despite the Clean Air Programme and EU environmental directives being put into effect. Poland's air quality continues to fall short of WHO and EU standards. Peak smog occurrences have been linked to higher rates of cancer, strokes, cardiovascular and respiratory disorders, and other public health issues. The hazards are exacerbated by toxic emissions, such as carcinogens and heavy metals, necessitating more thorough interventions and stronger controls (Burchard-Dziubińska, 2019).

around the past, a lot of cities had very high pollution levels around the middle of the 20th century, but in areas with strong pollution management, these levels have considerably dropped. However, a lot of developing and transitional economies still have high levels of PM, SO_2, NO_2, and O_3 emissions, which are mostly caused by burning fossil fuels. Strong correlations between exposure to pollution and an increased risk of cardiovascular and respiratory disorders have been confirmed by research. Interestingly, there is no safe level at which contaminants stop being harmful to human health. The health effects of even little increases in pollutant concentrations are quantifiable, especially for vulnerable groups such as children, the elderly, and people with underlying medical disorders (Chen & Kan, 2008).

While chronic exposure leads to long-term diseases and noticeably higher mortality rates from heart and lung disorders, short-term exposure to air pollution increases daily mortality and hospital admissions. Reduced pollution has been shown to improve public health. Interventions in Hong Kong and Dublin, for instance, showed that quick drops in air pollution were associated with instantaneous gains in population health. In order to promote incremental improvements in developing nations, the WHO created air quality recommendations with intermediate targets after realizing these hazards (Chen & Kan, 2008).

Research on air pollution covers a wide range of topics, including regulatory evaluations, health effects, emission inventories, and pollution control. Emission inventories (EIs), which provide thorough records of pollutant emissions over particular geographies and time periods, are essential tools in this discipline. Models of air quality, policy formulation, and compliance strategies are all influenced by these inventories. For instance, the U.S. Environmental Protection Agency uses activity rates and emission factors to measure pollution levels in its standardized emissions data reporting systems. Pollution has a wide range of negative health effects, from higher mortality rates from cardiovascular disease to decreased sperm quality and increased childhood eczema. It is crucial to assess the socioeconomic costs of air pollution, quantifying the effects on industry, public health, and agriculture using techniques including engineering cost estimates and market value methods. Another research topic is pollution control efficiency, where methods such as data envelopment analysis are employed to evaluate how well different industries lower emissions. In order to notify the public of pollution spikes and start emergency responses, early warning systems that combine weather forecasts with air quality data are essential. In order to simulate the dispersion of pollutants in complicated terrains more accurately, modern air pollution modeling has moved from two-dimensional to three-dimensional systems. To improve forecasting accuracy, hybrid models integrate multiple forecasting techniques, such as geographic information systems and artificial intelligence. In order to provide a more comprehensive understanding of the dynamics of air quality, models such as BlueSky and WRF-Chem are made to predict emissions from natural occurrences like volcanic activity and wildfires (Bai et al., 2018).

When coal, gasoline, fuel, and wood are burned, particulate matter a significant contributor to urban air pollution is released into the atmosphere. PM may also emerge from secondary chemical reactions in the atmosphere involving nitrogen oxides and Volatile Organic Compounds (VOC). Size, makeup, length of exposure, and personal vulnerability all affect its health effects. According to Sierra-vargas and Teran (2012), symptoms can include coughing, eye irritation, worsening of chronic respiratory conditions, and early death. The interaction of sunlight with organic molecules and nitrogen oxides produces ozone (O_3), which has a negative impact on lung function and makes disorders like asthma worse. Particularly for those who are sensitive, physical activity during times of high ozone can exacerbate these effects (Sierra-vargas & Teran, 2012).

Because their immune systems are still developing and their breathing rates are faster, children are more susceptible to air pollution. Pollutants can exacerbate respiratory conditions and allergic reactions. For example, it has been demonstrated that diesel exhaust particles interact with allergens and exacerbate inflammation of the airways. Oxidative stress, in which reactive oxygen and nitrogen species damage cellular proteins and tissues, is one of the main mechanisms of pollution-induced damage. According to Sierra-vargas & Teran (2012), this procedure raises the risk of cancer, chronic inflammation, and other degenerative disorders. Reductions in PM exposure have been linked to better respiratory outcomes and longer life expectancies, especially in susceptible populations, according to encouraging research (Sierra-vargas & Teran, 2012).

Coordination between individual and governmental efforts is necessary to reduce air pollution. Personal tactics include enhancing indoor air circulation, particularly in homes that use biomass fuels, and minimizing outdoor activities during pollution peaks. Government initiatives like the implementation of renewable energy, green infrastructure, and more stringent automobile pollution regulations are essential. For instance, brief increases in air quality during the 2008 Beijing Olympics were associated with fewer hospitalizations for asthma. By escalating pollution concentrations through rising temperatures and changed weather patterns, climate change makes air quality even more problematic. Therefore, comprehensive climate and pollution regulations are required, but they are still difficult to police and execute (Sierra-vargas & Teran, 2012).

The study of pollution dispersion from point sources, particularly industrial stacks, has benefited greatly from models such as the Gaussian Plume Model. The model, which was created using empirical observations, helps establish stack heights for regulatory compliance and simulates how pollutants spread under different atmospheric conditions. Lagrangian and Eulerian models, which follow pollutants along wind paths or inside predetermined spatial grids, respectively, were later developments. Global simulations, long-distance transport studies, and the creation of hybrid methodologies have all been made possible by these models (Daly & Zannetti, 2007).

These days, models like AERMOD and CALPUFF are frequently utilized. CALPUFF mimics the behavior of pollutants in complicated terrains and under non-steady situations, whereas AERMOD is a steady-state Gaussian model that incorporates meteorological data. In order to mimic pollutant transformations over various geographical and temporal scales, advanced photochemical models like CMAQ and CAMx employ mathematical representations of atmospheric chemistry. These simulations are improved by meteorological inputs from models such as MM5 and CALMET. Furthermore, statistical and particle-based models can explain dispersion, particularly in situations where it is challenging to establish clear cause-and-effect links (Daly & Zannetti, 2007).

Reducing air pollution has been repeatedly linked to major benefits in public health, particularly for vulnerable groups like children, the elderly, and people with pre-existing medical illnesses. Research indicates that improved respiratory function and longer life expectancy are linked to reduced particulate matter concentrations. Therefore, governments have a crucial role to play in managing air quality. Expanding green urban design, switching to cleaner energy sources, enforcing strict emission regulations, and assisting with public awareness campaigns are examples of effective policy actions. Outpatient visits for respiratory illnesses significantly decreased during the Beijing Olympics, demonstrating the practical benefits of short-term air quality measures (Sierra-vargas & Teran, 2012).

To lessen the negative health effects of air pollution, comprehensive methods that incorporate community involvement and state regulation are required. Individuals can experience modest but significant respite from smog by taking simple steps like wearing protective masks during smog events, reducing exposure to heavy traffic corridors, and using indoor air purifiers. In order to support these initiatives, policymakers should enforce cleaner fuel standards, encourage public transit, and provide subsidies for eco-friendly home appliances. As pollution and climate change continue to interact, these initiatives become increasingly more critical because, particularly in warmer weather circumstances, more frequent photochemical smog occurrences could degrade air quality. Countries like Mexico have enacted climate laws to cut carbon emissions and encourage renewable energy, but there are still significant obstacles due to enforcement issues and inadequate urban planning (Sierra-vargas & Teran, 2012).

Models such as the Gaussian Plume Model, which was initially developed to estimate the spread of contaminants from industrial stack emissions, have significantly advanced scientific understanding of air pollution dispersion. In order to help regulators choose the right stack heights to reduce ground-level concentrations, this model computes the horizontal and vertical spread of contaminants. The model simulates real-world dispersion conditions by using weather and atmospheric stability data. It developed over time to support line and area sources and incorporate multiple reflections. The scientific community had moved away from regional pollution control and toward acknowledging air pollution as a worldwide issue by the 1970s. The significance of long-range air transport was highlighted when it was shown that tall emission stacks, which were previously believed to lessen local effects, actually contributed to regional acidification (Daly & Zannetti, 2007).

The creation of increasingly complex modeling systems was spurred by this realization. Lagrangian approaches track pollutant packages as they travel, whereas Eulerian models split a region into fixed grid cells to examine pollution within each segment. Accuracy has improved in both urban and regional settings as a result of the combination of these methods. In regulatory and planning applications, contemporary models like CALPUFF, a non-steady-state model that works well in changeable weather circumstances, and AERMOD, a steady-state Gaussian dispersion tool, have become indispensable. The intricate chemistry of atmospheric processes is taken into consideration by photochemical models such as CAMx and CMAQ (Community Multiscale Air Quality), which model pollutant interactions and transformations throughout time (Daly & Zannetti, 2007).

The atmospheric characteristics required for these dispersion models are provided by meteorological models like MM5 and CALMET. To provide more accurate models of pollution plumes across different terrains, many additionally incorporate modules for particle behavior and plume ascent. Particularly for localized air quality assessments, statistical models that use ambient concentration data without presuming causation are frequently used (Daly & Zannetti, 2007).

Research conducted in China in recent years has brought attention to the substantial impact that rising air pollution places on public health. Since 2011, haze outbreaks in cities like Jinan have alarmingly increased, mostly due to rising levels of PM2.5, PM10, SO_2, and NO_2. In the past, the accuracy of health effect assessments was constrained by a lack of high-resolution PM2.5 data. Recent research, however, has shown definite relationships between pollution levels and mortality rates utilizing better monitoring and longitudinal data. Rising PM10 and NO_2 concentrations were linked to a considerable increase in daily death rates, according to the Jinan study, which was carried out between 2011 and 2015. In winter, when smog seasons were marked by stagnant air and high emissions, the health effects were particularly pronounced. The most vulnerable groups were determined to be women and the elderly, with the greatest impact on cardiovascular and respiratory mortality. Crucially, this study demonstrated the severe impact that smog outbreaks have on public health, as mortality increases happen only a few days after pollution peaks. The thorough five-year dataset confirmed the validity of these conclusions and emphasized the significance of local data in developing successful air quality regulations (Zhang et al., 2017).

A common tool for informing the public about pollution levels is the air quality index (AQI). These indices combine information from different pollutants, including CO, NO_2, PM2.5, and PM10, to give a simplified picture of air quality. To direct public health responses, regulatory agencies such as the U.S. Environmental Protection Agency establish pollution thresholds under laws like the Clean Air Act. For instance, in Pakistan, places like Abbottabad have attained improved air standards, but Lahore continues to rank among the most polluted cities in the world because of industrial and vehicle emissions (Isha et al., 2024).

Bangladesh has the worst levels of smog in the world, and nations like China, India, Pakistan, Indonesia, and Bangladesh continue to have disproportionately high rates of air pollution-related deaths. The United States, New Zealand, and Canada, on the other hand, record much lower levels of pollutants than countries with stricter regulations. Industrial intensity, transportation regulations, and waste management effectiveness are some of the factors influencing these disparities (Isha et al., 2024). Smog is a mixture of dangerous gasses and fine particulates that can seriously harm people's health. It increases cardiovascular events, worsens respiratory conditions, and may even have an impact on reproductive health. Smog particles can increase the risk of antibiotic-resistant bacteria by entering the lungs, according to recent research. New information about the connections between air quality and health was made available by the COVID-19 pandemic. High levels of air pollution have been linked in studies to higher incidence of postpartum depression, respiratory discomfort, and gestational diabetes in expectant mothers. Furthermore, a large number of industrial chemicals that the EPA has determined are carcinogenic, and exposure to these airborne pollutants has been linked to COVID-19 issues on multiple occasions (Isha et al., 2024).

Biological remediation techniques are emerging remedies for air pollution. Pseudomonas putida and other microorganisms are used in bacterial bioremediation to convert nitrogen oxides and hydrocarbons into less toxic byproducts. Although these bacteria are susceptible to environmental restrictions and may be blocked by high pollutant concentrations, they have shown efficacy in degrading pollutants under particular conditions. Additionally, fungi have intriguing remedial properties. Fungal enzymes are used in mycoremediation processes to break down complex contaminants, such as heavy metals and hydrocarbons. White-rot fungi have demonstrated remarkable efficacy in breaking down a wide range of organic substances. Although these tactics can be used both in situ and out of situ, their effectiveness frequently depends on certain environmental factors (Isha et al., 2024).

An other frontier in pollution control is nanotechnology. By capturing or neutralizing a wide range of contaminants, engineered nanoparticles can improve on conventional bioremediation methods. That being said, some nanomaterials are harmful and persistent in the environment, which calls for more investigation and regulation. A plant-based method for air purification is called phytoremediation, which uses species that have the ability to absorb pollutants. Cost-effective and adaptable to urban landscaping, this approach's effectiveness is contingent upon biological and climatic conditions. Despite the potential of these novel techniques, obstacles like the complexity of pollutants, environmental unpredictability, and possible secondary consequences prevent their general adoption. Integration with regulatory frameworks, urban planning, and continuing genetic engineering and nanoscience research are necessary to overcome these obstacles (Isha et al., 2024).

Although bioremediation techniques show great promise, a number of obstacles still prevent their widespread use. The effectiveness of biological decomposition is limited by the trace amounts of many contaminants. Moreover, specific environmental parameters like temperature, humidity, and pollution load are frequently necessary for bioremediation to be successful. During the degrading process, there is also a chance of producing hazardous secondary byproducts. These drawbacks highlight the necessity of interdisciplinary integration to improve the practical efficacy of bioremediation by fusing biotechnology, nanoscience, urban planning, and environmental policy. Innovations in genetic engineering and nanotechnology are probably going to be very important in enhancing these biological systems' efficiency, specificity, and robustness in both urban and rural settings (Isha et al., 2024).

Global ecosystems and human well-being are under risk due to the interconnected environmental problems of air pollution and climate change. The burning of fossil fuels is one of the main anthropogenic activity that contributes to greenhouse gas emissions, which raise the levels of carbon dioxide, methane, and nitrous oxide in the atmosphere. Sea level rise, biodiversity loss, weather pattern disturbances, and challenges to food and water security are all consequences of these gases' acceleration of climatic volatility and driving of global warming. By lowering air quality and increasing the overall environmental burden, pollution, especially in the form of airborne gases and particles, makes these problems worse. Although the main causes are automobile exhaust and industrial pollution, natural occurrences like dust storms and volcano eruptions also contribute (Usman et al., 2023).

One of the most dangerous forms of air pollution is smog, which is usually a dense yellow or brown haze made up of particulate matter and ground-level ozone. It is caused by sunlight interacting with nitrogen oxides and volatile organic chemicals, and it is particularly common in crowded urban environments. Exposure to smog can cause everything from minor irritations like eye, nose, and throat pain to serious respiratory and cardiovascular conditions, reduced lung function, and even early mortality. Smog has a direct impact on ecological and economic sustainability since it impairs photosynthesis, lowers crop output, and deteriorates vision. The WHO has released thorough air quality standards that include safe concentration limits for main pollutants like PM2.5, PM10, CO, NO_2, SO_2, and O_3 in order to combat these effects. Governments use these recommendations as a baseline for creating national policies and public health initiatives. The recent tightening of PM2.5 regulations to lower exposure and associated mortality is only one example of how updated limits frequently take into account new scientific knowledge about the dangers of pollution. Since ground-level ozone is a strong oxidant created mostly by industrial and vehicle emissions in direct sunlight, it continues to be a crucial area of concern (Usman et al., 2023).

Air quality issues have gotten worse due to rapid urbanization and the growth of megacities, which are urban agglomerations with more than 10 million residents. Megacities increased from 10 to 28 between 1990 and 2014, and by 2030, there are expected to be 41 of them, with a combined population of nearly 730 million. In Asia, where cities like Delhi, Beijing, Dhaka, and Karachi routinely report PM2.5 concentrations far beyond global health limits, the environmental pressure from this growth is particularly noticeable. Millions of people die each year from pollution-related causes in these areas, where the majority of urban dwellers are exposed to dangerously high levels of air pollution on a daily basis (Marlier et al., 2016).

Globally, pollution trends differ. Other places, particularly in South and East Asia, have seen rising pollution levels as a result of industrial expansion and inadequate enforcement, while other regions, such as the eastern United States, have successfully decreased PM2.5 levels by regulatory initiatives between 1998 and 2012. Vehicle emissions, industrial operations, burning in agriculture, and the use of fossil fuels are some of the main causes. To monitor pollutant concentrations, assess the efficacy of policies, and predict pollution extremes, satellite data, ground-based monitoring stations, and atmospheric modeling methods are combined (Marlier et al., 2016).

The effects of urban air pollution on health are widespread. Chronic exposure is associated with developmental abnormalities in children, reduced labor productivity, cognitive impairment, and cardiovascular and respiratory ailments. Exposure during pregnancy or early childhood, for example, may have long-term consequences on intellectual development and lung capacity. Despite efforts to reduce emissions, urban centers like Beijing and Shanghai often surpass international pollution regulations. On the other hand, stricter enforcement of regulations, cleaner transportation, and effective urban planning have allowed places like Tokyo to preserve comparatively cleaner air (Marlier et al., 2016).

Countries' efforts to reduce urban pollution differ greatly from one another. To cut emissions, some governments have put in place legislative frameworks, environmental fees, and infrastructural improvements like wider metro systems and automobile limitations. For instance, Delhi's odd-even car rule and Beijing's alternate-day driving program have had varying degrees of effectiveness. However, such actions are not necessarily guaranteed to be equitable. Richer people may frequently avoid regulations or protect themselves from pollution, which lessens public pressure for systemic change. Therefore, to guarantee broad support for air quality measures, inclusive and socially equitable strategies are required (Marlier et al., 2016).

Interestingly, air pollution mitigation also contributes to climate change adaptation. Numerous contaminants are also climate forcers, including ozone precursors and black carbon. By lowering their concentrations, global warming is moderated and public health is enhanced. These interactions, though, can be intricate. Certain actions taken to lower one pollutant may unintentionally increase another or change the dynamics of the atmosphere in unforeseen ways. For example, non-linear chemical processes may cause aggressive NOx reductions to occasionally raise ozone levels. These interdependencies must therefore be carefully taken into account when designing policies (Marlier et al., 2016).

One of the main causes of air pollution worldwide is transportation. Since World War II, the number of vehicles produced annually has increased dramatically from 5 million to about 70 million, with nations such as China currently dominating the manufacture of both cars and motorcycles. Personal automobiles have been more and more in demand as the world's population has grown, especially in urban areas. Over 1 billion vehicles were in use worldwide by 2002, and this number has been rising at a rate of 27 million new cars per year (Global Trends in Motor Vehicle Pollution Control: A 2011 Update, 2011).

PM, NOx, CO, SO_2, and VOCs are among the dangerous pollutants that are released by motor vehicles. Smog, respiratory ailments, heart disease, and climate change are all exacerbated by these. Because fine particulate matter (PM2.5 and PM10) can enter the bloodstream and lungs deeply, it presents unique hazards. There is no threshold below which these particles are deemed harmless, according to the WHO. Similarly, respiratory discomfort and asthma episodes are linked to ground-level ozone produced by NOx and VOCs. Disproportionately impacted are children, the elderly, and people with pre-existing ailments (Global Trends in Motor Vehicle Pollution Control: A 2011 Update, 2011).

In addition to their negative health impacts, automobile emissions lower crop yields, decrease visibility, and hasten structure degradation. Acid rain is caused by pollutants like SO_2 and NOx, which harm infrastructure and ecosystems. In the meantime, global warming is intensified by greenhouse gases like CO_2 and black carbon. Rapid motorization is causing emissions to rise in poorer countries, despite the fact that wealthy countries have implemented cleaner technologies and more stringent laws. International collaboration, technological advancement, and strict enforcement are necessary to meet these difficulties (Global Trends in Motor Vehicle Pollution Control: A 2011 Update, 2011).

No doubt that air pollution has not influence direct health consequences, but also imposes substantial socio-economic burdens over the country's economy. For instance, ozone levels correlates with increased healthcare expenditures, reduced labour productivity, and heightened social inequality. Accordingly, in low- and middle-income countries such as Asian and Latin American countries where vulnerable populations lack access to protective measures like air purifiers or quality healthcare. To give a clear example, Delhi and Lahore, school and workplace absenteeism rises during smog episodes, while informal sector workers—unable to work remotely—face disproportionate exposure. All of this deep disequilibrium underlines how air pollution exacerbates existing socio-economic divides, turning environmental degradation into a matter of environmental justice.

The health risks of uncontrolled emissions are highlighted by past pollution incidents. Extremely high levels of SO_2 and suspended particles during the 1985 smog crisis in North Rhine-Westphalia, West Germany, prompted emergency responses. During the haze period, there was a significant increase in hospital admissions, ambulance calls, and cardiovascular fatalities, particularly among people who already had heart issues (Wichmann et al., 1989). Comparably, the Great London Smog of 1952 sparked the creation of contemporary air quality laws and caused over 4,000 immediate deaths. Current research supports the effects of air pollution on the cardiovascular system. Even brief exposure to PM2.5 increases the risk of arrhythmias, heart attacks, and strokes. Smokers, diabetics, and the elderly are the groups most impacted. Atherosclerosis and endothelial dysfunction are caused by fine particles that enter the bloodstream and cause systemic inflammation. Risk can be decreased by taking precautions like wearing masks, utilizing air purifiers, and remaining indoors during smog episodes, but more extensive mitigation calls for stricter laws and better technology (Mishra, 2017).

A global public health emergency is the unequal burden of air pollution in low- and middle-income nations. Even while many affluent countries have cut their emissions, more than 90% of people on Earth still live in places with pollution levels higher than those recommended by the WHO. Accurately assessing exposure is a major obstacle in resolving this issue. Population density is not taken into consideration by conventional measures that rely on national averages. A fuller picture of the true health hazards is provided by a more equitable approach called population-weighted exposure (Wolf et al., 2022).

This method was used in a recent study that assessed exposure to six pollutants in more than 200 countries: PM2.5, O_3, NOx, SO_2, CO, and VOCs. The findings showed glaring differences. While legislative measures have led to advancements in certain countries, such as China and India, others continue to face increased exposure. The study also emphasized intricate relationships between pollutants, including how atmospheric feedback loops may cause decreases in NOx and VOCs to inadvertently raise ozone levels. These results assist governments anticipate trade-offs and improve their environmental plans by informing policy design. Importantly, countries can monitor progress and enact more fair, efficient air quality regulations when routinely updated, population-relevant exposure data is available (Wolf et al., 2022).

The supplied bibliometric map, which was created with VOSviewer and is based on data taken from PubMed, provides a thorough visual representation of the state of research on smog and air pollution control worldwide, as shown in Fig. 1. Each node in the map represents a keyword or idea that is commonly used in the literature, and the relationships between them show co-authorship or a thematic association. This map illustrates keyword co-occurrence networks. While colors indicate various subject groupings or research topics, node size indicates the frequency of occurrence.

Figure 1. VOSviewer-generated keyword co-occurrence network based on PubMed Database on Globa; Trends in SMOG and Air Pollution Control

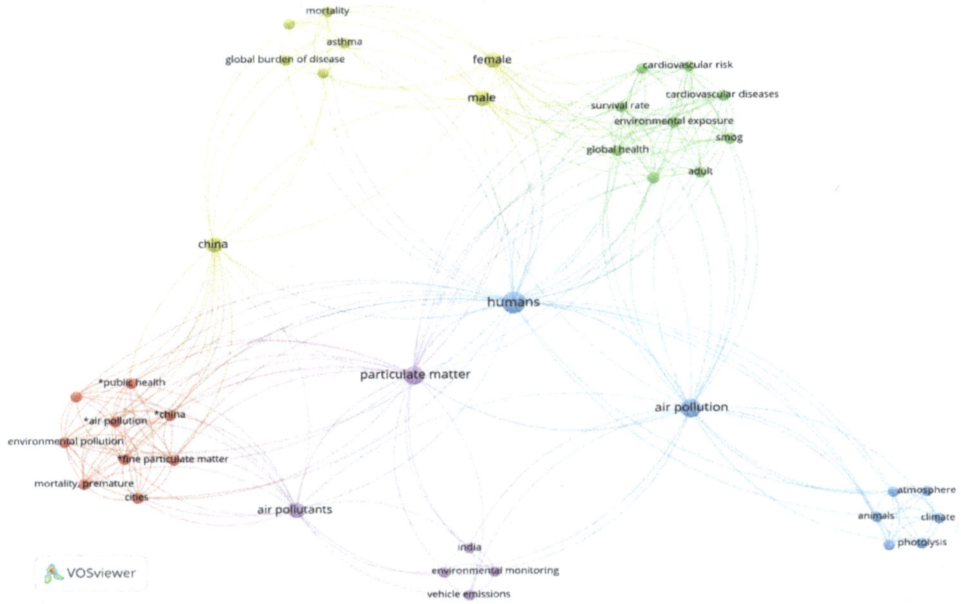

The phrase "humans," which serves as a link between different subject groups, is located at the middle of the map. This is in line with the literature's heavy emphasis on human health in relation to smog and air pollution. As evidenced by its association with key phrases such as "air pollution," "particulate matter," and "public health,"

a significant amount of the study is focused on the effects of pollution on human populations. Clusters pertaining to health outcomes are closely related, especially the green cluster, which includes terms such as "global health," "cardiovascular risk," "environmental exposure," "cardiovascular diseases," and "smog." These phrases draw attention to the rising worry over chronic diseases brought on by pollution, particularly those that impact the heart and lungs.

Terms like "female," "male," "mortality," "asthma," and "global burden of disease" are included in another important theme area, which is highlighted in yellow. This implies that attention should be paid to variations in pollution exposure and sensitivity based on demographic and gender factors. Research on how some groups, women, children, and the elderly, are disproportionately impacted by pollution and fine particulates is becoming more and more popular. Within this cluster, mortality-related research seems to be the main focus, especially early mortality brought on by long-term exposure to contaminants.

"Public health," "fine particulate matter," "cities," and "premature mortality," which stand for the urban health aspect of air pollution, are at the core of the red cluster. It implies that a great deal of academic research is focused on the ways that urban pollution is causing higher death rates and more general public health issues. The inclusion of "China" as a strongly connected node in this cluster highlights the nation's pivotal position in the global conversation about air pollution, probably as a result of its high emissions, fast urbanization, and substantial smog-related health burden.

Among the environmental and ecological issues covered by a separate blue cluster are "climate," "photolysis," "atmosphere," and "animals." This suggests that there is a strong body of study examining the effects of air pollution on atmospheric chemistry and ecological systems in addition to health-related issues. Relationships between "air pollution" and "climate" demonstrate how scientists are becoming more aware of the dangers posed by both pollution and global warming.

Vehicle emissions are an important research topic, and the purple cluster, which includes the terms "air pollutants," "vehicle emissions," "India," and "environmental monitoring," illustrates ongoing concerns in South Asia. The inclusion of India implies that it is a focus of health and monitoring studies related to pollution, just like China.

The research field is geographically and disciplinary varied, according to this bibliometric analysis. With significant regional emphasis on China and India, it covers epidemiology, environmental science, public health, and climate studies. The prevalence of health-related terms confirms that air pollution is a serious global health catastrophe in addition to an environmental problem.

Recognising fast and robust development in technology and community engagement, there are more optimist for air quality management. For instance, low-cost air sensors now empower citizens and local governments to monitor hyperlocal pollution in real time, enabling targeted interventions in data-scarce regions. On top of that, green infrastructure, including urban forests, green roofs, and permeable pavements, play key role for filtering particulates and mitigating urban heat islands. Such actions cause indirectly reducing ozone formation. From the sicentificat point of view, biological approaches like phytoremediation and mycoremediation offer sustainable, low-cost alternatives for pollutant absorption, particularly in resource-limited settings. All these promising development have combined with AI-enhanced forecasting and public awareness campaigns. As a result, these innovations represent a shift toward decentralized, participatory air quality governance that complements top-down regulation.

CONCLUSION

One of the most complicated and urgent environmental issues facing the modern world is still air pollution. It still has a significant impact on the environment and human health in spite of growing awareness and scientific breakthroughs. The historical evolution of air pollution issues has been outlined in this chapter, showing how initial worries about smoke and soot gave way to thorough knowledge of airborne chemical elements, atmospheric behavior, and their extensive effects.

Rapidly urbanizing and industrializing countries bear a disproportionate amount of the cost of air pollution, especially in Asia, where megacities like Delhi, Beijing, Dhaka, and Lahore frequently endure extremely high pollution levels. The information provided unequivocally shows that the main pollutants, PM2.5, ozone, nitrogen dioxide, and sulfur dioxide, are to blame for a variety of diseases, such as heart attacks, strokes, asthma, chronic obstructive pulmonary disease, and even neurological and reproductive issues. With mortality and morbidity rates strongly correlated with pollution exposure, vulnerable populations, including children, the elderly, and people with pre-existing conditions, are particularly at danger.

Both classical and photochemical smog have become obvious indicators of urban environmental deterioration. Because photochemical smog is linked to ground-level ozone, a highly reactive and harmful gas, it is especially deadly. Photochemical smog is caused by industrial activities and automobile emissions. Repeated smog crises in nations like China and Pakistan have prompted both grassroots calls for cleaner air and official involvement.

Technologies for modeling and forecasting, such CMAQ, CALPUFF, and AERMOD, provide useful resources for comprehending the dispersion of pollutants and anticipating pollution occurrences. The accuracy and utility of air quality prediction have significantly improved with the addition of AI and machine learning to forecasting algorithms. Policymakers are now able to make more informed decisions on pollution control because to these models, enhanced emission inventories, and real-time monitoring.

On the policy front, nations have adopted a range of tactics, from fuel switching and automobile limits to public transit spending and the promotion of renewable energy. Enforcement is still uneven, though, and in many areas, environmental preservation is still subordinated to economic expansion. Addressing the environmental and socioeconomic aspects of air pollution calls for a more equitable and coordinated strategy.

Promising avenues for the future are presented by novel biological techniques including bioremediation, phytoremediation, and pollution control based on nanotechnology. Even though there are still obstacles to overcome, like environmental constraints and the intricacy of pollutant combinations, ongoing research and interdisciplinary cooperation can be helpful. Short-term protection can also be obtained through individual behavioral modifications, including as wearing face masks, enhancing indoor air quality, and eating a diet high in antioxidants, particularly in areas with high exposure levels.

Combating air pollution necessitates a multifaceted strategy that incorporates public involvement, science, policy, and technology. Collaborative international action is necessary due to the global character of air pollution, including its transboundary movement, climate effects, and economic impacts. Significant improvement is achievable, as this chapter has demonstrated, but in order to guarantee cleaner air and healthier communities in the next decades, persistent dedication, creativity, and equity-driven initiatives will be essential.

REFERENCES

Bai, L., Wang, J., Ma, X., & Lu, H. (2018). Air Pollution Forecasts: An Overview. *International Journal of Environmental Research and Public Health, 15*(4), 780. DOI: 10.3390/ijerph15040780 PMID: 29673227

Burchard-Dziubińska, M. (2019). Air pollution and health in Poland: anti-smog movement in the most polluted polish cities. *Economics and Environment, 69*(2), 15. https://ekonomiaisrodowisko.pl/journal/article/view/85

Butt, M. U., Waseef, R. F., & Ahmed, H. (2018). Perception about the factors associated with smog among medical students [Article]. *Biomedica, 34–34*(4). https://www.researchgate.net/publication/358615094

Chen, B., & Kan, H. (2008). Air pollution and population health: A global challenge. *Environmental Health and Preventive Medicine, 13*(2), 94–101. DOI: 10.1007/s12199-007-0018-5 PMID: 19568887

Daly, A., & Zannetti, P. (2007). Air Pollution Modeling – An Overview. In P. Zannetti, D. Al-Ajmi, & S. Al-Rashied (Eds.), *AMBIENT AIR POLLUTION*. The Arab School for Science and Technology (ASST). https://home.iitk.ac.in/~anubha/Modeling.pdf

Fowler, D., Brimblecombe, P., Burrows, J., Heal, M. R., Grennfelt, P., Stevenson, D. S., Jowett, A., Nemitz, E., Coyle, M., Liu, X., Chang, Y., Fuller, G. W., Sutton, M. A., Klimont, Z., Unsworth, M. H., & Vieno, M. (2020). A chronology of global air quality. *Philosophical Transactions. Series A, Mathematical, Physical, and Engineering Sciences, 378*(2183), 20190314. DOI: 10.1098/rsta.2019.0314 PMID: 32981430

Global trends in motor vehicle pollution control: a 2011 update. (2011). In *COMBUSTION ENGINES* (pp. 106–107).

Isha, N., Ali, S., Khalid, A., Naseer, I. A., Raza, H., & Chang, Y. (2024). Bioremediation of smog: Current trends and future perspectives. *Processes (Basel, Switzerland), 12*(10), 2266. DOI: 10.3390/pr12102266

Marlier, M. E., Jina, A. S., Kinney, P. L., & DeFries, R. S. (2016). Extreme air pollution in global megacities. *Current Climate Change Reports, 2*(1), 15–27. DOI: 10.1007/s40641-016-0032-z

Mishra, S. (2017). Is smog innocuous? Air pollution and cardiovascular disease. *Indian Heart Journal, 69*(4), 425–429. DOI: 10.1016/j.ihj.2017.07.016 PMID: 28822504

Rani, B., Singh, U., Chuhan, A. K., Sharma, D., & Maheshwari, R. (2011). Photochemical smog pollution and its mitigation measures. In *Journal of Advanced Scientific Research*. http://www.sciensage.info/jasr

Sierra-vargas, M. P., & Teran, L. M. (2012). Air pollution: Impact and prevention. *Respirology (Carlton, Vic.)*, *17*(7), 1031–1038. DOI: 10.1111/j.1440-1843.2012.02213.x PMID: 22726103

Usman, N. M., Amjad, N. S., & Khan, N. A. (2023). Clearing the Air: Legal Strategies for Combating Smog and Pollution. *Journal of Strategic Policy and Global Affairs*, *04*(01), 15–21. DOI: 10.58669/jspga.v04.i01.02

Wichmann, H. E., Mueller, W., Allhoff, P., Beckmann, M., Bocter, N., Csicsaky, M. J., Jung, M., Molik, B., & Schoeneberg, G. (1989). Health effects during a smog episode in West Germany in 1985. *Environmental Health Perspectives*, *79*, 89–99. DOI: 10.1289/ehp.897989 PMID: 2495934

Wolf, M. J., Esty, D. C., Kim, H., Bell, M. L., Brigham, S., Nortonsmith, Q., Zaharieva, S., Wendling, Z. A., De Sherbinin, A., & Emerson, J. W. (2022). New insights for tracking global and local trends in exposure to air pollutants. *Environmental Science & Technology*, *56*(7), 3984–3996. DOI: 10.1021/acs.est.1c08080 PMID: 35255208

Zhang, J., Liu, Y., Cui, L., Liu, S., Yin, X., & Li, H. (2017). Ambient air pollution, smog episodes and mortality in Jinan, China. *Scientific Reports*, *7*(1), 11209. Advance online publication. DOI: 10.1038/s41598-017-11338-2 PMID: 28894179

ADDITIONAL READING

Haagen-Smit, A. J. (1964). The Control of Air Pollution. *Scientific American*, *210*(1), 24–31. https://www.jstor.org/stable/24935982. DOI: 10.1038/scientificamerican0164-24 PMID: 14088561

Haagen-Smit, A. J. (1970). A Lesson from the Smog Capital of the World. *Proceedings of the National Academy of Sciences of the United States of America*, *67*(2), 887–897. DOI: 10.1073/pnas.67.2.887 PMID: 4943183

Hu, D., & Jiang, J. (2013). A Study of Smog Issues and PM_{2.5} Pollutant Control Strategies in China. *Journal of Environmental Protection*, *04*(07), 746–752. DOI: 10.4236/jep.2013.47086

Wang, Y., Li, M., Wang, L., Wang, H., Zeng, M., Zeng, B., Qiu, F., & Sun, C. (2019). Can remotely delivered electricity really alleviate smog? An assessment of China's use of ultra-high voltage transmission for air pollution prevention and control. *Journal of Cleaner Production*, *242*, 118430. DOI: 10.1016/j.jclepro.2019.118430

Watkins, A., Danilewitz, M., Kusha, M., Massé, S., Urch, B., Quadros, K., Spears, D., Farid, T., & Nanthakumar, K. (2012). Air pollution and arrhythmic risk: The smog is yet to clear. *The Canadian Journal of Cardiology*, *29*(6), 734–741. DOI: 10.1016/j.cjca.2012.09.005 PMID: 23219609

KEY TERMS AND DEFINITIONS

Air Pollution: The existence of dangerous compounds in the air that endanger both the environment and human health.

Air Quality Modelling: Using computer models to forecast the movement and changes of pollutants in the atmosphere

Emission Sources: The places where pollutants are emitted into the atmosphere, including industry, powcr plants, and automobilcs.

Environmental Policy: Laws and rules aimed at lowering pollution and safeguarding public health and air quality.

Health Effects: The detrimental effects that air pollution has on people's health, such as heart and respiratory conditions.

Particulate Matter: Dust and soot are examples of particulate matter, which are tiny solid or liquid airborne particles that can enter the lungs and cause major health issues.

Photochemical SMOG: It is a form of air pollution created when sunlight combines with pollutants such as car exhaust to create ozone and other compounds.

Smog: It is a thick, frequently observable type of air pollution that is produced by a combination of chemical pollutants, smoke, and fog.

Chapter 13

Health Implications of Global Trends in Smog and Air Pollution Control:

Strategies for Cleaner Air and Public Well-Being

Praveen Katiyar
https://orcid.org/0000-0001-5902-8073
CSJM University, Kanpur, India

Shivansu Sachan
https://orcid.org/0000-0002-9893-8018
CSJM University, Kanpur, India

ABSTRACT

Air pollution and smog remain critical global health challenges, contributing significantly to respiratory illnesses, cardiovascular diseases, and premature mortality. Despite advancements in air quality monitoring and control strategies, urbanization, industrialization, and vehicular emissions continue to exacerbate pollution levels in many regions. This paper examines the health implications of global trends in smog and air pollution, highlighting the interconnectedness between environmental degradation and public health outcomes. It further evaluates contemporary strategies for cleaner air, including technological innovations, regulatory frameworks, and sustainable urban planning, as well as their effectiveness in reducing exposure-related health risks. The study underscores the necessity for integrated approaches that combine policy intervention, public awareness, and cross-sectoral collaboration to achieve long-term improvements in air quality and public well-being.

DOI: 10.4018/979-8-3373-2469-2.ch013

INTRODUCTION

Air pollution has emerged as one of the most pressing environmental and public health challenges of the 21st century. According to the *World Health Organization (WHO)*, approximately *7 million premature deaths occur each year due to exposure to polluted air*, making it one of the leading risk factors for global morbidity and mortality (WHO, 2022). Among the various forms of air pollution, *smog*—a dense mixture of smoke and fog—stands out as one of the most dangerous phenomena, particularly in urban and industrial regions. Smog formation is primarily driven by the interaction of pollutants such as nitrogen oxides (NOx), volatile organic compounds (VOCs), and particulate matter (PM) under specific meteorological conditions, often aggravated by temperature inversions.

The implications of air pollution for human health are extensive and profound. It affects multiple physiological systems, particularly the *respiratory and cardiovascular systems*, while also contributing to chronic conditions such as asthma, chronic obstructive pulmonary disease (COPD), and ischemic heart disease. Emerging evidence suggests that air pollution may even have neurological effects, increasing the risk of *cognitive decline and neurodegenerative disorders*. Beyond health, the economic burden associated with air pollution is staggering, encompassing increased healthcare expenditures, productivity losses, and environmental degradation.

The *rapid pace of urbanization and industrialization* has exacerbated air quality concerns globally. Mega-cities in Asia, such as *Delhi, Beijing, and Dhaka*, consistently record *Air Quality Index (AQI)* values that far exceed recommended thresholds. These regions experience prolonged episodes of hazardous smog, primarily during winter months when atmospheric inversions trap pollutants close to the surface. In contrast, urban centers in *North America and Europe* face episodic smog events often associated with traffic congestion and industrial emissions, although stringent regulatory frameworks have mitigated the severity of these occurrences. Rural areas are not immune; practices such as *biomass burning, agricultural residue burning, and natural dust storms* significantly contribute to particulate matter concentrations in the atmosphere, affecting both rural populations and nearby urban settlements.

Compounding this issue is the *climate change–air pollution nexus*, where rising global temperatures and altered atmospheric circulation patterns intensify the formation and persistence of smog. Warmer temperatures accelerate *photochemical reactions*, increasing ground-level ozone concentrations, while changing precipitation patterns influence pollutant dispersion and deposition. This creates a *feedback loop*, where greenhouse gas emissions contribute to climate change, which in turn worsens air quality, leading to additional health and environmental stressors.

Despite increasing awareness of the hazards associated with air pollution, mitigation efforts remain fragmented and uneven across regions. *High-income countries*

(HICs) have achieved significant progress through stringent environmental regulations, technological innovations, and robust enforcement mechanisms. Conversely, *low- and middle-income countries (LMICs)* continue to grapple with limited financial resources, inadequate institutional capacity, and conflicting priorities between economic growth and environmental sustainability. These disparities highlight the need for *globally coordinated strategies* that integrate policy, technology, and community engagement.

This chapter aims to comprehensively examine the *health implications of global trends in smog and air pollution*, evaluate existing control measures, and propose strategic pathways for achieving cleaner air and improved public well-being. The discussion begins with an exploration of *global trends and drivers of air pollution*, followed by an in-depth analysis of its health impacts, including respiratory, cardiovascular, and neurological outcomes. It then addresses the *social and economic costs* associated with air pollution before reviewing contemporary strategies and presenting illustrative case studies of successful interventions. Finally, the chapter identifies persistent challenges in policy implementation and outlines forward-looking recommendations that leverage technological innovation, regulatory harmonization, and community-driven solutions.

The importance of this discourse cannot be overstated. *Air quality management is no longer an environmental issue alone—it is a public health imperative and a socioeconomic priority*. It forms a critical pillar for achieving the *United Nations Sustainable Development Goals (SDGs)*, particularly *SDG 3 (Good Health and Well-being) and SDG 13 (Climate Action)*. Effective air pollution mitigation is essential for protecting ecosystems, reducing healthcare costs, ensuring health equity, and safeguarding the productivity and resilience of future generations.

Air pollution is a *pervasive global challenge*, intrinsically tied to the structural characteristics of industrialization, urbanization, energy consumption, and transportation systems. Despite notable successes in reducing emissions in certain high-income regions, the global scenario remains highly uneven, with *developing economies bearing the brunt of pollution-related health and environmental impacts*. This section explores the prevailing trends, regional variations, key pollutants, underlying drivers, and the compounding influence of climate change on air quality.

KEY POLLUTANTS AND THEIR SOURCES

Air pollution consists of a wide range of harmful substances released into the atmosphere, which have significant implications for human health, ecosystems, and the climate. Among these, particulate matter, nitrogen oxides, volatile organic compounds, and sulfur dioxide are considered the most critical due to their role in smog formation and their severe health impacts.

Particulate Matter (PM)

Particulate matter (PM) refers to tiny particles suspended in the air, which vary in size and chemical composition. The most dangerous forms are $PM_{2.5}$ (particles less than 2.5 micrometers in diameter) and PM_{10} (particles less than 10 micrometers). $PM_{2.5}$ is particularly harmful as it penetrates deep into the alveolar regions of the lungs, leading to cardiovascular and respiratory diseases (WHO, 2021).

Major Sources of PM

1. **Combustion of Fossil Fuels**: Coal-fired power plants, oil refineries, and natural gas processing units emit substantial quantities of fine particles.
2. **Vehicular Emissions**: Internal combustion engines release carbonaceous soot, nitrates, and sulfates that form PM.
3. **Biomass Burning**: Burning of crop residue and wood for cooking contributes heavily to PM emissions, especially in developing regions.
4. **Industrial Processes**: Cement production, metal smelting, and chemical manufacturing release both primary and secondary particles into the atmosphere.

Recent studies suggest that long-term exposure to $PM_{2.5}$ is linked to increased mortality from cardiovascular and respiratory conditions as well as lung cancer (Lelieveld et al., 2019). Global exposure to $PM_{2.5}$ has been estimated to reduce life expectancy by up to 1.8 years in heavily polluted regions such as South Asia (IQAir, 2023).

Nitrogen Oxides (NO_x) and Volatile Organic Compounds (VOCs)

Nitrogen oxides, primarily NO and NO_2, are significant precursors to tropospheric ozone, a major component of photochemical smog. They also contribute to the formation of secondary particulate matter. NO_X emissions originate mainly from:

- **Vehicle Exhaust**: Cars and trucks emit large amounts of nitrogen oxides due to high-temperature combustion.
- **Power Plants**: Fossil fuel-based power generation is a dominant source of NO_X emissions globally.

- **Industrial Solvents and Chemicals**: VOCs from paints, coatings, and cleaning agents react with NO_X in the presence of sunlight to form ozone.

Exposure to high concentrations of ozone and nitrogen oxides leads to respiratory irritation, reduced lung function, and exacerbation of asthma (US EPA, 2022).

Sulfur Dioxide (SO_2)

Sulfur dioxide is primarily emitted from coal combustion in power plants and various industrial processes, including metal smelting and oil refining. While SO_2 itself is an irritant, its greater impact comes from its transformation into *sulfate aerosols*, a major component of fine particulate matter that affects both health and visibility. Additionally, SO_2 contributes to *acid rain*, which damages crops, aquatic ecosystems, and infrastructure (Kampa & Castanas, 2008).

Drivers of Air Pollution

Despite decades of air quality regulation in certain regions, several systemic factors continue to drive air pollution globally. These drivers are deeply embedded in socioeconomic development pathways, making air quality management a complex challenge.

Rapid Urbanization

The *United Nations Department of Economic and Social Affairs (UN DESA)* projects that *urban populations will reach 68% of the global total by 2050*, up from 55% in 2018 (UN DESA, 2018). This rapid urbanization concentrates sources of emissions such as vehicles, power generation, and residential heating in densely populated areas. Urban environments also experience the *urban heat island effect*, which enhances photochemical smog formation.For instance, megacities like Delhi, Mexico City, and Jakarta face severe smog episodes during winter months due to a combination of traffic emissions, industrial activity, and stagnant atmospheric conditions.Economic growth in emerging economies such as India and China remains heavily reliant on *coal-based energy*. Coal combustion accounts for over *40% of global electricity generation* and is the largest single source of both SO_2 and CO_2 emissions (IEA, 2023). While developed nations have made significant progress in transitioning to cleaner energy, industrial growth in low- and middle-income countries often prioritizes economic output over environmental sustainability.

Agricultural Practices

Agricultural emissions are a major contributor to episodic air quality crises, especially in South Asia. The practice of *stubble burning*—burning crop residue after harvest—emits enormous amounts of particulate matter, black carbon, and greenhouse gases. This phenomenon is particularly prevalent in northern India, contributing significantly to Delhi's severe winter smog episodes (Venkataraman et al., 2018).

Similarly, open biomass burning in parts of Africa and Southeast Asia releases large quantities of fine particles and methane, a potent greenhouse gas, exacerbating regional and global air quality concerns.The growth of personal vehicles, especially in rapidly developing nations, contributes significantly to urban air pollution. The lack of robust public transportation systems, coupled with rising incomes, leads to higher private vehicle usage. As a result, nitrogen oxides, carbon monoxide, and particulate emissions from road transport continue to rise globally.While many countries have established air quality standards, enforcement remains weak in several low- and middle-income nations due to *institutional capacity constraints*. Corruption, lack of monitoring infrastructure, and competing policy priorities further undermine compliance.The relationship between climate change and air pollution is *bidirectional* and complex. Air pollutants like black carbon, tropospheric ozone, and methane contribute to climate warming, while climate change alters meteorological conditions that govern pollutant dispersion and chemical transformations. Rising global temperatures accelerate *photochemical reactions*, leading to increased formation of ground-level ozone. Ozone episodes are particularly pronounced during heatwaves, which are becoming more frequent under climate change scenarios (Jacob & Winner, 2009). Elevated ozone levels exacerbate respiratory conditions and contribute to higher mortality rates during extreme heat events.

Drought and Wildfire Events

Drought conditions, increasingly common under climate change, intensify *wildfire frequency and severity*. Wildfires release vast amounts of particulate matter, carbon monoxide, and volatile organic compounds, creating transboundary haze episodes that affect air quality over large regions. For example, the *2019–2020 Australian bushfires* contributed to hazardous $PM_{2.5}$ levels across multiple continents (Johnston et al., 2021).

CHANGING WIND PATTERNS AND POLLUTANT DISPERSION

Climate change is also altering *atmospheric circulation patterns*, which affects the transport and deposition of air pollutants. For instance, weaker winds and temperature inversions can trap pollutants near the surface, leading to prolonged smog episodes in urban valleys such as Los Angeles and Beijing (Seinfeld & Pandis, 2016).

1. Regional Variations in Air Pollution Burden

The burden of air pollution is not uniformly distributed; instead, it is shaped by *geographic, economic, and sociopolitical factors*. The *Global Burden of Disease (GBD) Study* reveals stark disparities in air pollution exposure between developed and developing nations:

- **South and East Asia**: Countries like *India, China, Pakistan, and Bangladesh* experience some of the world's highest concentrations of $PM_{2.5}$, often exceeding WHO guidelines by more than tenfold. Seasonal factors, such as *crop residue burning* in northern India and *coal-based heating* in northern China, amplify pollution during winter months, resulting in recurrent smog crises.
- **Africa**: Rapid urbanization combined with reliance on *biomass fuels for cooking and heating* has elevated indoor and outdoor air pollution levels. In many sub-Saharan countries, the absence of robust monitoring networks obscures the true magnitude of the problem.
- **Latin America**: Urban centers such as *Mexico City* and *São Paulo* grapple with vehicular emissions and industrial activities. Despite improvements in fuel standards, persistent congestion contributes to ozone and particulate pollution.
- **North America and Europe**: These regions have achieved significant air quality improvements through regulatory frameworks like the *Clean Air Act (USA)* and *EU Air Quality Directives*. However, *episodic smog events* still occur, often triggered by traffic congestion and wildfires, as observed in *California* and *Southern Europe* during summer months.

Regional Patterns of Air Pollution

Air pollution manifests differently across geographical regions, shaped by economic structures, industrial practices, population density, and environmental policies.

- **Asia-Pacific**: The Asia-Pacific region, particularly South and East Asia, bears the heaviest burden of air pollution globally. Rapid industrialization, coal-based energy systems, vehicular growth, and agricultural residue burning contribute to hazardous levels of particulate matter (PM2.5 and PM10). Countries like India, Pakistan, and Bangladesh consistently feature among the most polluted nations, with major cities such as Delhi, Lahore, and Dhaka recording Air Quality Index (AQI) values that exceed WHO's safe limits by several folds. Beijing and other Chinese cities, once notorious for smog, have made progress through stringent regulations and technological interventions, although challenges persist due to high energy demand and economic growth pressures.
- **Europe**: European nations have implemented stringent air quality standards through frameworks such as the *European Union Ambient Air Quality Directive*, leading to notable improvements over the past two decades. However, Eastern Europe continues to struggle with coal-dependent energy systems, while Western Europe faces episodic pollution spikes driven by traffic congestion and industrial emissions. London's Ultra-Low Emission Zone (ULEZ) and Germany's promotion of electric mobility exemplify proactive measures adopted to reduce urban smog.
- **North America**: The United States and Canada have achieved significant reductions in major pollutants through the *Clean Air Act* and associated programs. Nevertheless, challenges persist in regions reliant on fossil fuel industries and areas affected by wildfires, which have become increasingly frequent due to climate change. Wildfire smoke contributes significantly to episodic spikes in particulate matter levels, impacting both urban and rural communities.
- **Africa**: Air quality data for Africa remains limited, but emerging evidence suggests worsening pollution trends in major urban centers such as Lagos, Cairo, and Nairobi. Key contributors include unregulated industrial activities, biomass burning, and vehicle emissions from aging fleets. The health burden in Africa is compounded by limited healthcare infrastructure and weak enforcement of environmental regulations.
- **Latin America**: Cities such as São Paulo and Mexico City face persistent smog issues driven by vehicular emissions, industrialization, and geographical factors that trap pollutants. While some progress has been made through public transport modernization and stricter fuel standards, urban growth continues to strain mitigation efforts.

Major Pollutants and Their Sources

Air pollution consists of a complex mix of particulate matter and gaseous pollutants, each with distinct sources and health implications:

- **Particulate Matter (PM2.5 and PM10)**: Fine particles primarily originate from combustion processes, industrial activities, and vehicular emissions. PM2.5 is of particular concern due to its ability to penetrate deep into the respiratory system, posing severe health risks.
- **Nitrogen Oxides (NOx)**: Emitted mainly by motor vehicles and power plants, NOx contributes to ground-level ozone formation and respiratory disorders.
- **Sulfur Dioxide (SO2)**: Produced from coal combustion and industrial processes, SO2 irritates the respiratory tract and contributes to acid rain.
- **Ozone (O3)**: Unlike other pollutants, ozone at ground level is not directly emitted but forms through chemical reactions between NOx and volatile organic compounds (VOCs) in sunlight.
- **Carbon Monoxide (CO)**: Predominantly from incomplete combustion of fuels, CO impairs oxygen delivery in the bloodstream, causing cardiovascular stress.

Drivers of Air Pollution

Several interconnected factors drive global air pollution trends:

- **Industrialization**: Energy-intensive industries in emerging economies rely heavily on coal and fossil fuels, contributing to persistent emissions.
- **Urbanization**: Rapid urban expansion increases energy demand, traffic congestion, and waste generation, amplifying air quality challenges.
- **Transportation**: Motor vehicles remain a significant source of NOx, CO, and PM emissions. The lack of robust public transportation exacerbates reliance on private vehicles in many cities.
- **Agricultural Practices**: Crop residue burning, fertilizer application, and livestock farming emit pollutants that worsen seasonal smog episodes.
- **Climate Change**: Rising temperatures and altered wind patterns influence pollutant dispersion, while increased frequency of wildfires adds to particulate matter concentrations.

CLIMATE CHANGE LINKAGES

Air pollution and climate change share a bidirectional relationship. Many air pollutants, such as black carbon and methane, are also short-lived climate pollutants (SLCPs) that contribute to global warming. Conversely, climate change affects atmospheric conditions, influencing the dispersion and concentration of pollutants. For example, heatwaves and stagnant air masses intensify smog formation, while droughts increase wildfire risks, leading to spikes in particulate pollution. Integrated strategies addressing both climate and air quality objectives are therefore essential for long-term sustainability.

Health Implications of Air Pollution

Air pollution is not merely an environmental concern; it is one of the leading risk factors for global morbidity and mortality. Exposure to pollutants such as particulate matter (PM2.5 and PM10), nitrogen oxides (NOx), sulfur dioxide (SO_2), ozone (O_3), and volatile organic compounds (VOCs) triggers a wide range of health problems, affecting multiple organ systems. According to the *World Health Organization (WHO, 2022)*, approximately *99% of the global population breathes air that exceeds WHO air quality guidelines*, leading to *7 million premature deaths annually*. This section provides a comprehensive overview of the health implications associated with air pollution, covering respiratory, cardiovascular, neurological, and systemic effects, with emphasis on vulnerable populations.

Respiratory Diseases

The respiratory system serves as the primary point of entry for airborne pollutants. Chronic exposure to fine particulate matter (PM2.5) and gaseous pollutants can damage the respiratory tract, impair lung function, and exacerbate pre-existing respiratory conditions.

- **Asthma**: Airborne allergens and irritants such as ozone and nitrogen dioxide are known triggers for asthma attacks. Studies indicate that children living in high-traffic areas are at significantly greater risk of developing asthma and experiencing severe episodes.
- **Chronic Obstructive Pulmonary Disease (COPD)**: Long-term exposure to particulate matter is strongly associated with the development and progression of COPD. This condition leads to persistent airflow limitation, chronic bronchitis, and emphysema, which severely impair quality of life.

- **Lung Cancer**: According to the *International Agency for Research on Cancer (IARC)*, outdoor air pollution, particularly PM2.5, is classified as a Group 1 carcinogen, meaning it is a confirmed cause of lung cancer in humans. Long-term exposure increases cancer risk even among non-smokers.

Cardiovascular Disorders

The relationship between air pollution and cardiovascular health has been extensively documented. Fine particulate matter can penetrate the alveolar barrier and enter the bloodstream, triggering systemic inflammation and oxidative stress, which are key mechanisms underlying cardiovascular diseases.

- **Hypertension and Atherosclerosis**: Chronic exposure to pollutants accelerates arterial stiffening and plaque formation, increasing the risk of hypertension and atherosclerosis.
- **Ischemic Heart Disease and Stroke**: Numerous epidemiological studies have demonstrated a strong link between PM2.5 exposure and ischemic heart disease, myocardial infarction, and stroke. Short-term spikes in air pollution are associated with increased hospital admissions and mortality from cardiovascular events.

Neurological and Cognitive Effects

Emerging research suggests that air pollution adversely affects the central nervous system. Ultrafine particles (less than 0.1 microns) can cross the blood-brain barrier, causing neuroinflammation and oxidative damage.

- **Neurodegenerative Diseases**: Chronic exposure to particulate matter has been linked to higher risks of Alzheimer's disease and Parkinson's disease.
- **Cognitive Decline**: Studies involving older adults indicate that long-term exposure to polluted air correlates with accelerated cognitive decline and an increased incidence of dementia.
- **Children's Neurodevelopment**: Air pollution exposure during pregnancy and early childhood can impair brain development, resulting in reduced cognitive performance and behavioral disorders.

Maternal and Child Health

Pregnant women and children represent particularly vulnerable groups due to physiological and developmental factors.

- **Low Birth Weight and Preterm Birth**: Exposure to high levels of air pollution during pregnancy increases the risk of low birth weight, preterm delivery, and stillbirth.
- **Infant Mortality**: Neonatal deaths have been associated with elevated PM2.5 concentrations, especially in low-income regions where healthcare systems are weak.

Mental Health and Well-being

Recent studies highlight the psychological impact of prolonged exposure to polluted air.

- **Depression and Anxiety**: Chronic exposure to smog has been associated with higher prevalence rates of depression and generalized anxiety disorders.
- **Stress and Quality of Life**: Residents of highly polluted cities report higher stress levels, which can contribute to mental fatigue and decreased overall well-being.

Vulnerable Populations

The health impacts of air pollution are not evenly distributed. Vulnerability is shaped by age, pre-existing health conditions, socioeconomic status, and geographic location.

- **Children**: Developing respiratory systems make children more susceptible to pollutant exposure.
- **Elderly**: Older adults with compromised immune and cardiovascular systems are at heightened risk of severe health outcomes.
- **Low-income Communities**: Poorer populations often live in areas with higher pollution levels and have limited access to healthcare services, exacerbating health inequalities.

GLOBAL BURDEN OF DISEASE

The *Global Burden of Disease (GBD) Study* consistently identifies air pollution as one of the top environmental risk factors worldwide. In 2019, air pollution accounted for *6.7 million deaths globally*, making it the fourth leading cause of death, surpassing risks associated with malnutrition and unsafe water.

Health Impacts Beyond Humans

While the primary focus of this chapter is on human health, it is essential to acknowledge that air pollution also affects animal health and ecosystem stability. Pollutants harm livestock productivity, contaminate food chains, and reduce biodiversity, indirectly influencing human nutrition and livelihoods.

Economic and Social Costs of Air Pollution

Air pollution not only undermines human health but also imposes significant economic and social burdens on societies worldwide. These costs manifest through increased healthcare expenditures, productivity losses, environmental degradation, and socioeconomic disparities. According to a World Bank report (2022), the global economic cost of air pollution is estimated at *over $8 trillion annually*, equivalent to nearly *6.1% of the global GDP*, making it a critical development challenge.

Healthcare Expenditures

The health consequences of air pollution drive substantial healthcare costs, particularly in managing chronic conditions such as asthma, chronic obstructive pulmonary disease (COPD), and cardiovascular disorders. Treatment of these illnesses requires frequent hospitalizations, medication, and long-term care, straining public health systems.

- In *low- and middle-income countries (LMICs)*, where healthcare infrastructure is often inadequate, the costs are disproportionately high relative to national income.
- For instance, in India, air pollution-related diseases cost the economy approximately *$37 billion annually*, representing nearly *1.36% of the GDP* (Lancet Planetary Health, 2020).

Productivity Loss

Air pollution reduces workforce productivity through increased absenteeism, decreased work capacity, and premature mortality. A polluted environment leads to:

- Higher rates of sick leave in urban centers with persistent smog.
- Impaired cognitive function, reducing efficiency in sectors reliant on mental alertness.

- Reduced outdoor labor output due to extreme pollution episodes, particularly in agriculture and construction.

A study by the OECD (2021) suggests that air pollution-related productivity losses account for *up to 1% of GDP annually* in highly polluted economies.

Economic Impact of Premature Deaths

Premature mortality from air pollution has profound economic implications. The *Value of Statistical Life (VSL)* approach used by economists estimates the cost of lost life years and productivity. The World Bank (2020) estimates that *7 million premature deaths annually* translate to trillions of dollars in lost economic value globally.

Impact on Education

Children exposed to polluted environments often suffer from frequent respiratory illnesses, leading to school absenteeism and poor academic performance. Long-term cognitive impacts due to air pollution further impair learning outcomes, perpetuating cycles of poverty and inequality.

Social Inequality and Environmental Justice

Air pollution exacerbates social disparities, disproportionately affecting marginalized populations who often live in areas with higher pollution levels. Low-income communities:

- Reside near industrial zones or congested traffic areas.
- Lack resources to access clean technologies or relocate to less polluted regions.
- Bear a disproportionate health and economic burden, contributing to environmental injustice.

ENVIRONMENTAL COSTS AND GLOBAL STRATEGIES FOR SMOG CONTROL

Environmental Costs of Air Pollution

Air pollution is not merely a health crisis; it also imposes significant environmental costs that cascade into economic and social burdens. These costs stem from widespread ecological damage and the need for adaptive interventions to mitigate environmental degradation. Several critical environmental implications include:

Impact on Agriculture and Crop Yields

Ground-level ozone, a by-product of vehicular and industrial emissions, has a detrimental effect on agricultural productivity. According to studies by the Food and Agriculture Organization (FAO), global crop losses attributable to ozone exposure range between *4% and 15% annually*, with staple crops like wheat, soybeans, and rice being most affected. For instance, in India, estimated wheat yield losses due to tropospheric ozone were approximately *6 million tons per year*, translating to significant economic and food security challenges. Reduced agricultural output not only jeopardizes farmer incomes but also intensifies global hunger and nutritional deficiencies, especially in low-income regions.

Damage to Ecosystems and Biodiversity

Airborne pollutants such as nitrogen oxides (NOx), sulfur dioxide (SO_2), and particulate matter (PM) contribute to soil acidification, water contamination, and forest degradation. Acid rain, primarily caused by SO_2 and NOx emissions, disrupts the pH balance of freshwater ecosystems, leading to the decline of sensitive aquatic species. Forest ecosystems suffer as well; exposure to ozone damages leaf tissues, reduces photosynthesis, and stunts plant growth. The cumulative effect is biodiversity loss, which destabilizes ecosystem services such as carbon sequestration and water purification.

Increased Environmental Restoration and Adaptation Costs

The economic implications extend to restoration projects and climate adaptation initiatives. Governments and local authorities must allocate significant resources to rehabilitate polluted rivers, restore degraded forests, and implement urban greening programs to counteract smog impacts. These efforts require continuous funding, specialized expertise, and long-term monitoring, creating financial strain on national and municipal budgets.

Global and Regional Smog Control Strategies

Addressing smog and air pollution necessitates a comprehensive approach encompassing *regulatory frameworks*, *technological interventions*, and *community-driven initiatives*. While some regions have achieved substantial progress, disparities persist due to varying economic capabilities, governance structures, and societal engagement. This section outlines major global and regional strategies for combating air pollution.

INTERNATIONAL FRAMEWORKS AND GUIDELINES

WHO Air Quality Guidelines

The World Health Organization (WHO) serves as a global authority in setting scientific benchmarks for air quality. The *2021 update of the WHO Air Quality Guidelines* introduced more stringent standards, lowering the recommended annual mean concentration for $PM_{2.5}$ *from 10 µg/m*^{3} *to 5 µg/m*^{3} and for NO_2 *from 40 µg/m*^{3} *to 10 µg/m*^{3}. These reductions were based on epidemiological evidence linking even low-level exposure to increased cardiovascular and respiratory morbidity. However, *compliance remains a global challenge*, especially in low- and middle-income countries (LMICs), where pollutant levels often exceed the new limits by several folds.

Integration with Sustainable Development Goals (SDGs)

Air quality improvement intersects with multiple United Nations Sustainable Development Goals (SDGs). *SDG 3 (Good Health and Well-being)* underscores reducing illness from hazardous chemicals and air contamination. *SDG 11 (Sustainable Cities and Communities)* promotes clean urban environments, while *SDG 13 (Climate Action)* advocates for emissions reduction and resilience strategies. Synergizing air quality initiatives with climate policies amplifies both health and environmental benefits. For example, transitioning to renewable energy not only reduces carbon emissions but also decreases SO_2, NOx, and PM levels, delivering *dual benefits*.

Regulatory Approaches at National and Regional Levels

United States – The Clean Air Act (CAA)

The U.S. *Clean Air Act (CAA)*, enacted in *1970*, remains one of the most robust and influential air quality legislations worldwide. It introduced *National Ambient Air Quality Standards (NAAQS)*, regulating pollutants such as ozone, particulate matter, and sulfur dioxide. Moreover, the Act mandated emission controls for vehicles and established mechanisms for state and federal collaboration. According to the *U.S. Environmental Protection Agency (EPA)*, from 1970 to 2020, emissions of key pollutants decreased by *77%*, while the economy expanded by over *270%*, demonstrating the feasibility of decoupling economic growth from environmental degradation.

European Union – Air Quality Directives

The *European Union (EU)* has enforced strict air quality directives, including the *Ambient Air Quality Directive (2008/50/EC)*. This legislation requires member states to monitor pollutant levels and report data to the European Commission. Complementary initiatives such as the *European Green Deal aim for zero pollution by 2050*, integrating air quality targets with climate neutrality objectives. Notably, *Low Emission Zones (LEZs)* implemented in cities like London, Berlin, and Milan restrict high-pollution vehicles, significantly lowering NO_2 concentrations. For instance, *London's Ultra-Low Emission Zone (ULEZ)* achieved a *44% reduction in roadside* NO_2 within its first year of implementation.

China – Air Pollution Prevention and Control Action Plan

China's *Air Pollution Prevention and Control Action Plan*, launched in *2013*, represents an aggressive nationwide approach to combating smog. Key measures included:

- *Closure of coal-fired power plants* in major urban centers.
- *Vehicle ownership restrictions* in cities like Beijing and Shanghai.
- Expansion of *public transportation networks.*

These interventions yielded measurable success; between 2013 and 2017, $PM_{2.5}$ *concentrations in Beijing declined by 35%*. Despite these gains, China continues to face challenges related to industrial growth and rural coal dependency.

India – National Clean Air Programme (NCAP)

India introduced the *National Clean Air Programme (NCAP)* in *2019*, targeting a *20–30% reduction in* $PM_{2.5}$ *and* PM_{10} *levels by 2024* across 132 cities. Strategies include:

- Promoting *electric vehicle (EV) adoption* under the *FAME II scheme.*
- Banning *agricultural stubble burning.*
- Installing *advanced air filtration systems* in highly polluted zones.

However, rapid urbanization, weak enforcement, and reliance on coal-based energy remain barriers to achieving NCAP targets.

TECHNOLOGICAL INNOVATIONS FOR SMOG REDUCTION

Renewable Energy Transition

Transitioning from coal to *renewable energy sources* such as solar, wind, and hydro power is a cornerstone of air pollution mitigation. Countries like *Germany* and *Denmark* are leaders in renewable integration, aligning air quality and climate objectives. According to the *International Renewable Energy Agency (IRENA)*, achieving global energy transition goals could reduce premature deaths caused by air pollution by *40% by 2050.*

Electrification of Transport

Transportation contributes significantly to urban smog through vehicular emissions. *Electric vehicles (EVs)* offer a sustainable alternative. China dominates global EV markets, with Europe and the U.S. rapidly scaling charging infrastructure. India's *FAME II program* incentivizes EV adoption by offering subsidies to manufacturers and buyers. The cumulative impact of these programs is substantial, reducing tailpipe emissions of NOx and PM.

Smart Air Monitoring Systems

The integration of *Internet of Things (IoT)* and *satellite-based monitoring* enables real-time pollution tracking. Platforms like *AirVisual* and *BreezoMeter* provide accurate, localized air quality data, empowering both policymakers and citizens. Advanced *predictive analytics*, driven by *artificial intelligence (AI)*, allow for early warnings and proactive interventions, reducing smog episodes in major metropolitan areas.

Industrial Emission Controls

Technological solutions for industrial emissions include:

- **Electrostatic Precipitators (ESPs)** for capturing particulate matter.
- **Flue Gas Desulfurization (FGD)** systems to reduce SO_2 emissions.
- **Selective Catalytic Reduction (SCR)** for NOx removal.

These technologies are increasingly mandated in thermal power plants and heavy industries across Asia, Europe, and North America.

URBAN PLANNING AND GREEN INFRASTRUCTURE

Green Belts and Urban Forests

Urban greenery plays a vital role in mitigating air pollution. Trees absorb gaseous pollutants, intercept particulate matter, and cool urban heat islands. Programs like *Mexico City's Green Roof Initiative* encourage residents to install rooftop gardens, reducing smog and enhancing biodiversity in urban landscapes.

Low Emission Zones (LEZs)

European cities have pioneered *Low Emission Zones (LEZs)* to curb vehicular emissions. London's *ULEZ*, for instance, achieved a *44% reduction in NO_2 concentrations* and decreased traffic congestion. These zones not only improve air quality but also encourage the transition toward cleaner transportation alternatives such as EVs and cycling.

COMMUNITY ENGAGEMENT AND BEHAVIORAL CHANGE

Public Awareness Campaigns

Behavioral change is indispensable for sustainable air quality management. Public campaigns raise awareness about practices such as *avoiding open waste burning, reducing vehicle idling, and using cleaner fuels*. Educational programs targeting schools and community groups foster a culture of environmental responsibility.

Citizen Science Initiatives

Innovative approaches like *low-cost air sensors* enable citizens to monitor local air quality and contribute to open-data platforms. These initiatives democratize environmental monitoring, enhance accountability, and promote policy advocacy.

Case Example – Delhi's "Odd-Even" Scheme

In Delhi, the *Odd-Even traffic rationing scheme* restricts car use on alternate days based on license plate numbers. Though temporary, the program successfully reduced vehicular emissions during high-pollution periods, highlighting the effectiveness of behavioral interventions in smog control.

Role of International Cooperation

Cross-border pollution challenges necessitate global cooperation. Examples include:

- **Convention on Long-Range Transboundary Air Pollution (CLRTAP)** in Europe.
- **ASEAN Agreement on Transboundary Haze Pollution** in Southeast Asia.

Such frameworks promote information sharing, joint research, and coordinated policy responses.

6. CASE STUDIES: COMPARATIVE ANALYSIS OF GLOBAL AIR POLLUTION CONTROL EFFORTS

Case studies provide critical insights into the practical challenges and successes of air pollution control strategies across different geographic, socio-economic, and political contexts. This section examines four key regions—Delhi (India), Beijing (China), Los Angeles (USA), and selected European Union (EU) cities—highlighting their historical air pollution patterns, health consequences, interventions, and policy frameworks. These case studies underscore the role of governance, technology, and public participation in shaping air quality outcomes.

6.1 Delhi, India

Air Quality Trends and Sources of Pollution

Delhi consistently ranks among the world's most polluted megacities, with Air Quality Index (AQI) values frequently surpassing the "hazardous" threshold of 400 during winter months (IQAir, 2022). The city's air quality crisis is driven by multiple factors:

- **Vehicular Emissions:** Delhi has over 11 million registered vehicles, contributing approximately 40% of PM2.5 emissions in the city (CPCB, 2022).
- **Industrial Activities:** Small-scale industries and power plants burning coal exacerbate pollution levels.
- **Agricultural Residue Burning:** Seasonal stubble burning in neighboring states (Punjab, Haryana) significantly worsens smog conditions in October–November.
- **Geographic and Meteorological Conditions:** The Indo-Gangetic plains' topography and winter temperature inversion trap pollutants close to the ground, preventing dispersion.

According to the State of Global Air Report (2020), Delhi's annual average PM2.5 concentration exceeds 110 $\mu g/m^3$—more than ten times the WHO guideline of 10 $\mu g/m^3$.

Health Impacts

Delhi's chronic smog episodes have far-reaching public health consequences. Studies estimate that long-term residents experience a reduction in life expectancy by approximately *10 years* (Greenstone & Fan, 2019). The city has witnessed:

- **Respiratory Illnesses:** Increased incidence of asthma, bronchitis, and chronic obstructive pulmonary disease (COPD).
- **Cardiovascular Diseases:** Elevated risk of heart attacks and strokes, linked to prolonged PM exposure.
- **Child Health:** Children exhibit impaired lung development, with a significant rise in pediatric respiratory hospital admissions during high-pollution months (Lal et al., 2021).
- **Maternal and Neonatal Risks:** Air pollution correlates with premature births and low birth weights.

WHO (2021) attributes approximately 54,000 premature deaths annually in Delhi to air pollution-related illnesses.

Governmental and Policy Interventions

The Delhi government has implemented multiple measures to mitigate smog:

1. **Odd-Even Traffic Rationing Scheme:** Vehicles with odd/even-numbered plates operate on alternate days to reduce congestion and emissions.

2. **CNG Transition:** Mandatory conversion of public buses and auto-rickshaws to compressed natural gas (CNG).
3. **Ban on Firecrackers and Construction Dust:** Temporary prohibitions during critical pollution episodes.
4. **Introduction of Electric Buses and Air Purifiers:** Adoption of electric mobility and air filtration towers in high-traffic zones.
5. **Graded Response Action Plan (GRAP):** A tiered intervention system triggered by AQI thresholds.

Limitations and Challenges

Despite these efforts, the interventions are *seasonal and reactive* rather than preventive. Weak enforcement, high vehicular dependency, and socio-political complexities surrounding crop residue burning remain major obstacles. Furthermore, limited inter-state coordination hampers comprehensive regional solutions.

6.2 Beijing, China

Historical Context and Pollution Dynamics

In the early 2010s, Beijing faced global attention for its *"airpocalypse"* events, with PM2.5 levels exceeding 500 $\mu g/m^3$—nearly 20 times the WHO limit (Chen et al., 2018). Key pollution sources included:

- **Coal Combustion:** Coal-fired power plants and industrial boilers accounted for the majority of SO_2 and PM emissions.
- **Vehicular Growth:** Rapid motorization increased NO_X emissions significantly.
- **Urban Expansion:** Dust from construction and lack of vegetation worsened particulate matter levels.

Meteorological conditions, such as low wind speeds and temperature inversions, exacerbated pollution accumulation.

Health and Social Consequences

The health impacts of Beijing's pollution crisis included a surge in respiratory infections, asthma, cardiovascular disorders, and pregnancy complications. Public dissatisfaction with hazardous air conditions escalated into *social unrest and policy pressure*, making air quality a central governance priority.

Control Strategies and Measurable Success

Beijing adopted a *comprehensive, multi-sectoral approach under the Air Pollution Prevention and Control Action Plan (2013–2017)*:

- **Coal Phase-Out:** Closure of coal plants and transition to natural gas and renewable energy.
- **Industrial Relocation:** Moving polluting industries out of urban zones.
- **Vehicle Restrictions:** Implementation of Euro VI-equivalent emission standards and car ownership limits.
- **Green Urbanization:** Large-scale tree planting and development of green belts.

Outcomes

Between 2013 and 2020, Beijing reduced PM2.5 concentrations by *over 35%*, marking one of the most dramatic improvements globally (World Bank, 2021). The city's success demonstrates that *political will, strict enforcement, and energy diversification* are pivotal for sustainable air quality improvements.

6.3 Los Angeles, USA

Historical Background

In the mid-20th century, Los Angeles became infamous for *photochemical smog*, primarily caused by vehicular emissions and industrial activity under sunny climatic conditions conducive to ozone formation. By the 1970s, smog posed severe threats to visibility, human health, and environmental quality.

Policy Interventions and Technological Innovations

The turning point came with the *Clean Air Act (1970)* and the establishment of the *Environmental Protection Agency (EPA)*. California implemented *stringent vehicle emission standards*, compelling automobile manufacturers to introduce *catalytic converters* and low-emission technologies. Additionally:

- **Fuel Reforms:** Phasing out leaded gasoline.
- **Inspection Programs:** Regular vehicle emission checks.
- **Public Transit Investments:** Expansion of metro and bus networks.

Current Scenario

Although Los Angeles continues to experience ozone exceedances due to geography and traffic density, PM2.5 concentrations have declined by over *60%* since 1980 (EPA, 2022). The region exemplifies the effectiveness of *long-term regulatory frameworks, innovation incentives, and monitoring systems*.

6.4 European Union Cities

Historical Lessons

European cities such as *London, Paris, and Berlin* experienced severe air pollution episodes in the mid-20th century, notably the *Great Smog of London (1952)*, which caused over *12,000 excess deaths* (Bell & Davis, 2001). These historical tragedies catalyzed environmental reforms across Europe.

Modern Policy Framework

The European Union adopted *stringent air quality directives* aligned with WHO guidelines, setting strict limits on PM and NO_2. Notable strategies include:

- **Low-Emission Zones (LEZs):** Restricting diesel vehicles in city centers.
- **Public Transport Investments:** Expansion of metro systems, cycling lanes, and pedestrian-friendly infrastructure.
- **Green Urban Planning:** Integration of parks, green roofs, and carbon sinks.
- **Energy Transition:** Commitment to *net-zero emissions by 2050* under the European Green Deal.

Achievements and Persistent Challenges

Many EU cities have achieved significant air quality improvements, yet challenges persist:

- **Diesel Dependency:** Despite bans, older diesel fleets remain prevalent.
- **Urban Hotspots:** Industrial hubs and traffic corridors exhibit recurring exceedances.
- **Climate Policy Integration:** Synchronizing air quality and climate action requires enhanced coordination.

Comparative Insights

- **Governance Matters:** Beijing and Los Angeles illustrate the role of *strong regulatory frameworks and enforcement.*
- **Public Engagement is Crucial:** EU cities demonstrate the impact of community-driven sustainability initiatives.
- **Technological Shifts:** Transitioning to clean energy and electric mobility remains a common success factor.
- **Persistent Challenges:** Rapid urbanization, political resistance, and transboundary pollution complicate solutions in Delhi and other developing regions.

7. CHALLENGES AND IMPLEMENTATION GAPS IN GLOBAL AIR QUALITY MANAGEMENT

Despite growing recognition of the health and environmental risks posed by air pollution, many regions—particularly low- and middle-income countries (LMICs)—face persistent barriers to achieving sustainable air quality improvements. These challenges arise from structural, economic, political, and technological limitations that hinder effective implementation of pollution control strategies. Below is a comprehensive discussion of these critical challenges.

7.1 Regulatory Weakness in Developing Countries

One of the most significant barriers to effective air quality management is the *absence of stringent legal frameworks and weak institutional capacity* in many LMICs. While countries like the United States and members of the European Union have implemented comprehensive air quality standards and monitoring systems, several developing economies operate under outdated or fragmented regulatory regimes (Gulia et al., 2021).

- **Limited Monitoring Infrastructure:** In India, for example, only a fraction of cities have real-time air quality monitoring stations. Many rural and peri-urban areas remain unmonitored, leading to underreporting of pollution levels (CPCB, 2022).
- **Lack of Enforcement:** Even when regulations exist, weak enforcement mechanisms undermine compliance. Industries often operate without adequate emission controls due to corruption, lack of inspections, and insufficient penalties.

- **Judicial and Administrative Delays:** Legal actions against violators frequently face prolonged litigation, reducing the deterrent effect of environmental laws.

The lack of institutional capacity further exacerbates these challenges. Environmental agencies in resource-constrained regions are often underfunded and understaffed, limiting their ability to conduct inspections and enforce compliance effectively.

7.2 Economic Prioritization Over Environmental Protection

Emerging economies often prioritize *rapid industrialization and economic growth* over environmental sustainability. This results in policy trade-offs that favor polluting sectors, such as manufacturing, construction, and coal-based energy production.

- **Industrial Expansion:** In countries like India, China, and parts of Africa, industrial clusters contribute heavily to ambient particulate matter concentrations (World Bank, 2021).
- **Energy Dependence on Fossil Fuels:** Many developing nations still rely on coal for electricity generation due to its affordability and availability. This reliance undermines efforts to reduce sulfur dioxide and particulate matter emissions.
- **Political Pressure:** Policymakers often face competing pressures to deliver employment and economic development, making stringent environmental regulations politically challenging to enforce.

The short-term economic benefits of industrial growth are weighed against the long-term health and social costs of air pollution, often resulting in *reactive rather than preventive measures.*

7.3 Technological Barriers and Financial Constraints

Advanced pollution control technologies—such as catalytic converters, electrostatic precipitators, and real-time monitoring systems—are costly to deploy and maintain. This creates a technological gap between developed and developing nations.

- **High Capital Costs:** Retrofitting existing power plants and industries with emission control equipment requires significant investment.

- **Limited Access to Clean Technologies:** In LMICs, the adoption of electric vehicles (EVs) and renewable energy technologies remains slow due to high upfront costs and inadequate infrastructure.
- **Insufficient Financing Mechanisms:** International climate funds and green financing instruments are insufficient to cover the global demand for clean technology adoption in air quality management.

These economic and technological constraints perpetuate dependence on outdated, polluting technologies, particularly in small-scale industries.

7.4 Transboundary Pollution

Air pollution is a *transboundary problem*, as pollutants travel across borders through atmospheric circulation. This complicates national-level policy responses.

- **Regional Pollution Hotspots:** For example, South Asia experiences widespread air quality deterioration during winter months due to combined emissions from multiple countries (UNEP, 2019).
- **Forest Fires and Agricultural Practices:** Seasonal forest fires in Southeast Asia and stubble burning in the Indo-Gangetic plains contribute to regional haze episodes that affect multiple nations.
- **Weak Cross-Border Governance:** Unlike climate change agreements, there are few binding international treaties for cooperative air quality management.

The lack of regional coordination leads to *policy fragmentation* and limits the effectiveness of unilateral interventions.

7.5 Public Awareness and Behavioral Challenges

Community engagement plays a vital role in the success of air quality interventions. However, *limited public awareness* about the health risks of air pollution and the importance of compliance with regulations remains a persistent challenge.

- **Low Risk Perception:** Many urban residents underestimate the health consequences of chronic exposure to polluted air.
- **Resistance to Behavior Change:** Interventions like vehicle rationing or bans on firecrackers during festivals often face public resistance due to cultural and lifestyle preferences.

- **Insufficient Health Education:** Lack of targeted health campaigns limits the adoption of protective measures, such as using masks during high-pollution episodes.

Public non-cooperation reduces the effectiveness of even well-designed policy interventions.

7.6 Climate Change Feedback Loops

Climate change exacerbates air pollution by altering atmospheric dynamics. Rising global temperatures intensify *heat waves*, increasing ground-level ozone formation. Changes in wind patterns and precipitation reduce pollutant dispersion, while extreme weather events like wildfires contribute additional particulate matter.

This *feedback loop* complicates policy planning because air quality and climate change require integrated strategies. Addressing one without considering the other leads to suboptimal results.

Summary of Challenges:

These gaps underscore the need for *multi-level governance models*, enhanced financing mechanisms, technological transfer programs, and integrated climate-air quality strategies to achieve long-term improvements in global air quality.

8. EMERGING SOLUTIONS AND FUTURE DIRECTIONS FOR CLEANER AIR

The future of global air quality management lies in *innovative, multi-pronged strategies* that combine technology, policy reforms, and community engagement. This section explores cutting-edge solutions and forward-looking strategies that can address existing gaps.

8.1 Technological Innovations in Air Quality Management

Technological advancement is pivotal in monitoring, predicting, and mitigating air pollution. The following innovations represent the next frontier in air quality management:

8.1.1 AI-Driven Air Quality Monitoring

Artificial Intelligence (AI) and Internet of Things (IoT) technologies have revolutionized environmental monitoring. AI-powered models enable:

- **Real-Time Air Quality Forecasting:** Predictive analytics can forecast pollution episodes days in advance, allowing governments to implement early interventions.
- **Data Integration:** AI systems combine satellite imagery, meteorological data, and ground-based monitoring for comprehensive analysis.
- **Low-Cost Sensors:** Deployment of affordable sensors connected via IoT networks allows granular monitoring, even in remote regions.

8.1.2 Electrification of Transportation

Transportation accounts for a significant share of urban air pollution. The rapid deployment of *electric vehicles (EVs)* offers transformative potential:

- **Government Incentives:** Countries like Norway and China lead global EV adoption through tax breaks and subsidies.
- **Charging Infrastructure:** Investments in fast-charging networks are essential for scaling EV usage.
- **Shared Mobility Solutions:** Electric ride-sharing fleets reduce congestion and emissions simultaneously.

8.1.3 Green Infrastructure and Urban Design

Integrating *nature-based solutions* into urban planning mitigates air pollution and urban heat islands:

- **Urban Forests and Green Belts:** Trees act as carbon sinks and particulate matter filters.
- **Vertical Gardens and Green Roofs:** These solutions improve local air quality in densely populated areas.
- **Smart Cities:** Incorporating clean energy, sustainable transport, and efficient waste management into urban design.

8.2 Policy Innovations and Governance Mechanisms

8.2.1 Global Air Quality Governance

Unlike climate change, there is no global treaty dedicated to air pollution control. Future strategies should include:

- **International Agreements:** A framework similar to the Paris Agreement, setting binding targets for air quality improvement.
- **Regional Cooperation:** Mechanisms like ASEAN Haze Agreement provide models for managing transboundary pollution.

8.2.2 Economic Instruments

Market-based tools can incentivize emission reductions:

- **Carbon Pricing and Pollution Taxes:** Charging polluters creates financial incentives for cleaner technologies.
- **Green Subsidies:** Supporting industries that adopt low-emission practices.

8.2.3 Integration with Development Planning

Air quality considerations must be embedded in *urban planning, transportation policy, and industrial strategy* to ensure sustainability.

8.3 Community Engagement and Behavioral Interventions

8.3.1 Citizen Science Initiatives

Low-cost sensor networks allow citizens to monitor local air quality, empowering communities to demand policy action.

8.3.2 Public Health Campaigns

Awareness campaigns can promote:

- **Behavioral Changes:** Reducing private vehicle use, avoiding open burning.
- **Protective Measures:** Encouraging the use of N95 masks and indoor air purifiers during pollution episodes.

8.4 Integration with Climate Action

Air pollution and climate change share common sources (fossil fuel combustion), creating opportunities for *co-benefit strategies*:

- **Renewable Energy Transition:** Wind, solar, and hydropower reduce greenhouse gas and pollutant emissions simultaneously.
- **Energy Efficiency:** Improving industrial and residential energy efficiency lowers both carbon and particulate emissions.
- **Nature-Based Solutions:** Reforestation mitigates carbon dioxide and acts as a pollutant sink.

9. CONCLUSION

Air pollution remains a *pressing global challenge*, threatening public health, economic stability, and environmental sustainability. While success stories from *Beijing, Los Angeles, and European cities* illustrate that *comprehensive policy frameworks, technological innovations, and community participation* can significantly improve air quality, many regions still struggle with regulatory gaps, financial constraints, and transboundary challenges.Looking forward, *integrated governance*, leveraging *emerging technologies*, and aligning air quality management with *climate goals* are essential. Cleaner air is not only an environmental necessity but a *public health imperative*, requiring collective action at *local, national, and global levels.* Achieving this vision is fundamental to meeting the *United Nations Sustainable Development Goals*, safeguarding human well-being, and ensuring a sustainable future for generations to come. Air pollution represents an existential challenge at the intersection of environmental sustainability, public health, and economic development. Despite decades of scientific research, technological innovation, and policy interventions, the global community continues to grapple with uneven progress and persistent disparities. While *high-income countries* have demonstrated the feasibility of significant air quality improvements, *developing economies* remain locked in a struggle between industrial growth and environmental stewardship.

The health burden of air pollution—reflected in millions of premature deaths, widespread respiratory and cardiovascular diseases, and rising healthcare costs—underscores the urgency of adopting *integrated, multi-sectoral strategies*. Addressing this crisis demands:

- **Stronger regulatory frameworks**, tailored to national contexts but aligned with global best practices.

- **Technological innovations**, such as renewable energy deployment, electric mobility, and advanced air quality monitoring.
- **Community engagement and behavioral change**, ensuring that citizens become active participants in pollution control efforts.
- **Synergies between air quality management and climate action**, leveraging co-benefits to maximize impact.

Air pollution is no longer an isolated environmental issue—it is a *global health emergency and an economic risk multiplier*. To achieve cleaner air and safeguard public well-being, the international community must embrace *collaborative governance*, *policy harmonization*, and *equitable technology transfer mechanisms*. The path forward lies in *holistic solutions* that integrate science, policy, and society in a shared commitment to sustainability.

REFERENCES

Bell, M. L., & Davis, D. L. (2001). Reassessment of the lethal London fog of 1952: Novel indicators of acute and chronic consequences of acute exposure to air pollution. *Environmental Health Perspectives*, *109*(Supplement 3), 389–394. DOI: 10.1289/ehp.01109s3389 PMID: 11427388

Central Pollution Control Board (CPCB). (2022). *National ambient air quality status and trends in India – 2022*. Government of India.

Chen, Z., Wang, J.-N., Ma, G.-X., & Zhang, Y.-S. (2018). China tackles the health effects of air pollution. *Lancet*, *382*(9909), 1959–1960. DOI: 10.1016/S0140-6736(13)62064-4 PMID: 24332069

Greenstone, M., & Fan, C. (2019). *Clean air and its impact on life expectancy in India.* Energy Policy Institute at the University of Chicago (EPIC).

Gulia, S., Nagendra, S. M. S., Khare, M., & Khanna, I. (2021). Urban air quality management—A review. *Atmospheric Environment*, *246*, 118093. DOI: 10.1016/j.atmosenv.2020.118093

International Energy Agency (IEA). (2023). *World energy outlook 2023*. IEA.

IQAir. (2022). *World air quality report 2022.* IQAir.

IQAir. (2023). *World air quality report 2023.* IQAir.

Jacob, D. J., & Winner, D. A. (2009). Effect of climate change on air quality. *Atmospheric Environment*, *43*(1), 51–63. DOI: 10.1016/j.atmosenv.2008.09.051

Johnston, F. H., Borchers-Arriagada, N., Morgan, G. G., Jalaludin, B., Palmer, A. J., Williamson, G. J., & Bowman, D. M. J. S. (2021). Unprecedented health costs of smoke-related PM2.5 from the 2019–2020 Australian bushfires. *Nature Sustainability*, *4*(1), 42–47. DOI: 10.1038/s41893-020-00610-5

Kampa, M., & Castanas, E. (2008). Human health effects of air pollution. *Environmental Pollution*, *151*(2), 362–367. DOI: 10.1016/j.envpol.2007.06.012 PMID: 17646040

Lal, R., Kumar, P., & Banerjee, T. (2021). Air pollution exposure and children's health in India: A review. *Environmental Research*, *194*, 110700. DOI: 10.1016/j.envres.2020.110700

Lelieveld, J., Klingmüller, K., Pozzer, A., Pöschl, U., Fnais, M., Daiber, A., & Münzel, T. (2019). Cardiovascular disease burden from ambient air pollution in Europe reassessed using novel hazard ratio functions. *European Heart Journal, 40*(20), 1590–1596. DOI: 10.1093/eurheartj/ehz135 PMID: 30860255

Seinfeld, J. H., & Pandis, S. N. (2016). *Atmospheric chemistry and physics: From air pollution to climate change* (3rd ed.). Wiley.

United Nations Department of Economic and Social Affairs (UN DESA). (2018). *World urbanization prospects: The 2018 revision.* UN.

United Nations Environment Programme (UNEP). (2019). *Air pollution in Asia and the Pacific: Science-based solutions*. UNEP.

United States Environmental Protection Agency (US EPA). (2022). *Integrated science assessment (ISA) for oxides of nitrogen, oxides of sulfur, and particulate matter ecological criteria.* EPA.

Venkataraman, C., Brauer, M., Tibrewal, K., Sadavarte, P., Mahtta, R., Zhang, Q., & Wang, Y. (2018). Source influence on emission pathways and ambient PM2.5 pollution over India (2015–2050). *Atmospheric Chemistry and Physics, 18*(11), 8017–8039. DOI: 10.5194/acp-18-8017-2018 PMID: 33679902

World Bank. (2020). *The cost of air pollution: Strengthening the economic case for action*. World Bank Group.

World Bank. (2021). *Air pollution and development: Strengthening policy responses in developing countries*. World Bank Group.

World Health Organization (WHO). (2021). *WHO global air quality guidelines: Particulate matter (PM2.5 and PM10), ozone, nitrogen dioxide, sulfur dioxide, and carbon monoxide*. WHO.

World Health Organization (WHO). (2022). *Air pollution and health: Summary report*. WHO.

Chapter 14

The Impact of Corporate Sustainability Performance on Cost of Debt:

Is Debt Pricing a Pathway to Achieving Long-Term Sustainability?

Witness Siwela

https://orcid.org/0009-0000-4414-2332

University of South Africa, South Africa

ABSTRACT

South Africa is arguably one of the major emitters experiencing high levels of air pollution. The blessing of natural ground resources come with a price as major mining activities contribute to seasonal air pollution and smog issues. The primary objective of the study is to investigate if corporate sustainability performance influences debt pricing. The study employs quantitative research and adopts a positivist perspective. Archival data were employed as extracted from companies constituted in the FTSE/JSE Responsible Investment Index selected primarily for their environmental impact. A short panel data set of 19 FTSE/JSE Responsible Investment Index companies is employed for statistical analysis. The chapter adopts first differenced econometric models. The results posit that financiers such as banks are becoming more sensitive to corporate sustainability performance. Limitation

DOI: 10.4018/979-8-3373-2469-2.ch014

of this study manifest itself in the sampled companies that are environmentally responsible. Future, studies may expand the scope of this study to include companies that are not part of the FTSE/JSE RII.

1. INTRODUCTION AND BACKGROUND

Smog is a significant concern in the Republic of South Africa. In the South African context, smog is driven by air pollution especially in large Cities such as Johannesburg. Besides, South Africa is arguably one of the major emitters experiencing high levels of air pollution which often exceeds the World Health Organisation's recommended standards (Kyaw, Olugbode & Petracci, 2021). Moreover, Steyn (2014) postulates that the blessing of natural ground resources come with a price as major mining activities contribute to seasonal smog issues which poses a health threat. This study seeks to investigate if corporate sustainability performance influences cost of debt within the context of green financing. Thus, corporate sustainability performance in this study is represented by key sustainability indicators such as water consumption, energy consumption and carbon emissions, which are at the centre of smog formation. The findings of the study render insights in the South African context, if debt pricing plays a role in mitigating smog and air pollution issues.

Africa is not an exception in as far as calamities brought by the climate crisis are concerned. Africa exist as an integral part of the global village. Thus, Africa receives its fair share of scrutiny as far as pushing the Environmental, Social and Governance (ESG) agenda is concerned (Hector, 2021). Hector (2021) argues that in countries where ESG agenda is fostered as an imperative, smog formation and air pollution related problems are supressed. Undoubtedly, Africa has made strides in pushing for sustainability reporting to ensure that companies consider the triple bottom line to consolidate the elements of ESG in reporting. Bala and Ibrahim (2022) posit that the nexus of the global village has been about pushing companies to formulate ESG strategies in conjunction with initiatives of the UN Global Compact and European Union.

In Nigeria, Bala and Ibrahim (2022) examined the extent to which ESG imperatives drive the accounting performance of companies listed on the Nigerian Stock Exchange (NSE). Results show that the environmental and governance dimensions have no impact on the accounting performance of these companies. This indicates that advocacy groups and various lobby groups still have much work to do to conscientising all players about the importance of embracing ESG practices. Furthermore, Bala and Ibrahim (2022) suggested that only the social dimension positively affect the accounting performance of those listed companies. This is evidence that communities are also viewed as stakeholders that value ESG practices of those listed companies. Consequently, the study recommended a policy shift to favour the governance and the environmental imperatives to promote good practices in Nigeria amidst growing calls to deal with environmental concerns.

While the push is on for companies to adopt ESG practices and assist in combating the climate crisis, the climate investment challenge remains a topical issue. It will take green investments for the world to realise a rapid low-carbon transition and ultimately the goals of the Paris Agreement. Ameli, Dessens, Winning, Cronin, Chenet, Drummond, Calzadilla, Anandarajah and Grubb (2021) contends that the geographical distribution of low carbon finance is unequal, with Africa on the receiving end of lack of access to such funding. This means that regions that are developed, are by far the largest recipients, while developing economies in Africa are allocated a small portion. These imbalances further exacerbate environmental problems such as smog and air pollution in the Southern hemisphere, with South Africa included. However, where green finances are present, financiers have leverage somewhat to punish the borrowers who exhibit poor environmental performance in a bid to control further adverse environmental impact.

In tandem to these assertions, Ameli et al. (2021) revealed how modelled decarbonisation pathways differ substantially between different regions across the globe. Thus, it is evident that while finance is crucial for the anticipated green energy transition, there is uneven access to lower cost of finance with developing markets like Africa remaining disadvantaged. Moreover, Salisu, Omoke and Fadiya (2022) comment that climate related funding is chronically insufficient, hence the scramble which leaves poor developing economies like Africa on a climate investment trap. Thus, Ameli et al. (2021) suggest that policy interventions will play a crucial role in putting disadvantaged regions on a trajectory to reach net zero emissions within the stipulated period of time if not earlier.

Klusak, Agarwala, Burke, Kraemer and Mohaddes (2021) speculate that financial decision makers do not have adequate information to make sound decisions in as far as greening the global financial system is concerned. However, it is evident that financial resources must be distributed tactfully across regions to curb uncontrollable levels of climate change. These financial resources assist in curbing environmental issues in that capital funders can cascade such resources on strict terms that pushes companies to comply with ESG demands. Thus, such compliance yield results as environmental degradation, air pollution and smog formation begins to dwindle. In a global study which incorporated African countries, Klusak et al. (2021) simulated climate change effect on sovereign credit ratings for 108 countries. The study revealed evidence of climate-induced sovereign downgrades. This means that poor performance at country level with regards to the climate crisis affects the country's creditworthiness. Consequently, such countries face challenges in accessing finances at affordable interest rates. The current study will thus show, at country level, if debt pricing plays a significant role in curbing the environmental impact of companies operating in South Africa.

In one study, Buhr, Murinde, Donovan, Pullin, Kling, Volz and Lo, (2018) revealed several insightful findings on climate change and its effect on the cost of debt with particular focus on developing countries. Firstly, the study suggests that climate vulnerability has increased the cost of debt in developing countries. Secondly, developing countries have already incurred substantial amounts towards interest payments both in the private and public sectors, and these additional interest payments which are attributable to climate vulnerability are expected to increase in the foreseeable future (Buhr et al., 2018). Thirdly, Buhr et al. (2018) further posited that only investments in social preparedness can mitigate partly, the impact of climate vulnerability on sovereign interest rates given that such investments can increase the countries' economic and social resilience. Thus, the current study seeks to empirically establish if this phenomenon manifest itself at prime lending rates of South African financiers.

Notwithstanding a few studies that have produced negative and positive relationships between environmental performance and cost of debt financing, some studies have taken a neutral stance in explaining the association between these variables. In one such study, Bussman, Tanda and Yu (2022) investigated the role of ESG disclosure and its behaviour on the cost of capital. It should be noted that cost of capital comprises two components namely cost of equity and cost of debt. Thus, the two components represent the cost of the two streams of finance for companies, i.e. debt finance and equity finance. Bussman, Tanda and Yu (2022) contends that while debt pricing is paramount in resolving companies' environmental impact, government policy and regulation has to be deliberate in targeting companies that are perceived to be a nuisance in the environment. Furthermore, some studies have provided evidence in favour of the importance of ESG on the perceived riskiness of investment, either by equity investors or capital lenders (Sultana, Zainal & Zulkifli, 2017; Kozak, 2021; Kuo et al., 2021; Xu et al., 2021; Kordsachia, Focke & Velte, 2022). However, some contrasting results are emerging from studies that have considered different countries. Bussman, Tanda and Yu (2022) claims that country variables intertwined with some key fundamental variables other ESG elements maintain their crucial role in predicting cost of capital. Thus, while ESG factors are not key drivers of cost of capital, such variables are becoming more imperative and some features were found to have a role to play in the prediction of cost of capital for large companies listed in the stock exchange.

2. THE THEORETICAL FOUNDATIONS OF THE STUDY

The study focusses on the nexus between companies and capital lenders of money within the context of growing environmental concerns, particularly air pollution and smog formation in South Africa. Capital lenders of money or financiers are crucial stakeholders for business especially in the advent of green financing where finance injections towards the greening agenda are peaking pace globally. A mutually rewarding relationship between companies and banks can be leveraged in dealing with a collage of environmental concerns. As such, this study finds footing on the stakeholders' theory which stipulates that the success of the company depends on the management of the nexus between them and other stakeholders (Lakhani & Herbert, 2022). Thus, stakeholders' theory indicates that companies ought to respond to the needs and expectations of stakeholders. Such responses should be in the form of exhibiting their performance, both financial and nonfinancial, through sound disclosures following the guidelines set out by the Global Reporting Initiative (GRI) (Fuhrmann *et al.*, 2017). According to Raimo *et al.*, (2021) stakeholders represent groups or an entity that has vested interest in the activities of the business. Stakeholders' support is perceived to be indispensable for the existence of the company. According to the normative view of the stakeholders' theory, stakeholders have a right to be treated with fairness in the bi-directional relations they have with the company. However, the status quo depicts that various external stakeholders are on the receiving end given the worrisome levels of environmental degradation, air pollution and smog formation. Through the lances of the stakeholder theory, the current study agrees that financiers as a crucial stakeholder that provides capital to the business can leverage that power to champion companies towards a path of long-term sustainability.

3. AIR POLLUTION CRISIS IN SOUTH AFRICA

Africa is developing at a faster pace than before, with projections indicating that by the year 2060 about 65 percent of the population would have migrated to live in urban cities (Muttakin *et al.*, 2020). Notwithstanding the economic and social benefits that may emanate from this phenomenon, increased urbanisation comes at a price when it comes to environmental pollution. Most recent reports pinpoints that South Africa is a significant polluter on the African continent (Lakhani & Herbert, 2022). The country is seized with industries such as mining that releases hazardous pollutants into the air while performing their economic activities. Moikwatlhai and Yasseen (2019) comments that while there is economic beneficiation on these minerals that are extracted and exported, the downside things are increased environmental risks and human health crisis. Thus, air pollution and smog formation remains a challenge in areas like the Mpumalanga province and densely populated cities like Johannesburg. In 2019 alone, the country recorded 25 800 immature deaths which were linked to air pollution to confirm this phenomenon as the second largest health risk (Dube & Maroun, 2017). Consequently, the effects of air pollution in the country are long-lasting given reduced productivity and loss of foreign direct investment (FDI). Moreover, the tourism sector also suffers dwindling revenue generation given land disturbance and general environmental degradation emanating company operations in the country.

Besides, South Africa is ranked 12th in the list of largest emitters of greenhouse gases in the world (Dubihlela & Ngxukumeshe, 2016). The major culprit for such a performance at a global stage is the energy sector. Drawing the statistics stand-point, South Africa houses six out of 10 hotspots of the largest nitrogen dioxide emissions globally. Additionally, in the African continent, South Africa hosts nine out of 10 point sources of the continent's largest nitrogen dioxide emissions. Moreover, the country also has the largest sulphur dioxide emission hotspots. Dilotsotlhe and Duh (2021) postulates that the population density in the big cities such as Johannesburg has been a culprit to black carbon emissions emanating from residential fuel combustion for cooking and heating. At a micro level, such practices of air pollution cause immediate health impact which further strains an already dilapidated health care system in the country. Literature in the public domain shows that the production of sulphur dioxide and nitrogen dioxide from the high-temperature combustion of fossil fuel have caused serious illnesses and deaths in some cases (Marx & van Dyk, 2011).

Apart from the mining industry, another culprit in South African is the transport industry where in the City of Johannesburg alone there is a record number of over five million cars releasing hazardous exhaust fumes and gases into the air every day. To add to this problem is the increased use of private vehicles as opposed to cultured use of public transport. As a result, massive traffic congestions on the roads of cities such Johannesburg does not make air pollution any better resulting in high levels of emissions that are associated with traffic volumes on the roads. Furthermore, the industrial hubs in the Mpumalanga province as well as Vaal Triangle also adds to the smog and air pollution related problems given that these industries rely on coal-fired power stations.

3.1 Sustainability Concerns and Cost of Debt From Prior Literature

Prior studies have demonstrated that environmental performance impact on debt pricing in the context of increasing interest toward ESG activities. For instance, a Nigerian study analysed the effect of CSR on cost of capital for listed companies in the Nigerian Securities Exchange (NSE) (Ofogbe, Nnamani, Anisiuba & Ezuwore-Obodoekwe, 2021). The study used cost of equity and cost of debt as the measure for cost of capital given that these are two main components that make up the capital structure. Ofogbe et al. (2021) revealed a significant correlation between cost of equity and CSR disclosure. Moreover, despite differences in the individual performance of the ESG dimensions, the study provided strong evidence that favourable environmental performance leads to easy access to capital at reduced interest rate (Ofogbe et al., 2021). The study is plausible for spurring those companies that neglect ESG activities to adopt these as part of their daily operations for their own good. However, Ofogbe et al. (2021) did not use extensive data to measure companies' ESG imperatives, as such the amount of data which can be accessed from the IARs or KLD Index was ignored.

To complement these findings in Nigeria, Abba, Said and Mahat (2017) examined performance effects on costs of capital structure financing. While capital structure financing involves equity finance, we note findings postulated for the cost of debt financing. Following a statistical analysis of data obtained from companies listed on the NSE, results demonstrated that negative effects exist between environmental performance and costs of capital structure financing (Abba et al., 2017). This means that favourable environmental performance reduces the cost of not only equity, but debt financing as well. However, the study employed cross-sectional data which is deemed unstable for a study of this magnitude. Thus, current studies have headroom to employ panel data for a longitudinal period to arrive at more accurate results.

In support of these findings, Ali and Abdel ghaffar (2022) investigated the effect of CSR practices on the debt finance of Egyptian companies. Using an extensive sample of 120 companies that are listed on the Egyptian Exchange, findings indicated that CSR activities have a significant negative association with debt finance (Ali et al., 2022). Moreover, this effect was found to be more pronounced in companies that have high conservatism. The negative relationship entails easy access to debt finance by companies that have high CSR ratings. Furthermore, easy access to finance implies that companies obtain finance at reduced cost because of their green practices. Though insightful the study may be, the study failed to investigate individual dimensions of ESG, with poor quality of ESG disclosure due to lack of data.

According to (Kiron, 2012), the advent of IR as advocated by its proponents brings along IR incentives that are of great benefit to the reporting companies. In one study, the researcher examined whether IR is in anyway associated with cost of debt financing and interrogated if it moderates the relationship between reporting quality and cost of debt (Muttakin, Mihret, Lemma & Khan, 2020: 517). In conjunction with Ali and Abdel ghaffar's (2022) approach, the researchers utilised companies that are listed on the JSE for the period 2009 to 2015 to analyse observations for the purposes of statistical analysis.

Similarly, (Ofogbe et al., 2021; Ali & Abdel ghaffar, 2022), Muttakin et al. (2020) demonstrated that companies who produce IRs tend to benefit from reduced banks' loan prices vis a vis those companies that have not embraced IR. Moreover, the researchers report an inverse association between the quality of the reports and the cost of debt financing. Thus, capital lenders find information provided to IRs to be value relevant when debt contracting. The study is commended for accentuating, beyond financial data, the need for ESG related nonfinancial information when making economic decisions.

In tandem to the findings by Buhr et al. (2018), Kling, Volz, Murinde and Ayas (2021) cascaded the study to firm-level by performing a systematic investigation of the impact of climate vulnerability on companies' cost of capital and access to finance with a focus on the developing countries. Kling et al. (2021) demonstrated that companies with the highest exposure to climate risks are financially constrained in as far as debt funding is concerned. Moreover, the study posited that climate vulnerability heighten the cost of debt financing. Jung, Herbohn and Clarkson (2015) extended the above findings by examining whether capital lenders of money integrate companies' exposure to carbon related risks into lending decisions through the cost of debt. Moreover, the study sought to understand if demonstrating awareness of their carbon risk could help mitigate debt penalties. Findings indicate that there is a positive relationship between carbon risk and cost of debt financing for companies that do not comply with the carbon disclosure project (Jung, Herbohn & Clarkson, 2018). This entails that these companies pay more to access finance while firms that are compliant with the carbon disclosure project exhibit a negative relationship.

Though informative, Jung, Herbohn and Clarkson's (2018) study acknowledges criticisms levelled against it, thereby opening opportunities for new research areas for the current study. First, the study did not consider other contractual mechanisms for debt contracting. Secondly, the empirical proxy utilised for cost of debt is questionable, current studies may employ proxies that may render more conclusive results. However, despite these criticisms, this robust is applauded for shedding light on the importance of carbon awareness as a strategy for firms who are high emitters.

3.2 Adverse Impact of Sustainability Performance on Cost of Debt From Prior Studies

Several African studies demonstrate that environmental performance does not influence cost of debt financing. In one such study in Ghana, Boachie and Tetteh (2021) examined the relationship between CSR disclosure and cost of debt financing in an attempt to determine if companies with higher CSR disclosure scores have reduced cost of debt financing. The study is of relevance to this research because of the variables being investigated and population employed, that is, listed companies on the Ghana Stock Exchange (GSE). Therefore, if the methodology employed is concrete, the results may be generalisable to those companies listed on the JSE as investigated in the current study.

Boachie and Tetteh (2021) found that the cost of debt finance is positively associated with the CSR disclosure scores. Thus, these findings rejected the anticipated hypothesis which was based on prior studies that have shown a negative relationship suggesting that high CSR disclosure score lowers cost of debt finance (Ofogbe et al., 2021; Ali & Abdel ghaffar, 2022). In contrast to this assertion, a positive association suggests that there is no inverse relationship whatsoever between the variables being studied. This means that higher CSR disclosure score does not translate into reduced cost of debt. Moreover, results demonstrated that the levels of long-term debt increased proportionately with CSR disclosures in those high risk industries (Boachie & Tetteh, 2021).

Although plausible for focusing on the little explored relationship between these variables, Boachie and Tetteh's (2021) study is criticised for a narrowed sample which was only limited to three industries in Ghana. Besides, some control variables were neglected which renders the results of the study inconclusive. Additionally, the study excluded the quality dimension of the CSR information. As a result, the quality of the results is compromised.

In a recent landmark study which utilised company-level data from 51 countries, Srivastava, Sampath and Gopalakrishnan (2022) examined if ESG is the key to unlock debt financing during the period of the Covid 19 crisis. The study emphasised the importance of maintaining good relations with different stakeholders. Findings pinpointed that companies with greater stakeholder engagement obtained higher debt financing during the Covid 19 crisis (Srivastava et al., 2022). Moreover, higher debt financing was found to be greatly facilitated by stakeholder engagement especially in emerging economies. Thus, the ability to obtain finance and negotiate for cheaper cost of debt was not necessarily influenced by ESG practices, rather it was influenced by the companies' stakeholder engagement practices. Similarly, Azmi, Hassan, Houston and Karim (2021) examined the relationship between ESG practices and bank value using a sample of 251 banks in the emerging economies. The results demonstrated that there is no relationship between ESG practices and bank value indicating that low levels of ESG practices positively affect bank value (Azmi et al., 2021). Moreover, while ESG practices were positively linked to both cash flows and efficiency, these showed a negative relationship with the cost of equity with no effect whatsoever on the cost of debt financing.

4. METHODOLOGICAL APPROACH AND DATA ANALYSIS

According to Meissner, Creswell, Klassen, Ann Plano, Smith and Katherine (2011), a researcher must set out the philosophical assumptions underlying a study. In this context, the researcher adopts positivism to frame the study (Morgan, 2007). This paradigm is preferred because of its association with quantitative techniques, where variables are empirically tested through observation and measurement (Crowe, Cresswell, Robertson, Huby, Avery & Sheikh, 2011). As a philosophical stance of the natural scientist, positivism focuses on observable and measurable facts (Saunders & Lewis, 2018). The quantitative nature of the positivist paradigm therefore makes it suitable for this study. According to Groenewald (2004), quantitative research is used when researchers are interested in testing the relationship between the variables. This approach is suitable for this study, as it employs a deductive approach to explain the relationship between two or more variables using numerical data and statistical methods (Mertler & Reinhart, 2016; Saunders & Lewis, 2018). Moreover, given the nature of the current study, the researcher adopted a multiple case study design to investigate the underlying phenomenon in the study. A multiple case study approach allows the researcher to select a number of cases to enhance analytical generalisation and make comparisons across several cases (Crowe et al., 2011). This design is deemed a suitable fit for the current study that seeks to scrutinise the cases of companies in the FTSE/JSE Responsible Investment Index.

4.1 Population, Sampling Technique and Research Sample

The population for the current study constitute the aggregate of all the known objects, members or subjects that conform to the set of specifications (Leedy & Ormorod, 2010; Kitsikopoulos, Schwaibold & Taylor, 2018). The JSE introduced the Socially Responsible Index (SRI) in 2004, comprising 57 companies, following the introduction of sustainability initiatives globally and the King Code in South Africa (Greyvenstein, 2010). Companies appearing in the JSE All Share Index were reviewed annually against the ESG concerns for inclusion in the SRI. The SRI grew to 82 companies by the end of December 2014 which marked the last review under the SRI banner. However, SRI was subsequently replaced by the FTSE/JSE Responsible Investment Index (RII) which was launched in October 2015 to continually promote sustainable development and foster good corporate citizenship. Therefore, the population of this current study consist of all companies appearing on the FTSE/JSE RII from 2016 to 2021. This implies that the population will comprise of 60 companies listed on the FTSE/JSE RII as of 28 June 2021. Furthermore, Hui, Foxcroft, Richardson and Macfadyen (2011) refer to a sample is a subsection of a population chosen to participate in the study. Thus, the current study employed judgemental sampling technique. This technique is a non-probability sampling method employed when the researcher's judgement is used to select elements for the sample based on a range of premises (Etikan, 2016). The researcher collected environmental performance data of companies that have reported the environmental performance indicators consistently for a period of six years. Consequently, the sample comprises 19 companies drawn from the 60 FTSE/JSE SRI in the 2021 fiscal year. This represents 32% of the population, which is deemed acceptable for the study given the environmental impact and reporting practices of companies sampled purposively. The table below shows the list of companies selected to form part of the sample for this study.

Table 1. Sampled companies and their respective industry sectors19

Firm	Sector
African Rainbow Minerals Ltd	Mining and Resources
Anglo American	Mining and Resources
Anglo American Platinum	Mining and Resources
Anglogold Ashanti	Mining and Resources
Anheuser-Busch InBev	Industrial and Consumption
Barloworld	Industrial and Consumption
BHP Group Plc	Mining and Resources
British American Tobacco PLC	Industrial and Consumption
Exxaro Resources	Mining and Resources
Glencore	Mining and Resources
Gold Fields	Mining and Resources
Harmony Gold Mining company Ltd	Mining and Resources
Impala Platinum Hlds	Mining and Resources
Kumba Iron Ore	Mining and Resources
Mondi Plc	Industrial and Consumption
Remgro	Industrial and Consumption
Sappi	Industrial and Consumption
Sasol	Industrial and Consumption
Sibanye Stillwater	Mining and Resources

Source: Aurthor's compilation, 2021

In Table 1 above, a clear computation of companies sampled is presented to show their respective industry sectors. Thus, companies that are included in the sample are mainly those in mining and resources as well as industrial and consumption. Most of these companies are part of the JSE main board when ranked according to market capitalisation. The companies are worthy utilising in the sample considering the impact they have on the environment. Furthermore, the inclusion of such companies guarantees more conclusive results given the nature of the explanatory variables utilised in the study.

4.2 Research Data and Collection Procedure

Secondary data are data that have already been collected from other sources and are readily available for use by the researcher. Boslaugh (2009) argues that while this data is collected for a purpose other than the research at hand, it helps the researcher to have a better understanding of a phenomena. Secondary data is suitable for its employability in exploratory, correlational, descriptive and evaluative studies (Hox & Boeije, 2004). Given the benefits of using secondary data, this study utilised panel data retrieved from the Integrated Annual Reports (IARs) of the top FTSE/JSE SRI companies. These companies will be investigated for the six financial years from 2016 to 2021. The choice of these financial years is justifiable given that the FTSE/JSE SRI was launched in 2015. Moreover, reporting on corporates' environmental performance only became mandatory for listed companies in 2010 (Kamala, Wingard & Cronje, 2015; JSE, 2016). Thus, the fiscal years 2016 to 2021 are considered most recent to be employable given that markets and environmental sustainability concerns have evolved sharply at global stage level calling for a wholistic ESG criterion.

Furthermore, this study employed content analysis to collect data needed to meet the objective of the study. The researcher employed this approach to sift through archival documents to identify the needed data for the phenomena being studied (Guthrie, Petty, Yongvanich & Ricceri, 2004). Data were collected from the documents sourced and captured on Microsoft Excel spreadsheets in preparation for the data analysis process using chosen statistical software packages. Where a variable needed calculation, such calculations were done on the Excel spreadsheet. Based on the sampled companies, it was anticipated that some companies may be computing their reports using foreign currency. In such cases, the researcher converted all foreign currency to the local one for the purposes of consistency in the analysis processes. Additionally, where the variable measurement method employed by reporting companies differs, conversions were done on an Excel spreadsheet for standardisation purposes. Where the researcher encountered missing data, the interpolation (also known as adjustment) technique was employed to fill in the missing data points. Its applicability is justified as the interpolation technique is widely employed in most econometrics and statistics research (Church, 2002; Boslaugh, 2009).

4.3 Operationalisation of Main Variables for the Study

The table (Table 2) below shows how the data that relate to independent variable and control variables as extracted from the IARs and SRs for the purposes of analysis.

Table 2. Measurement of independent and control variables

Variable	Variable sub-name	Measurement method in IARs and SRs
Independent variable	Energy consumption (ConsE)	Giga-joules (Gj) or Mega-watt hours (Mwh)
	Water consumption (CoW)	Million cubic metres (Mm^3) and kilolitres (Kl).
	Greenhouse gases emissions (EmGHG)	Metric tonnes of CO2 equivalent (Mt CO2e)
Control variables	Leverage (LEV))	Total assets to total liabilities
	Total monetary value of collateral (MvCOL)	Total tangible fixed assets reflected in the IARs
	Total earnings (Te)	Total of Net Profit after tax
	Company size (CoSZ)	Company's total assets

Source: Author's compilation, 2024

The environmental performance indicators presented in Table 2 above are informed by the Global Reporting Initiative (GRI), Eco-Management and Audit Scheme (EMAS), and European Commission's guidelines. The reporting companies are guided by these guidelines to produce standardised reports for non-financial information. Companies presents such information in their sustainability reports, ESG reports, and or IARs. These indicators are presented as a panel data set of the sampled companies to statistically analyse if they influence stakeholders' support in the context of the current study.

4.4 Operationalisation of Dependent Variables

Prior research renders a plethora of incentives financial sector entities can use to make strides towards playing a key role in stopping climate change (Fahim & Mahadi, 2022; Macey, Harris & Macey, 2022; Petelczyc, 2022). Degryse et al. (2021) contends that numerous largest banks are still visible in carbon heavy industries and finance fossil fuel companies to contradict the prevailing agenda of going green. The current study speculates that debt pricing may render insights if capital lenders of money such as banks are responsive to corporate environmental performance.

Traditionally, the control variables included below (Table 3) renders a premise for accessing finance, as such the results are interpreted in the lenses of these variables.

Table 3. Measurement of the dependent variable

Variable		Variable sub-name	Measurement method in IARs
Dependent variable		Debt pricing (Dp)	Interest paid on bank's loan

Source: Author's compilation, 2021

In Table 3 above measurement of banks' loan pricing is presented together with the measurement of control variables associated with this variable. These proxies are adapted from previous studies (Aifuwa et al., 2019; Chatzitheodorou et al., 2021; Anagnostopoulos et al., 2018; Cui et al., 2018; Serrano-Cinca, Gutiérrez-Nieto & Reyes, 2016).

4.5 Panel Data Analysis Procedures

The study involves a balanced short panel data set of all the companies depicted in the sample (see Table 1 above). Frees (2004) explains that panel data analysis techniques are useful tools for dealing with cross-section time series data. Thus, such techniques have been widely used by researchers in the field of economics (Hsiao, 2007). To this end, panel data analysis is preferred by most researchers when investigating numerous research questions in diverse fields of research. Furthermore, panel data analysis tools have been employed by researchers such as Nervlove (2002), Arellano (2003), and Baltagi (2001). Baltagi (2001) argues through econometrics that panel data is preferred in that it supplies the researcher with highly informative data that tend to be robust in yielding more conclusive results on the phenomena being investigated.

The data utilised using this statistical tool provides greater variability and less collinearity between research variables. Moreover, the data provides more degrees of freedom and efficiency (Baltagi, 2001). To support this assertion, Nervlove (2002) indicates that the advancement in technology has aided panel data to be utilised in a more comprehensive manner where more econometric relationships are analysed. In addition, panel data uses both cross-sectional and times series data sets in a repetitive manner for a given number of observations (Arellano, 2003). Therefore, the researcher employed panel data analysis given that the current study uses variables presented in time series cross-sectional data sets. The study employed two powerful software packages for data analysis, i.e. STATA and Eviews 13. Descriptive statistical analysis, pairwise correlation, Panel Vector Autoregressions for Random-effects Generalized Least Squares (GLS), Random-effects Maximum

Likelihood Estimator (MLE), and Fixed-effects estimations were mainly conducted using STATA for a modest presentation of results. To complement STATA, Eviews 13 was utilised to run some statistical diagnostic tests.

Test of Hypothesis

Therefore, the adopted research design and methodological approach employed in this study is deemed adequate to test the null hypotheses stated below as:

H1: *Corporate sustainability performance does not influence debt pricing.*

Table 3. The main decision rules

Assumptions	Decision rule
Pearson pairwise correlation matrix	If $P < 0.05$, where $\alpha = 0.05$ *(Significance level)*, then Reject the null hypothesis of 'no correlation', accept the alternative.
Hausman test for Random vs Fixed effects	If $P < 0.05$, where $\alpha = 0.05$ *(Significance level)*, then Reject the null hypothesis that 'Random effects model' is consistent, accept the alternative in favour of Fixed effects.
Breusch and Pagan Lagrangian multiplier (LM) test for Random effects	If $P < 0.05$, where $\alpha = 0.05$ *(Significance level)*, then Reject the null hypothesis favouring 'simple ordinary least squares (OLS) regression', accept the alternative in favour of Random effects Generalised Least Squares (GLS) regression

4.6 Descriptive Statistics and Pearson Pairwise Correlation

This study employed the statistical software STATA to generate both descriptive statistics and pairwise correlation matrix. Fisher and Marshall (2009) argues that in descriptive statistics, data can be organised, analysed, and presented in a meaningful manner. The current study envisaged presenting data such as mean, minimum, maximum, observations, and standard deviation in a table format. According to Abba et al. (2019), descriptive statistics is a useful tool for non-parametric and parametric tests. Moreover, it is also argued that descriptive statistics is imperative in both quantitative and qualitative research, hence it should be performed to give the researcher a glimpse of the dataset utilised in the study (None & Kumar Datta, 2011; Ionescu, 2021).

Furthermore, descriptive statistics is of paramount importance as it unravels normality in the distribution of the sample employed in the study and gives a clear depiction of any outliers in the dataset (Gill et al., 2009; Fisher, Murray & Marshall,

2009; Marshall & Jonker, 2011). Most importantly, normality reveal the degree of sharpness and symmetry in the dataset being examined. Farrokhi and Mahmoudi-Hamidabad (2012) indicates that descriptive statistics is also adulated in research for providing information on the measures of dispersion and central tendency. Conversely, pairwise correlation is commendable for enabling the researcher to compare and explore the relationships between variables (Agresti & Finlay, 1998). The researcher pronounces that the research data utilised in this study fulfilled all relevant statistical diagnostic tests as expatiated in the next section.

4.7 Statistical diagnostic tests and Panel data analysis techniques

Prior to conducting the study, the researcher conducted several relevant preliminary diagnostic tests (Myers, 1990; Marshall & Jonker, 2011). This preceded the selection of the models employed in the study given such tests renders the selection criteria. Marshall and Jonker (2011) and Getahun (2014), suggests that model estimation process commences with the panel unit root test to check if the series is stationary at level – I(0) or at first difference – I(1). In the case of stationarity at I(0) the Pooled Ordinary Least Squares (POLS) model may be employed. However, in the latter the researcher proceeds to perform the cointegration test to check if long-run relationship exists between variables. The cointegration test may reveal that variables are not cointegrated which allows the research to perform the Hausman specification test to choose between the Panel Vector Autoregression (PVAR) Model for Fixed Effects or Random Effects tests. Conversely, for variables that are cointegrated, the Panel Vector Error Correction Model (PVECM) may be adopted to enable the researcher observe short-run and long-run dynamics among the variables. Accordingly, numerous statistical diagnostic tests performed comprised panel data tests for normality, serial correlation, multicollinearity, heteroscedasticity, and stationarity.

Variables for Kao Residual Cointegration test

This section provides details of the variables utilised in performing the cointegration test for respective hypothesis pursued by this study. The researcher employed Eviews 13 software to perform Kao Residual cointegration test on the variables for the study.

Table 4. Variables for Kao Residual Cointegration Test

Endogenous explanatory variables	**Endogenous outcome variable**
CoW(*x1*); ConsE(*x2*); EmGHG(*x3*); MvCOL(*x4*); Te(*x5*); LEV(*x6*); CoSZ(*x7*)	Dp(*y*)

Source: Aurthor's compilation, 2021

In tandem to these tests, the researcher applied varied statistical estimations to test the proposed hypothesis of the study. These estimations include Panel Vector Autoregression (PVAR) models for random effects GLS regression and subsequently random effects MLE, and fixed effects estimations to generate the main results of the study. The Pearson correlation matrix was computed to complement the results of the main model specified for this study. The next section deals with model specification and operationalisation for the purpose of statistical analysis.

4.8 Statistical modelling of the study

The researcher employed the techniques above in order investigate if corporates' environmental performance influence stakeholders' support. Getzmann, Lang and Spremann (2014) argue that statistical models in research can render intuitive visualisations that aid the researcher in identifying relationships between variables being studied and make predictions using such statistical models to raw data. As such, the next section presents composite variables and model specification adopted for the study.

Operationalisation of the basic panel data model

The basic panel data model can be presented as shown in Equation (1) below:

$$y_{i,t} = \alpha + x_{i,t}'\beta + c_i + \delta_t + \varepsilon_{i,t} \qquad (1)$$

where y and x denotes the dependent variable and vector of explanatory variables for firm i at time (year) t, respectively. The time trend t allows for a shift of the intercept (α) over time, capturing time effects such as technology and regulation changes among others. However, if the implicit assumption of a constant rate of change is strong, a set of time dummies is introduced (δ_t), one for each time-period except reference period. The subscript ε is the error term. The component c captures the variables responsible for unobserved heterogeneity (unobserved effect). It is usually a "nuisance" component of the model.

If the explanatory variables are so comprehensive that they capture all relevant attributes of i, then c_i can be dropped, and pooled ordinary least squares (OLS) can be used. However, this situation is highly unlikely. Generally, dropping c_i may lead to bias estimations resulting from missing variables problem. Thus, c_i is strongly assumed to be strictly exogenous to the conditional error term as shown in equation (2) below.

$$E[\varepsilon_{i,t}|x_{i,1}, x_{i,2}, \ldots, x_{i,T}, c_i] = 0, \ t = 1, \ldots, T \tag{2}$$

Under the strict exogeneity condition as highlighted in equation (2), equation (1) can be reproduced as follows:

$$E[y_{i,t}\varepsilon_{i,t}|x'_{i,t}, c_i] = \alpha + x_{i,t}'\beta + c_i + \delta_t \tag{3}$$

This implies that β' are partial effects holding c_i constant. However, to estimate β the relationship between $x_{i,t}$ and c_i must be clarified, giving rise to different estimators of Equation (3) such as the pooled, fixed effects, and random effects models. The pooled model's assumptions are often unrealistic, as previously mentioned. Hence, most researchers opt for either fixed or random effects model based on the results of the Hausman Test.

The fixed effects model (FEM) assumes that the unobserved effects (c_i) are correlated with the regressors (x'_i).

$$E[c_i|x_i] = g(x_i) = \alpha_i \tag{4}$$

Then,

$$y_{i,t} = \alpha_i + x_{i,t}'\beta + \varepsilon_{i,t} \tag{5}$$

As a result, the regression line can be shifted up and down by a fixed amount of each i. Equation (5) is consistently estimated using OLS.

The random effects model (REM) assumes that the differences between cross sectional units are random, drawn from a given distribution with constant parameters. Most importantly, it assumes that that the unobserved effects (c_i) and the regressors are uncorrelated.

$$E[c_i|x_i] = \mu \tag{6}$$

$$y_{i,t} = \mu^* + x_{i,t}'\beta + v_{i,t},$$

$$v_{i,t} = u_i + \varepsilon_{i,t} \tag{7}$$

The composed error term ($v_{i,t}$) introduces contemporaneous cross-correlations across the i group. While equation (7) can be consistently estimated using OLS, GLS and maximum likelihood estimator (MLE) may offer greater efficiency.

5. EMPIRICAL RESULTS, ANALYSIS AND DISCUSSIONS

5.1 Descriptive statistics of the panel data

The researcher commenced with statistical analysis of panel data by generating a summary of descriptive statistics. The data utilised were extracted from IARs and SRs of sampled companies. Notwithstanding the benefits of using any statistical software for data analysis, STATA and Eviews 13 were utilised to import data from MS Excel to perform the analysis. These software packages for statistical analysis are employed by most researchers for their easy to use interface, graphics quality, command line programming interface, sophisticated data management and powerful analytic tools (Onditi, 2016). Besides, these software packages simplify statistical analysis especially for the current study where panel data are used.

The researcher generated the descriptive statistics summary which was later exported to MS Excel for editing. Accordingly, the researcher edited the summary statistics table by eliminating sum square deviation and other post-estimations (Jarque-bera and probability) which were considered unfit for descriptive statistics analysis. Thus, mean, standard deviation, minimum and maximum values were presented for all composite variables employed in the study. Table 8.1 depicts positive mean values for all variables employed in the study. The minimum and maximum values assist the researcher understand how data are spread out in each variable. The range can also be calculated for each variable to check how wide the dataset is spread. Moreover, standard deviation abets the researcher in understanding how far observations in the dataset are from the mean values. Furthermore, the researcher also considered kurtosis values to understand the normality of the dataset being scrutinised. The data generated a positive kurtosis given that kurtosis values were higher than the normal distribution kurtosis of 3. This leptokurtic kurtosis suggests a peaked-curve for the given dataset with leverage (LEV) variable showing significant sharp-peakness of the curve with the kurtosis value above the mean.

In conjunction with consideration of the measures of normality, skewness values assisted to understand the degree of asymmetry in the dataset. Thus, energy consumption (ConsE) and suppliers' credit lending (SL) yielded skewness values which were indicative of normal skewness given that the distribution is symmetric around the mean values of 405755.6 and 10163.35 respectively (see Table 8.1). Conversely, bank loan pricing (BLP), consumer buying patronage (ConsBP), company size (CoSZ), water consumption (CoW), greenhouse gases emissions (EmGHG), investor share ownership (InvSO), leverage (LEV), level of profit (LvP) and market value of collateral (MvCOL) exhibited availability of outliers. The researcher employed truncation strategies to handle such outliers prior to the commencement of statistical estimations. The table below (Table 8.1) depicts the summary of descriptive statistics for the panel dataset employed in the study.

Table 5. Summary of descriptive statistics30

Variable		Mean	Std. Dev.	Min	Max	Observations
Water consumption (CoW)	overall	7896383	15200000	7583.725	52200000	N = 114
	between		15500000	9406.539	46200000	n = 19
	within		1751068	-5193284	13900000	T = 6
Greenhouse gas emission (EmGHG)	overall	5303.973	7476.846	65.571	35500	N = 114
	between		7575.85	67.054	29727	n = 19
	within		1024.331	-203.027	11076.97	T = 6
Energy consumption (ConsE)	overall	405755.6	1235945	14562.69	6220000	N = 114
	between		1258856	20846.33	5578333	n = 19
	within		113855.8	-412578	1047422	T = 6
Banks' loan pricing (BLP)	overall	1333.993	1729.952	68	7303	N = 114
	between		1567.22	89.5	6006.5	n = 19
	within		803.2652	-1526.01	6385.16	T = 6
Collateral value (MvCOL)	overall	68152.74	78000.32	634	388222	N = 114
	between		77996.11	3741	315318.5	n = 19
	within		16426.57	-15189.9	141056.2	T = 6
Total Earnings (Te)	overall	5790.132	15005.35	-91917	79021	N = 114
	between		7273.556	-4922.33	22948.67	n = 19
	within		13213.51	-81204.5	61862.46	T = 6
Leverage (LEV)	overall	0.72193	0.529764	0.15	3.79	N = 114
	between		0.386868	0.22	1.528333	n = 19
	within		0.370952	0.038597	3.630263	T = 6
Company size CoSZ)	overall	4.710789	0.543494	3.69	5.68	N = 114
	between		0.549595	3.74	5.623333	n = 19
	within		0.081833	4.254123	4.939123	T = 6

Source: Author's computation, 2024

In Table 5 above, all variables utilised in the study were presented to show measurements for dispersion and central tendency. It should be noted that '*N*' represents the number of observations in the panel dataset employed in the study. On the other hand, '*n*' represents the sample i.e. the number of companies whose panel data set is being examined. Most importantly, '*T*' represents the time-period being investigated (2016 to 2021) in the study. Moreover, the standard deviation shows how dispersed the data is relative to the mean values. The mean values depict the numeric quantity that represents the centre of a collection of numbers.

5.2 Pearson's pairwise correlation matrix

The pairwise correlation matrix is used to determine the strength and direction of the relationship between two variables. Notwithstanding the significance of data presented in Table 5 above, a pairwise correlation matrix was computed to understand how variables under study relate or depend on each other. Accordingly, data were exported from MS Excel to STATA 18 which is an advanced integrated software package used by researchers for all data science needs. This software package was preferred by the researcher for modest visualisation of pairwise correlation results. This implies that the correlation coefficient values, probability values and number of observations are clearly depicted in the correlation matrix. Moreover, STATA 18's automated reporting allows the researchers to present the pairwise correlation matrix with starred correlation coefficients where there is positive correlation which is significant at a predetermined level of significance.

Moreover, the estimated correlation coefficient of one (1), zero (0), and negative one (-1) represents a perfect positive, neutral, and perfect inverse relationship, respectively. In general, 0.5 benchmarks a strong and weak positive relationship, whereas -0.5 benchmarks a strong and weak negative relationship. Furthermore, the estimated correlation coefficients can be utilised to detect any potential multicollinearity between variables. An estimated correlation coefficient of at least 0.8 is considered an indication of the potential presence of multicollinearity between the variables (Baltagi, 2008). Table 6 reveals that there is no correlation coefficient above 0.8, indicating the absence of potential multicollinearity between the variables.

The results of the pairwise corelation matrix for the tested hypothesis produced mixed findings. The debt pricing (Dp) variable correlates negatively and positively with water consumption (CoW) and energy consumption (ConsE) variables respectively, and both relationships are statistically significant at 1%. Thus, Table 6 shows a starred correlation coefficient value of -0.258 between debt pricing (Dp) and water consumption (CoW). The relationships are, however, fragile given that the prevailing benchmark is 0.5 in distinguishing between weak and strong correlation between variables. Moreover, although weak, there is a positive correlation between debt pricing (Dp) and energy consumption (ConsE) as supported by the correlation coefficient of 0.252. This value is juxtaposed with a P value of 0.020 which is within the 0.05 level of significance. Thus, the P value suggests that the sample utilised has strong evidence for the existence of this said correlation between the concerned outcome and predictor variables, since at $P < 0.05$ we reject the null hypothesis of 'no correlation' to accept the alternative hypothesis. Therefore, the sample utilised provides stronger evidence that this correlation is true for the entire population of FTSE/JSE RII constituency.

Furthermore, Pearson's pairwise correlation probability values for greenhouse gases emissions (EmGHG) as an exposure variable also point to a positive correlation when paired with debt pricing (Dp), albeit weak. Consistent with prior correlation studies, for example Kozak (2021), Arora and Sharma (2022), and Kuo et al. (2021), this is indicative of a possible relationship between sustainability performance and cost of debt. The P value of 0.051 reflects evidence for the existence of correlation between debt pricing (Dp) and greenhouse gases emissions (EmGHG). Moreover, Table 6 shows a starred correlation coefficient value of 0.051 which is indicative of a positive correlation, albeit weak, given that it falls below the standard benchmark of 0.5. This result suggests that capital lenders or financiers are somewhat responsive to companies' sustainability performance regarding greenhouse gas emissions in the context of environmental degradation, smog formation and air pollution crisis.

Table 6. Pearson's pairwise correlations matrix

Variable	Water consumption (CoW)	Greenhouse gases emissions (EmGHG)	Energy consumption (ConsE)	Debt pricing (Dp)	Collateral value (MvCOL)	Total Earnings (Te)	Leverage (LEV)	Company size (CoSZ)
Water consumption (CoW)	1							
Greenhouse gas emission (EmGHG)	-0.15	1						
Energy consumption (ConsE)	-0.147	0.037	1					
Debt pricing (Dp)	-0.258***	0.051	0.252***	1				
Collateral value (MvCOL)	-0.218**	-0.002	0.013	0.711***	1			
Total Earnings (Te)	0.218**	-0.037	0.087	-0.164*	-0.007	1		
Leverage (LEV)	-0.121	0.182*	0.245***	0.396***	0.223**	-0.308***	1	
Company size (CoSZ)	-0.148	0.213**	0.133	0.618***	0.770***	0.194***	0.085	1

Note: *, **, and *** denote significance levels at 10%, 5%, and 1%, respectively.
Source: Author's computation, 2024

5.3 The statistical diagnostic tests conducted on panel dataset

The Cointegration test

The results of Kao Residual cointegration test was performed to find the Augmented Dickey-Fuller approach probability values. Thus, in conjunction with the panel unit test which confirmed that key variables employed in the study are stationary at first difference, the cointegration test was performed to ascertain the long-run relationship among variables (Getahun, 2014). Getahun (2014) further argues that cointegration test must be performed using variables on the level form such that raw data of such variables is considered for this purpose. Moreover, to decide whether variables are cointegrated or not, the decision criteria stipulates that the null hypothesis is rejected for $P < 0.05$. The researcher employed Eviews 13 software to perform Kao Residual cointegration test on the hypothesis variables for the study. In all cases, cointegration test indicated that all variables of interest in this study are not cointegrated. This demonstrates that although stationery at first difference, there is no evidence of long-run relationships among variables. Furthermore, all endogenous variables were selected with automatic selection for lag length to examine whether long-run relationships exists or not. Thus, the ADF P value of 0.1421 was generated. This entails accepting the null hypothesis of no cointegration among variables given that $P > 0.05$. Thus, there is no evidence of long-run relationship among variables computed for the tested hypothesis.

According to Xu, Hwan Lee and Ho Eom (2007), where cointegration tests indicates evidence of cointegration among variables, the Vector Error Correction Model (VECM) has to be adopted for further statistical analysis. However, Kaddoura and Westerlund (2023) reports that that Pooled Ordinary Least Squares Model (POLS) and Panel Vector Autoregression Model (PVAR) are common models for handling panel data where variables are not cointegrated. In tandem with the results of the cointegration test, the researcher employed the latter models to handle panel data presented for this study. Besides, Baltagi (2008) argues that these are more suitable where short run dynamics are being investigated given that this study does not render long time dimensions, rather only a six-year period is under scrutiny.

Table 6. Statistical diagnostic tests for panel data

Assumptions	Test employed	Decision rule	Remark
Normality	Jarque-bera tests	*Normal distribution, if P > 0.05*	*fulfilled*
Heteroscedasticity	Breusch Pagan-Godfrey heteroscedasticity test	*Homoscedasticity present, if P > 0.05*	*fulfilled*
Multicollinearity	Variance Inflation Factor (VIF)	*VIF values must be less than 10*	*fulfilled*
Serial correlation	LaGrange Multiplier (LM) test	*No serial correlation of any order, if P > 0.05*	*fulfilled*
Stationarity	Panel unit root test - "Fisher type" based on the Augmented Dickey-Fuller approach (ADF)	*Non-stationary, if P < 0.05*	*Fulfilled – I(1)*
Cointegration test	Kao Residual	*Not cointegrated, if P > 0.05*	*Fulfilled*

The Hausman test

The Hausman test is frequently used to determine whether the Fixed effects or Random effects models are better suited for panel data analysis. The test allows one to determine whether the Random effects assumption (that individual-specific effects are uncorrelated with the independent variables) is valid or if Fixed effects should be used instead. If the Hausman test yields a p-value of less than 5%, the null hypothesis that the Random effects model is consistent is rejected in favour of the Fixed effects model. If the p-value is greater than 5%, the null hypothesis is accepted, implying that the random effects model is more appropriate. Consequently, the results of the Hausman test indicated that a p-value of 0.060 for energy consumption. Therefore, the results show that the Random effects are the appropriate estimator for all the specifications of the energy consumption (ConsE) specification favours Random effects. This means that the specifications the energy consumption (ConsE) specification are estimated using Equation (7) stipulated above. Conversely, water consumption (CoW) and greenhouse gases emissions (EmGHG) specifications favour Fixed effects, with P-values of 0.017 and 0.022 respectively. Thus, the latter Fixed effects variables are estimated using equation (5) of the model specification for the study. In tandem to these results of the Hausman test, the study's main estimations are as expatiated in the next section.

5.4 Presentation of the main results

The main statistical diagnostics considered in the estimations are presented in Table 6 above. These included, among others, heteroscedasticity and the appropriateness of the estimation technique. Moreover, statistical diagnostic analysis for cross-sectional dependency and serial correlation tests were already fulfilled albeit not prioritized given that the study has a relatively short time frame i.e. 2016 to 2021. According to Baltagi (2008), cross-sectional dependency and serial correlation tests are typically applied to macro-panels with long time dimensions (at least 30 years). Moreover, the potential presence of multicollinearity, although fulfilled, is not prioritized given that the Pearson's pairwise correlation matrix presented in Table 5 also tackles this problem. Besides, the robust preliminary statistical diagnostics were presented deliberately so that the findings presented below are sufficiently consistent to be relied on for policy development.

The test to determine whether time-fixed effects are required when running Fixed effects regression is performed for water consumption (CoW) and greenhouse gases emission (EmGHG) specifications. Consequently, the results of Linear hypothesis indicated probability values of 0.343 and 0.399 for water consumption (CoW) and Greenhouse gases emissions (EmGHG) respectively. However, the need for Random effects GLS regression is confirmed for energy consumption (ConsE) specifications. This, follows that the LM test which produced a probability value less than 5% which implies that the null hypothesis in favour of pooled OLS regression is rejected. Thus, the LM test's p-value indicates that Random effects GLS regression are required. In contrast, the p-values of the linear hypothesis test are all greater than 5%, indicating that the null hypothesis that the coefficients for the years are jointly equal to zero cannot be rejected, and thus time fixed effects are unnecessary. Following these results, the estimates of the hypothesis being tested are provided in Table 7 below.

Table 7. Fixed effects and Random effects GLS/MLE results for the study

Dependent Variable: *ln*(Debt pricing)						
	Fixed effects using -*xtreg*-	**Random effects using-gls-**	**Fixed effects using -*xtreg*-**	**Using -*reghdfe*-**	**Using-mle-**	**Using -*reghdfe*-**
	(1)	**(2)**	**(3)**	**(1)**	**(2)**	**(3)**
Constant	*3.275*	*7.406****	*2.581****	*3.275*	*7.704****	*2.581****
	(5.869)	(0.866)	(5.824)	(5.869)	(1.317)	(5.824)
ln(Water consumption)	0.145			0.145		
	(0.115)			(0.115)		

Dependent Variable: *ln*(Debt pricing)						
	Fixed effects using -*xtreg*-	**Random effects using-gls-**	**Fixed effects using -*xtreg*-**	**Using -*reghdfe*-**	**Using-mle-**	**Using -*reghdfe*-**
ln(Energy Consumption)		0.084***			0.084***	
		(0.069)			(0.561)	
ln(Greenhouse gases emissions)			0.335***			0.335***
			(0.292)			(0.292)
ln(Collateral value)	-0.109	-0.066	-0.096	-0.109	-0.070	-0.096
	(0.101)	(0.145)	(0.087)	(0.101)	(0.121)	(0.087)
ln(Total Earnings)	-0.046***	-0.033***	-0.048***	-0.046***	- 0.031***	-0.048***
	(0.006)	(0.009)	(0.006)	(0.006)	(0.027)	(0.006)
ln(Company Size)	1.970	9.205***	1.888	1.970	9.425***	1.888
	(3.615)	(1.428)	(3.814)	(3.615)	(1.459)	(3.814)
ln(Leverage)	0.159	0.423***	0.145	0.159	0.444***	0.145
	(0.123)	(0.135)	(0.107)	(0.123)	(0.093)	(0.107)
Observations	*114*	*114*	*114*	*114*	*114*	*114*
R-Squared	*0.054*	*0.846*	*0.066*	*0.904*		*0.904*
Firm fixed effects	*YES*		*YES*	*YES*		*YES*
Number of firms	*19*	*19*	*19*	*19*	*19*	*19*
Prob > F	*0.000*	*0.000*	*0.000*	*0.000*	*0.000*	*0.000*
Prob >= chibar²					*0.000*	

Notes: *** denotes significant level at 1%. In parenthesis are heteroscedasticity-robust standard errors across all fixed effects regressions and random-effects GLS regression, while random-effects MLE standard errors are derived from asymptotic theory.

Source: Author's estimation using STATA, 2024

The classic Fixed effects regression estimates are tested for robustness using linear regression with multiple fixed effects, which is run with '*reghdfe*' command in the software STATA. The '*reghdfe*' command accounts for heteroscedasticity and correlation within panels. Thus, Random-effects GLS regression estimates are checked for robustness using Random effects MLE. Accordingly, the GLS regression was found to be fit for the data, as evidenced by very small F-statistics probability (<0.005) and relatively high R-Squared of 0.846. The accuracy of the results is envisaged so that the statistical analysis yields conclusive results whose modest contribution to literature enhances policy formulation.

The table above (Table 7) shows a marginal difference in magnitude between MLE and GLS regression estimates, while the status of the significance levels remains the same. Similarly, the 'extreg' and 'reghdfe' Fixed effects estimates concur. Imperatively, regarding Fixed effects regressions, it would be more interesting to confirm the consistency of the estimates using instrumental variable techniques, such as system GMM, which is robust to endogeneity. However, the period for this study (i.e.2016 to 2021) is relatively too short to implement instrumental variable techniques.

Nevertheless, Table 7 shows that the estimated coefficient of total earnings (Te) is negative and statistically insignificant at 1% for all specifications. This finding suggests that total earnings (Te) have a negative impact on debt (Dp), regardless of the corporate sustainability performance indicator used, at least for the indicators used in this study. In terms of control variables, the estimated coefficients of company size (CoSZ) and leverage (LEV) are positive and statistically significant at 1% in the energy consumption (ConsE) specification. This suggests that leverage (LEV) and company size (CoSZ) may significantly enhance debt pricing (Dp). However, this is only true if energy consumption (ConsE) is used as a proxy for corporate environmental performance, rather than water consumption (CoW) and greenhouse gas emissions (EmGHG) indicators.

Although the Fixed effects output for water consumption (CoW) is statistically insignificant, greenhouse gas emissions (EmGHG) produced a positive and statistically significant coefficient at 1% level of significance. Moreover, the Random effects GLS estimator also indicates that there is a positive and statistically significant effect of energy consumption (ConsE) on debt pricing (Dp). This revelation postulates that financiers are becoming more sensitive to firm's environmental performance. These results are congruent to the results emanating from Pearson's correlation matrix presented in Table 5. Interestingly, the estimated coefficient of the constant is also significant under the energy consumption (ConsE) specification alone.

5.5 Discussions and Conclusions

The statistical tests conducted for the tested hypothesis sought to determine if environmental performance influences debt pricing. Thus, the study investigated if debt pricing is a pathway for achieving long-term sustainability in the South African context. The estimations indicated mixed results for an assortment of endogenous variables utilised for corporate sustainability perormance. Thus, greenhouse gas emissions (EmGHG) produced a positive and statistically significant coefficient at 1% level of significance. Additionally, the random effects GLS estimator also indicates that there is a positive and statistically significant effect of energy consumption (ConsE) on debt pricing (Dp). Thus, results posit that capital lenders of money such as banks are becoming more sensitive to corporate's environmental performance. Such sensitivity comes as an effort to curb environmental crises such as air pollution and smog formation within the South African context. The findings are complemented by the Pearson's correlation matrix which yielded a positive association between the concerned variables. The debt pricing (Dp) variable correlates positively with energy consumption (ConsE) variable with the relationship statistically significant at 1%. Moreover, pairwise correlation probability values for greenhouse gases emissions (EmGHG), as an exposure variable for environmental performance employed in the study, also shows a positive correlation when paired with debt pricing (Dp) albeit weak. The results are consistent with prior studies. For example, Li, Hu and Hong (2022) notes a significantly positive association between environmental performance and cost of debt. Moreover, ESG compliant firms who were subjected to ESG disclosures were found to exhibit cheaper cost of debt and easy access to long term bank loans (Xu et al., 2021). Moreover, Kuo, Kuo and Chen (2021) also demonstrates that mandatory CSR assurance on CSR disclosure as provided by various accounting firms tend to reduce banks' loan pricing. Similarly, Ali et al. (2022) found a positive association between climate change actions and banks' loan pricing. The study argues that corporate climate change efforts as advocated by lobby groups and COP26 are rewarded by cheaper cost of debt financing. Furthermore, other claims also show that a company can reduce its cost of debt financing by reducing the intensity of CO2 emissions (Kozak, 2021). Raimo, Caragnano, Zito, Vitolla and Mariani, (2021)'s commendable in-depth study, where panel data of 919 companies was observed for a ten-year period, reports that ESG disclosures influence cost of debt financing. The study demonstrates that companies that are ESG compliant tend to have easy access to finance with better conditions from capital lenders.

Conversely, Fixed effects output for water consumption (CoW) was found to be statistically insignificant to explain the impact that corporate environmental performance has on debt pricing (Dp). Prior studies which concur with this assertion include Magnanelli and Izzo (2017) who presents findings of the linear regression model indicating that there is no significant relationship between CSP and cost of debt. The study utilised a data of 332 companies extracted for a five-year period. Results confirm evidence that ESG factors are not value drivers with an influence on cost of debt and the companies' risk profile (Magnanelli & Izzo, 2017). Moreover, Hamrouni, Uyar and Boussaada (2020) speculates that there is no inverse relationship whatsoever given that the social imperative is not a consideration for capital lenders of money. The study renders novel insights on the new tools that a key role player such as financiers can leverage in fostering environmental compliance by companies towards the greening agenda. Notwithstanding such a notable contribution, the limitation of this study manifest itself in the sampled companies that are environmentally responsible. Thus, future studies may expand the scope this study to include companies that are outside the FTSE/JSE RII for more conclusive results.

REFERENCES

Abba, A.. (2019). Supply Chain Sustainability Practices of Oil Servicing Firms in the Downstream Sector of Nigeria's Oil and Gas Industry. *Journal of Economic Info*, *6*(4), 11–14. DOI: 10.31580/jei.v6i4.1031

Abba, M., Said, R. M., & Mahat, F. (2017). Environment Operational Performance Effect on Costs of Capital Structure Financing of the Nigerian Listed Manufacturing Companies. *International Journal of Accounting and Financial Reporting*, *7*(1), 372–391. DOI: 10.5296/ijafr.v7i1.11381

Agresti, A., & Finlay, B. (1998). Statistical Methods for the Social Sciences. *Journal of the American Statistical Association*, *93*(1), 442–844. DOI: 10.2307/2670147

Aifuwa, H. O.. (2019). Accounting Information and Lending Decision: Does Sustainability Disclosure Matter? *Copernican Journal of Finance & Accounting*, *8*(4), 61–89.

Ali, K. M. and Abdel ghaffar, E. sayed (2022) 'The effect of corporate social responsibility on debt finance: The moderating effect of accounting conservatism', *Journal of Accounting, Business and Management (JABM)*, 29(1), pp. 01–67. .DOI: 10.31966/jabminternational.v29i1.821

Ameli, N., Dessens, O., Winning, M., Cronin, J., Chenet, H., Drummond, P., Calzadilla, A., Anandarajah, G., & Grubb, M. (2021). Higher cost of finance exacerbates a climate investment trap in developing economies. *Nature Communications*, *12*(1), 1–12. DOI: 10.1038/s41467-021-24305-3

Anagnostopoulos, T., Skouloudis, A., Khan, N., & Evangelinos, K. (2018). Incorporating sustainability considerations into lending decisions and the management of bad loans: Evidence from Greece. *Sustainability (Basel)*, *10*(12), 1012–4728. DOI: 10.3390/su10124728

Arellano, M. (2003). *Panel data econometrics*. OUP Oxford. DOI: 10.1093/0199245282.001.0001

Azmi, W., Hassan, M. K., Houston, R., & Karim, M. S. (2021). ESG activities and banking performance: International evidence from emerging economies. *Journal of International Financial Markets, Institutions and Money*, *70*(1), 101–277. DOI: 10.1016/j.intfin.2020.101277

Bala, S. A., & Ibrahim, M. (2022). Does the Disclosure of ESG Sustainability Matter to Selected Listed Firms Does the Disclosure of ESG Sustainability Matter to Selected Listed Firms Performance in Nigeria? *International Journal of Innovative Research in Accounting and Sustainability*, 7(1), 130–141.

Baltagi, B. H. (2001). *A companion to theoretical econometrics* (Baltagi, B. H., Ed.). Vol. 1). Blackwell.

Boachie, C., & Tetteh, J. E. (2021). Do creditors value corporate social responsibility disclosure? Evidence from Ghana. *International Journal of Ethics and Systems*, *37*(3), 466–485. DOI: 10.1108/IJOES-11-2020-0181

Boslaugh, S. (2009). *An Introduction to Secondary Data Analysis*. Secondary Data Sources for Public Health., DOI: 10.1017/CBO9780511618802.002

Buhr, B. *et al.* (2018) 'Climate Change and the Cost of Capital in Developing Countries', *United Nations*, 48(August), pp. 01–38.

Bussman, N., Tanda, A., & Yu, E. P. (2022) '*Can ESG shape cost of capital? A bibliometric review and empirical analysis through ML*', pp. 1–20. DOI: 10.2139/ssrn.4173890

Chatzitheodorou, K., Tsalis, T. A., Tsagarakis, K. P., Evangelos, G., & Ioannis, N. (2021). A new practical methodology for the banking sector to assess corporate sustainability risks with an application in the energy sector. *Sustainable Production and Consumption*, *27*, 1473–1487. DOI: 10.1016/j.spc.2021.03.005

Church, R. M. (2002). The effective use of secondary data. *Learning and Motivation*, *33*(1), 32–45. DOI: 10.1006/lmot.2001.1098

Crowe, S. *et al.* (2011) 'The case study approach', *BMC Medical Research Methodology*, 1(1), pp. 02–09.

Cui, Y., Geobey, S., Weber, O., & Lin, H. (2018). The impact of green lending on credit risk in China. *Sustainability (Basel)*, *10*(6), 1–16. DOI: 10.3390/su10062008

Dilotsotlhe, N., & Duh, H. I. (2021). 'Drivers of Middle-Class Consumers' Green Appliance Attitude and Purchase Behavior: A Multi-Theory Application'. *Social Marketing Quarterly*, *27*(2), 150–171. DOI: 10.1177/15245004211013737

Dube, S., & Maroun, W. (2017). Corporate social responsibility reporting by South African mining companies: Evidence of legitimacy theory. *South African Journal of Business Management*, *48*(1), 23–34. DOI: 10.4102/sajbm.v48i1.17

Dubihlela, J., & Ngxukumeshe, T. (2016). Customer Attributes and the Repurchase. *The International Business & Economics Research Journal, 15*(4), 163–175. DOI: 10.19030/iber.v15i4.9754

Farrokhi, F., & Mahmoudi-Hamidabad, A. (2012). Rethinking convenience sampling: Defining quality criteria. *Theory and Practice in Language Studies, 2*(4), 784–792. DOI: 10.4304/tpls.2.4.784-792

Fuhrmann, S., Ott, C., Looks, E., & Guenther, T. W. (2017). The contents of assurance statements for sustainability reports and information asymmetry. *Accounting and Business Research, 47*(4), 369–400. DOI: 10.1080/00014788.2016.1263550

Getahun, M. (2014) 'Determinants of capital structure and its impact on the performance of Ethiopian insurance industry'.

Getzmann, A., Lang, S., & Spremann, K. (2014). Target capital structure and adjustment speed in asia. *Asia-Pacific Journal of Financial Studies, 43*(1), 1–30. Advance online publication. DOI: 10.1111/ajfs.12038

Gill, A. *et al.* (2009) 'The Determinants of Capital Structure in the Service Industry : Evidence from United States'. .DOI: 10.2174/1874915100902010048

Greyvenstein, R. (2010) *Earnings management on the JSE before and after King II.*

Groenewald, T. (2004). A Phenomenological Research Design Illustrated. *International Journal of Qualitative Methods, 3*(1), 42–55. DOI: 10.1177/160940690400300104

Guthrie, J., Petty, R., Yongvanich, K., & Ricceri, F. (2004). Using content analysis as a research method to inquire into intellectual capital reporting. *Journal of Intellectual Capital, 5*(2), 282–293. DOI: 10.1108/14691930410533704

Hector, H. (2021). Understanding and development of sustainability investing in the recent history. *Humanum, 41*(2), 33–39.

Hox, J. J., & Boeije, H. R. (2004) 'Data Collection, Primary vs. Secondary', *Encyclopedia of Social Measurement*, pp. 593–599. DOI: 10.1016/B0-12-369398-5/00041-4

Hui, C., Foxcroft, L. C., Richardson, D. M., & MacFadyen, S. (2011). Defining optimal sampling effort for large-scale monitoring of invasive alien plants: A Bayesian method for estimating abundance and distribution. *Journal of Applied Ecology, 48*(3), 768–776. DOI: 10.1111/j.1365-2664.2011.01974.x

Ionescu, L. (2021). Corporate environmental performance, climate change mitigation, and green innovation behavior in sustainable finance. *Economics, Management, and Financial Markets, 16*(3), 94–106. DOI: 10.22381/emfm16320216

JSE. (2016). *JSE limited listings requirements*. Johannesburg Stock Exchange.

Jung, J., Herbohn, K., & Clarkson, P. (2018). Carbon Risk, Carbon Risk Awareness and the Cost of Debt Financing. *Journal of Business Ethics, 150*(4), 1151–1171. DOI: 10.1007/s10551-016-3207-6

Kaddoura, Y., & Westerlund, J. (2023). Estimation of Panel Data Models with Random Interactive Effects and Multiple Structural Breaks when T is Fixed. *Journal of Business & Economic Statistics, 41*(3), 778–790. DOI: 10.1080/07350015.2022.2067546

Kamala, P. N., Wingard, H. C., & Cronje, C. J. (2015). Evolution of corporate environmental reports in South Africa. *Journal of Accounting and Management, 5*(2), 31–34.

Kiron, D. (2012). Get Ready: Mandated Integrated Reporting Is The Future of Corporate Reporting. *MIT Sloan Management Review, 3*(53), 1–3.

Kitsikopoulos, C., Schwaibold, U., & Taylor, D. (2018). Limited progress in sustainable development: Factors influencing the environmental management and reporting of South African JSE-listed companies. *Business Strategy and the Environment, 27*(8), 1295–1301. DOI: 10.1002/bse.2176

Kling, G., Volz, U., Murinde, V., & Ayas, S. (2021). 'The impact of climate vulnerability on firms' cost of capital and access to finance'. *World Development, 137*(1), 105–131. DOI: 10.1016/j.worlddev.2020.105131

Klusak, P., ... (2021) *Rising temperatures, falling ratings: The effect of climate change on sovereign creditworthiness,* SSRN *Electronic Journal*. DOI: 10.2139/ssrn.3811958

Kordsachia, O., Focke, M., & Velte, P. (2022). 'Do sustainable institutional investors contribute to firms' environmental performance? Empirical evidence from Europe'. *Review of Managerial Science, 16*(5), 1409–1436. DOI: 10.1007/s11846-021-00484-7

Kozak, S. (2021) 'Will the reduction of co2 emissions lower the cost of debt financing? The case of eu countries', *Energies*, 14(24), pp. 01–13. .DOI: 10.3390/en14248361

Kuo, L., Kuo, P. W., & Chen, C. C. (2021). Mandatory CSR disclosure, CSR assurance, and the cost of debt capital: Evidence from Taiwan. *Sustainability (Basel), 13*(4), 1–19. DOI: 10.3390/su13041768

Kyaw, K., Olugbode, M., & Petracci, B. (2021). 'Stakeholder engagement: Investors' environmental risk aversion and corporate earnings'. *Business Strategy and the Environment, 1*(1), 1–12. DOI: 10.1002/bse.2951

Lakhani, L., & Herbert, S. L. (2022). Theoretical references applied in integrated reporting and supporting reporting research. *Suid-Afrikaanse Tydskrif vir Ekonomiese en Bestuurswetenskappe*, *25*(1), 1–12. DOI: 10.4102/sajems.v25i1.4427

Leedy, P. and Ormorod, J. (2010) *Practical Research: Planning and Design*.

Magnanelli, B. S., & Izzo, M. F. (2017). Corporate social performance and cost of debt: The relationship. *Social Responsibility Journal*, *13*(2), 250–265. DOI: 10.1108/SRJ-06-2016-0103

Marshall, G., & Jonker, L. (2011). An introduction to inferential statistics: A review and practical guide. *Radiography*, *17*(1), e1–e6. DOI: 10.1016/j.radi.2009.12.006

Marx, B., & van Dyk, V. (2011). Sustainability reporting and assurance. *Meditari Accountancy Research*, *19*(1/2), 39–55. DOI: 10.1108/10222521111178628

Meissner, H.. (2011). Best Practices for Mixed Methods Research in the Health Sciences. *Office of Behavioral and Social Sciences Research*, *29*(1), 1–39. DOI: 10.1002/cdq.12009

Mertler, C. A., & Reinhart, R. V. (2016) 'Advanced and multivariate statistical methods: Practical application and interpretation', in *A guide to multivariate techniques*, pp. 13–22. DOI: 10.4324/9781315266978

Moikwatlhai, K., & Yasseen, Y. (2019). ESG reporting and the institutional shareholder base: A quantitative study of listed companies on the JSE. *Southern African Journal of Accountability and Auditing Research*, *21*(1), 31–46.

Morgan, D. L. (2007). Paradigms Lost and Pragmatism Regained. *Journal of Mixed Methods Research*, *1*(1), 48–76. DOI: 10.1177/2345678906292462

Muttakin, M. B., Mihret, D., Lemma, T. T., & Khan, A. (2020). Integrated reporting, financial reporting quality and cost of debt. *International Journal of Accounting and Information Management*, *28*(3), 517–534. DOI: 10.1108/IJAIM-10-2019-0124

None, I., & Kumar Datta, S. (2011). Pro-environmental Concern Influencing Green Buying: A Study on Indian Consumers. *International Journal of Business and Management*, *6*(6), 124–133. DOI: 10.5539/ijbm.v6n6p124

Ofogbe, N. S., Nnamani, C., Anisiuba, C., & Ezuwore-Obodoekwe, C. (2021a). the Relationship Between the Corporate Social Responsibility and the Cost of the Capital of the Companies Listed on the Nigerian Stock Exchange. *Economic Horizons*, *23*(3), 277–292. DOI: 10.5937/ekonhor2103289O

Ofogbe, N. S., Nnamani, C., Anisiuba, C., & Ezuwore-Obodoekwe, C. (2021b). The Relationship Between the Corporate Social Responsibility and the Cost of the Capital of the Companies Listed on the Nigerian Stock Exchange. *Economic Horizons, 23*(3), 277–292. DOI: 10.5937/ekonhor2103289O

Onditi, A. (2016) 'Journal of Marketing and Consumer Research www.iiste.org ISSN', *An International Peer-reviewed Journal*, 29, pp. 37–45.

Raimo, N., Caragnano, A., Zito, M., Vitolla, F., & Mariani, M. (2021). Extending the benefits of ESG disclosure: The effect on the cost of debt financing. *Corporate Social Responsibility and Environmental Management, 28*(4), 1412–1421. DOI: 10.1002/csr.2134

Salisu, A., Omoke, P., & Fadiya, O. (2022). Climate Policy Uncertainty and Crude Oil Market Volatility. *Energy Research Letters, 3*(1), 1–5. DOI: 10.46557/001c.38781

Saunders, M., Lewis, P., & Thornhill, A. (2018). *Research Methods for Business Students* (8th ed.). Pearson Education.

Serrano-Cinca, C., Gutiérrez-Nieto, B., & Reyes, N. M. (2016). A social and environmental approach to microfinance credit scoring. *Journal of Cleaner Production, 112*, 3504–3513. DOI: 10.1016/j.jclepro.2015.09.103

Srivastava, J., Sampath, A., & Gopalakrishnan, B. (2022). Is ESG the key to unlock debt financing during the COVID-19 pandemic? International evidence. *Finance Research Letters, 49*(6), 103–125. DOI: 10.1016/j.frl.2022.103125

Steyn, M. (2014). 'Organisational benefits and implementation challenges of mandatory integrated reporting: Perspectives of senior executives at South African listed companies', *Sustainability Accounting. Management and Policy Journal, 5*(4), 476–503. DOI: 10.1108/SAMPJ-11-2013-0052

Sultana, S., Zainal, D., & Zulkifli, N. (2017). The influence of environmental, social and governance (ESG) on investment decisions: The Bangladesh perspective. *Pertanika Journal of Social Science & Humanities, 25*(11), 155–173.

Xu, H., & Lee, H. S. and Ho Eom, T. (2007) 'Introduction to Panel Data Analysis', *Handbook of Research Methods in Public Administration*, 1(1), pp. 571–587. DOI: 10.1201/9781420013276.ch32

Xu, H., Xu, X., & Yu, J. (2021). The Impact of Mandatory CSR Disclosure on the Cost of Debt Financing: Evidence from China. *Emerging Markets Finance & Trade, 57*(8), 2191–2205. DOI: 10.1080/1540496X.2019.1657401

Compilation of References

Abas, N., Saleem, M. S., Kalair, E., & Khan, N. (2019). Cooperative control of regional transboundary air pollutants. *Environmental Systems Research*, *8*(1), 1–14. DOI: 10.1186/s40068-019-0138-0

Abba, A.. (2019). Supply Chain Sustainability Practices of Oil Servicing Firms in the Downstream Sector of Nigeria's Oil and Gas Industry. *Journal of Economic Info*, *6*(4), 11–14. DOI: 10.31580/jei.v6i4.1031

Abba, M., Said, R. M., & Mahat, F. (2017). Environment Operational Performance Effect on Costs of Capital Structure Financing of the Nigerian Listed Manufacturing Companies. *International Journal of Accounting and Financial Reporting*, *7*(1), 372–391. DOI: 10.5296/ijafr.v7i1.11381

Abdelwahed, N. A. A., Al Doghan, M. A., Saraih, U. N., & Soomro, B. A. (2023). Green entrepreneurship in Saudi Arabia: Shaping the landscape of the greener economy. *Journal of Small Business and Enterprise Development*, *30*(7), 1352–1376. DOI: 10.1108/JSBED-05-2023-0239

Abildtrup, J., Jacobsen, J. B., Vedel, S. E., Mantau, U., Mavsar, R., Pettenella, D., Prokofieva, I., Schubert, F., Stenger, A., Varela, E., Vidale, E., & Thorsen, B. J. (2024). Preferences for climate change policies: The role of co-benefits. *Journal of Environmental Economics and Policy*, *13*(1), 110–128. DOI: 10.1080/21606544.2023.2223182

Acaroğlu, E., Kartal, C., & Fausto, P. G. M. (2023, May). Testing the environmental Kuznets curve hypothesis in terms of ecological footprint and CO_2 emissions through energy diversification for Turkey. *Environmental Science and Pollution Research International*, *30*(22), 63289–63304. DOI: 10.1007/s11356-023-26278-w

Afonso, T. L., Marques, A. C., & Fuinhas, J. A. (2021). Does energy efficiency and trade openness matter for energy transition? Empirical evidence for countries in the Organization for Economic Co-operation and Development. *Environment, Development and Sustainability*, *23*(9), 13569–13589. DOI: 10.1007/s10668-021-01228-z

Afzal, S. (2023). The Invisible Killer: Investigating the Lack of Research on Smog in Pakistan. *Annals of King Edward Medical University*, *29*(2), 73–74. DOI: 10.21649/akemu.v29i2.5423

Afzal, S., & Bajwa, N. (2024). Smog the New Unfortunate Reality: A Catastrophe Public Health Concern in Decreasing Life Spans of Pakistani Population. *Journal of Society of Prevention. Advocacy and Research KEMU*, *3*(1), 1–1.

Agresti, A., & Finlay, B. (1998). Statistical Methods for the Social Sciences. *Journal of the American Statistical Association*, *93*(1), 442–844. DOI: 10.2307/2670147

Ahmad, M. N., Büker, P., Khalid, S., Van Den Berg, L., Shah, H. U., Wahid, A., Emberson, L., Power, S. A., & Ashmore, M. (2013). Effects of ozone on crops in north-west Pakistan. *Environmental Pollution*, *174*, 244–249. DOI: 10.1016/j.envpol.2012.11.029 PMID: 23279903

Ahmed, I., Irshad, A., Zafar, S., Khan, B. A., Raza, M., & Ali, P. R. (2023). The role of environmental initiatives and green value co-creation as mediators: Promoting corporate entrepreneurship and green innovation. *SN Business & Economics*, *3*(4), 85. DOI: 10.1007/s43546-023-00465-w

Aifuwa, H. O.. (2019). Accounting Information and Lending Decision: Does Sustainability Disclosure Matter? *Copernican Journal of Finance & Accounting*, *8*(4), 61–89.

Ajmal, M., Tarar, M. A., Arshad, M. I., Gulshan, A. B., Iqbal, M. A., & Tanvir, F. (2016). Air pollution and its effect on human health: A case study in Dera Ghazi Khan urban areas, Pakistan. *Journal of Environment and Earth Science*, *6*, 87–93.

Akbar, A. (2020). *Theorizing the effect of smog on public health in Lahore*. DOI: 10.5539/ep.v9n2p14

Akbostancı, E., Türüt-Aşık, S., & Tunç, G. İ. (2009). The Relationship Between Income and Environment in Turkey: Is There an Environmental Kuznets Curve? *Energy Policy*, *37*(3), 861–867. DOI: 10.1016/j.enpol.2008.09.088

Akuraju, V., Pradhan, P., Haase, D., Kropp, J. P., & Rybski, D. (2020). Relating SDG11 indicators and urban scaling–An exploratory study. *Sustainable Cities and Society*, *52*, 101853. DOI: 10.1016/j.scs.2019.101853

Alamgir, W., & Shan, H. (2024). Air pollution: A challenge to public health and healthcare systems. *Life and science, 5* (1), 02-02. DOI: 10.37185/LnS.1.1.594

Albort-Morant, G., Henseler, J., Leal-Millán, A., & Cepeda-Carrión, G. (2017). Mapping the field: A bibliometric analysis of green innovation. *Sustainability (Basel), 9*(6), 1011. DOI: 10.3390/su9061011

Ali, K. M. and Abdel ghaffar, E. sayed (2022) 'The effect of corporate social responsibility on debt finance: The moderating effect of accounting conservatism', *Journal of Accounting, Business and Management (JABM)*, 29(1), pp. 01–67. .DOI: 10.31966/jabminternational.v29i1.821

Ali, M. I. (2023). Navigating smog: Legislation addressing children's right to health in india and pakistan. DOI: 10.56784/hrgs.v2i2.74

Ali, M., Siddique, I., & Abbas, S. (2022). Characterizing air pollution and its association with emission sources in Lahore: A guide to adaptation action plans to control pollution and smog. *Applied Sciences (Basel, Switzerland), 12*(10), 5102. DOI: 10.3390/app12105102

Ali, S., Niaz, H., Ahmad, S., & Khan, S. (2025). Investigating how Rapid Urbanization Contributes to Climate Change and the Social Challenges Cities Face in Mitigating its Effects. *Review of Applied Management and Social Sciences, 8*(1), 1–16. DOI: 10.47067/ramss.v8i1.429

Ali, Y., Razi, M., De Felice, F., Sabir, M., & Petrillo, A. (2019). A VIKOR based approach for assessing the social, environmental and economic effects of "smog" on human health. *The Science of the Total Environment, 650*, 2897–2905. DOI: 10.1016/j.scitotenv.2018.10.041 PMID: 30373066

Allam, Z., & Soomauroo, Z. (2024). How the green new deal can transoform urban landscapes and promote health equity. *Cities & health*null, 1-5. DOI: 10.1080/23748834.2024.2360301

Allaq, A. A. A., Mahid, H. H., Yahya, E. B., Saleh, A. M., Sidik, N. J., Abdulsamad, M. A., Saad, N. R., Elengoe, A., & Abdul-Aziz, A. (2023). Emerging drinking water borne diseases: A review on types, sources and health precaution. *Journal of Pharmaceutical Research International, 35*(31), 1–17. DOI: 10.9734/jpri/2023/v35i317462

Allen, J. C., & Barnes, F. D. (1985). The causes of deforestation in developing countries. *Annals of the Association of American Geographers, 75*(2), 163–184. DOI: 10.1111/j.1467-8306.1985.tb00079.x

Al-Thani, H. G., & Isaifan, R. J. (2024). Policies and regulations for sustainable clean air: an overview. *Sustainable Strategies for Air Pollution Mitigation: Development, Economics, and Technologies*, 409-437.

Ambreen, S., Wenbao, M., Khoso, W. M., Mangi, T. A., & Baloch, Z. A. (2025). Impact of socioeconomic and sociopolitical factors on environmental sustainability in Pakistan. *Environment, Development and Sustainability*, •••, 1–25. DOI: 10.1007/s10668-025-05989-9

Ameli, N., Dessens, O., Winning, M., Cronin, J., Chenet, H., Drummond, P., Calzadilla, A., Anandarajah, G., & Grubb, M. (2021). Higher cost of finance exacerbates a climate investment trap in developing economies. *Nature Communications*, *12*(1), 1–12. DOI: 10.1038/s41467-021-24305-3

American Psychological Association. (2020). The importance of social connection. https://www.apa.org/news/press/releases/2020/05/importance-social-connection

Anagnostopoulos, T., Skouloudis, A., Khan, N., & Evangelinos, K. (2018). Incorporating sustainability considerations into lending decisions and the management of bad loans: Evidence from Greece. *Sustainability (Basel)*, *10*(12), 1012–4728. DOI: 10.3390/su10124728

Andrews, R. N. L. (2020). *Managing the environment, managing ourselves: A history of American environmental policy* (3rd ed.). Yale University Press.

Anees, M., & Yan, W. (2019). An Overview about the Challenges of Urban Expansion on Environmental Health in Pakistan. *Journal of Energy and Environmental Policy Options*, *2*(3), 64–71.

Anjum, S., Zafar, M. M., & Kumari, A. (2024). A review of diseases attributed to air pollution and associated health issues: A case study of indian metropolitan cities. DOI: 10.1016/B978-0-443-16080-6.00004-5

Anjum, M. S., Ali, S. M., Subhani, M. A., Anwar, M. N., Nizami, A.-S., Ashraf, U., & Khokhar, M. F. (2021). An emerged challenge of air pollution and ever-increasing particulate matter in Pakistan; a critical review. *Journal of Hazardous Materials*, *402*, 123943. DOI: 10.1016/j.jhazmat.2020.123943 PMID: 33254830

Ansari, M. S. A. (2025). Innovation and green entrepreneurship as catalysts for sustainability in Oman. *Cleaner Waste Systems*, *11*, 100281. DOI: 10.1016/j.clwas.2025.100281

Antrop, M. (2004). Landscape change and the urbanization process in Europe. *Landscape and Urban Planning*, *67*(1-4), 9–26. DOI: 10.1016/S0169-2046(03)00026-4

Anwar, A., Sinha, A., Sharif, A., Siddique, M., Irshad, S., Anwar, W., & Malik, S. (2022). The nexus between urbanization, renewable energy consumption, financial development, and CO2 emissions: Evidence from selected Asian countries. *Environment, Development and Sustainability*, *24*(5), 6556–6576. DOI: 10.1007/s10668-021-01716-2

Appio, F. P., Lima, M., & Paroutis, S. (2019). Understanding Smart Cities: Innovation ecosystems, technological advancements, and societal challenges. *Technological Forecasting and Social Change*, *142*, 1–14. DOI: 10.1016/j.techfore.2018.12.018

Apte, J. S., & Manchanda, C. (2024). High-resolution urban air pollution mapping. *Science*, *385*(6707), 380–385. DOI: 10.1126/science.adq3678 PMID: 39052801

AQLI. (2022). *China's War on Air Pollution: The Results Are In*. Air Quality Life Index. https://aqli.epic.uchicago.edu

Arellano, M. (2003). *Panel data econometrics*. OUP Oxford. DOI: 10.1093/0199245282.001.0001

Aria, M., & Cuccurullo, C. (2017). Bibliometrix: An R-tool for comprehensive science mapping analysis. *Journal of Informetrics*, *11*(4), 959–975. DOI: 10.1016/j.joi.2017.08.007

Arif, M. I. U., & Mahsud, M. I. (2024). Disentangling Pakistan's Climate Change Governance Challenges: Trajectories and Underpinnings.

Arif, M., & Hassan, E. (2023). *Beyond the haze: a study of smog, its types, and mitigation*. Faisalabad Chamber of Commerce & Industry.

Asghar, N., Rehman, H., & Sultan, M. S. (2022). A nonlinear analysis of the relationship between urbanization and carbon dioxide emissions in Pakistan: A fresh evidence. *Review of Applied Management and Social Sciences*, *5*(4), 655–665. DOI: 10.47067/ramss.v5i4.276

Ashraf, F., (2019). Smog analysis and its effect on reported ocular surface diseases: A case study of 2016 smog event of Lahore, Atmos. Environ.

Ashraf, I., Ahmad, R. U., & Tareen, H. K. (2022). Worsening situation of smog in Pakistan: A tale of three cities. *Annals of Medicine and Surgery (London)*, *79*. Advance online publication. DOI: 10.1016/j.amsu.2022.103947 PMID: 35860153

Asif, M., Khan, M. I., & Pandey, A. (2024). Navigating the inclusive and sustainable energy transitions in South Asia: Progress, priorities and stakeholder perspectives. *Energy Conversion and Management*, *313*, 118589. DOI: 10.1016/j.enconman.2024.118589

Aslam, A. 2022. Impact of the built environment on climate change risk perception and psychological distancing: empirical evidence from Islamabad, Pakistan. Environ. Sci. Pol.

Astuti, R. D. P., & Rauf, A. U. (2024). Outdoor air pollution due to transportation, landfill, and. Health and Environmental Effects of Ambient Air Pollution: Volume 1: Air Pollution, Human Health, and the Environment, 257.

Aunan, K., & Pan, X. (2004). Exposure–response functions for health effects of ambient air pollution applicable for China—A meta-analysis. *The Science of the Total Environment*, *329*(1–3), 3–16. DOI: 10.1016/j.scitotenv.2004.03.008 PMID: 15262154

Awez, J., Khan, T., & Ali, M. (2024). Assessing air pollution crisis in South Asia: The health, economic, and environmental implications through case studies of Lahore, Delhi, and Dhaka. Pakistan Languages and Humanities Review, 8(2), 224–235.

Azhar, A.. (2019). The economic burden of outdoor air pollution-related morbidity and mortality in the cities of Lahore and Karachi, Pakistan. *Journal of Environmental Management*, *251*, 109557.

Azmi, W., Hassan, M. K., Houston, R., & Karim, M. S. (2021). ESG activities and banking performance: International evidence from emerging economies. *Journal of International Financial Markets, Institutions and Money*, *70*(1), 101–277. DOI: 10.1016/j.intfin.2020.101277

Bai, L., Wang, J., Ma, X., & Lu, H. (2018). Air Pollution Forecasts: An Overview. *International Journal of Environmental Research and Public Health*, *15*(4), 780. DOI: 10.3390/ijerph15040780

Bajwa, A. U., & Sheikh, H. A. (2023). Road transport's contribution to Pakistan's air Pollution in the urban environment. *Air*, *1*, 237–257. DOI: 10.3390/air1040018

Bala, S. A., & Ibrahim, M. (2022). Does the Disclosure of ESG Sustainability Matter to Selected Listed Firms Does the Disclosure of ESG Sustainability Matter to Selected Listed Firms Performance in Nigeria? *International Journal of Innovative Research in Accounting and Sustainability*, *7*(1), 130–141.

Baltagi, B. H. (2001). *A companion to theoretical econometrics* (Baltagi, B. H., Ed.). Vol. 1). Blackwell.

Balting, D. F., AghaKouchak, A., Lohmann, G., & Ionita, M. (2021). Northern Hemisphere drought risk in a warming climate. *NPJ Climate and Atmospheric Science*, *4*(1), 61. DOI: 10.1038/s41612-021-00218-2

Bansal, P., & Clelland, I. (2004). Talking trash: Legitimacy, impression management, and unsystematic risk in the context of the natural environment. *Academy of Management Journal, 47*(1), 93–103. DOI: 10.2307/20159562

Başar, S., & Sinan Temurlenk, M. (2007). Çevreye Uyarlanmış Kuznets Eğrisi: Türkiye Üzerine Bir Uygulama. *Atatürk Üniversitesi İktisadi Ve İdari Bilimler Dergisi, 21*(1), 1–12.

Beachey, W. (2022). *Respiratory care anatomy and physiology E-book: Foundations for clinical practice*. Elsevier Health Sciences.

Bell, M. L., & Davis, D. L. (2001). Reassessment of the lethal London fog of 1952: Novel indicators of acute and chronic consequences of acute exposure to air pollution. *Environmental Health Perspectives, 109*(Suppl 3), 389–394. DOI: 10.1289/ehp.01109s3389 PMID: 11427388

Belz, F. M., & Binder, J. K. (2017). Sustainable entrepreneurship: A convergent process model. *Business Strategy and the Environment, 26*(1), 1–17. DOI: 10.1002/bse.1887

Benis, A., Tamburis, O., Chronaki, C., & Moen, A. (2021). One digital health: A unified framework for future health ecosystems. *Journal of Medical Internet Research, 23*(2), e22189. DOI: 10.2196/22189 PMID: 33492240

Berardi, U., Civiero, P., Eicker, U., Gavalda, O., Montella, I., Ortiz, J., . . . Salom, J. (2025). Tools and protocols for cities/neighbourhood sustainability assessment. In Sustainability Certifications, Labels and Tools in the Built Environment (pp. 22-78). Routledge.

Bettencourt, L. M., Lobo, J., Helbing, D., Kühnert, C., & West, G. B. (2007). Growth, innovation, scaling, and the pace of life in cities. *Proceedings of the National Academy of Sciences of the United States of America, 104*(17), 7301–7306. DOI: 10.1073/pnas.0610172104 PMID: 17438298

Bhattarai, S. K., Dhakal, R., Acharya, S., & Arayal, S. (2024). Think global, act local: Developing an ebpm framework for air quality management in nepal. *Journal of water resources and pollution studies, 9*(3), 60-65. DOI: 10.46610/JoWRPS.2024.v09i03.005

Bich, H. N. T., & Dinh, T. L. (2024). Overcoming barriers to sustainable green transportation in ho chi minh city: A pathway toward achieving sdgs 11 and 13. *Sustainability (Basel), 16*(23), 10629–10629. DOI: 10.3390/su162310629

Biglari, S., Beiglary, S., & Arthanari, T. (2022). Achieving sustainable development goals: Fact or Fiction? *Journal of Cleaner Production, 332*, 130032. DOI: 10.1016/j.jclepro.2021.130032

Bilal, M., Hassan, M., Tahir, D. B. T., Iqbal, M. S., & Shahid, I. (2022). Understanding the role of atmospheric circulations and dispersion of air pollution associated with extreme smog events over South Asian megacity. *Environmental Monitoring and Assessment, 194*(2), 82. DOI: 10.1007/s10661-021-09674-y PMID: 35013892

Blair, M. J., Gagnon, B., Klain, A., & Kulišić, B. (2021). Contribution of biomass supply chains for bioenergy to sustainable development goals. *Land (Basel), 10*(2), 181. DOI: 10.3390/land10020181

Boachie, C., & Tetteh, J. E. (2021). Do creditors value corporate social responsibility disclosure? Evidence from Ghana. *International Journal of Ethics and Systems, 37*(3), 466–485. DOI: 10.1108/IJOES-11-2020-0181

Bölük, G., & Mert, M. (2015). The Renewable Energy, Growth and Environmental Kuznets Curve in Turkey: An ARDL Approach. *Renewable & Sustainable Energy Reviews, 52*(C), 587–595. DOI: 10.1016/j.rser.2015.07.138

Borenstein, M., Hedges, L. V., Higgins, J. P., & Rothstein, H. R. (2021). *Introduction to meta-analysis*. John wiley & sons. DOI: 10.1002/9781119558378

Boslaugh, S. (2009). *An Introduction to Secondary Data Analysis*. Secondary Data Sources for Public Health., DOI: 10.1017/CBO9780511618802.002

Bratman, G. N., Anderson, C. B., Berman, M. G., Cochran, B., de Vries, S., Flanders, J., Folke, C., Frumkin, H., Gross, J. J., Hartig, T., Kahn, P. H.Jr, Kuo, M., Lawler, J. J., Levin, P. S., Lindahl, T., Meyer-Lindenberg, A., Mitchell, R., Ouyang, Z., Roe, J., & Daily, G. C. (2019). Nature and mental health: An ecosystem service perspective. *Science Advances, 5*(7), eaax0903. Advance online publication. DOI: 10.1126/sciadv.aax0903

Brimblecombe, P. (2011). *The Big Smoke: A History of Air Pollution in London since Medieval Times*. Routledge.

Buhr, B. *et al.* (2018) 'Climate Change and the Cost of Capital in Developing Countries', *United Nations*, 48(August), pp. 01–38.

Büker, P., West, S. E., Bowyer, C., Apondo, W., Cinderby, S., Gray, C. M., Hahn, M., Lambe, F., Loh, M., Medcalf, A. J., Muhoza, C., Muindi, K., Njoora, T., Twigg, M., Waelde, C., Walnycki, A., Wainwright, M., Wendler, J., Wilson, M., & Price, H. D. (2024). Transdisciplinary research as a means of protecting human health, ecosystems and climate by engaging people to act on air pollution. DOI: 10.1079/onehealthcases.2024.0002

Burchard-Dziubińska, M. (2019). Air pollution and health in Poland: anti-smog movement in the most polluted polish cities. *Economics and Environment, 69*(2), 15. https://ekonomiaisrodowisko.pl/journal/article/view/85

Burchard-Dziubińska, M. (2019). Air pollution and health in Poland: Anti-smog movement in the most polluted polish cities. *Ecology & Environment, 69*(2), 15.

Burke, M., Driscoll, A., Heft-Neal, S., & Erhart, J. (2020). The changing risk and burden of wildfire in the United States. *Proceedings of the National Academy of Sciences of the United States of America, 117*(46), 24609–24617. DOI: 10.1073/pnas.2011048117 PMID: 33431571

Bussman, N., Tanda, A., & Yu, E. P. (2022) '*Can ESG shape cost of capital? A bibliometric review and empirical analysis through ML*', pp. 1–20. DOI: 10.2139/ssrn.4173890

Butt, M. U., Waseef, R. F., & Ahmed, H. (2018). Perception about the factors associated with smog among medical students [Article]. *Biomedica, 34–34*(4). https://www.researchgate.net/publication/358615094

Buttel, F. H. (2000). Ecological modernization as social theory. *Geoforum, 31*(1), 57–65. DOI: 10.1016/S0016-7185(99)00044-5

Butt, M. U., Waseef, R. F., & Ahmed, H. (2018). Perception about the factors associated with smog among medical students. *Biomedica, 34*(4).

Bye, B., Kaushal, K. R., Rosnes, O., Turner, K., & Yonezawa, H. (2023). The road to a low emission society: Costs of interacting climate regulations. *Environmental and Resource Economics, 86*(3), 565–603. DOI: 10.1007/s10640-023-00804-3

Caniglia, B. S., Vallée, M., & Frank, B. (Eds.). (2016). *Resilience, environmental justice and the city*. Routledge. DOI: 10.4324/9781315652054

Cao, S., Chen, L., Shankman, D., Wang, C., Wang, X., & Zhang, H. (2011). Excessive reliance on afforestation in China's arid and semi-arid regions: Lessons in ecological restoration. *Earth-Science Reviews, 104*(4), 240–245. DOI: 10.1016/j.earscirev.2010.11.002

Carta, F., Zidda, C., Putzu, M., Loru, D., Anedda, M., & Giusto, D. (2023). Advancements in forest fire prevention: A comprehensive survey. *Sensors (Basel), 23*(14), 6635. DOI: 10.3390/s23146635 PMID: 37514928

Ceder, A. (2021). Urban mobility and public transport: Future perspectives and review. *International Journal of Urban Sciences, 25*(4), 455–479. DOI: 10.1080/12265934.2020.1799846

Central Bank of the Republic of Turkey (TCMB). (2025). "Public Finance – Debt Stock (% of GDP)" table. Retrieved [02.04.2025] from https://www.tcmb.gov.tr/wps/wcm/connect/a26d8980-d468-4d79-beaf-b1b02c7fe4e3/6c11-1.pdf

Central Pollution Control Board (CPCB). (2020). *National Ambient Air Quality Status and Trends Report*. Ministry of Environment, Forest and Climate Change, Government of India.

Central Pollution Control Board (CPCB). (2022). *National ambient air quality status and trends in India – 2022*. Government of India.

Çetin, M., Eyyup, E., & Yucel, E. (2018). The impact of economic growth, energy consumption, trade openness, and financial development on carbon emissions: Empirical evidence from Turkey. *Environmental Science and Pollution Research International, 25*(36), 36589–36603. DOI: 10.1007/s11356-018-3526-5

Chandel, A. F. (2022). *Green entrepreneurship and sustainable development*. ResearchGate. https://www.researchgate.net/publication/362856766_Green_Entrepreneurship_and_Sustainable_Development

Chatzitheodorou, K., Tsalis, T. A., Tsagarakis, K. P., Evangelos, G., & Ioannis, N. (2021). A new practical methodology for the banking sector to assess corporate sustainability risks with an application in the energy sector. *Sustainable Production and Consumption, 27*, 1473–1487. DOI: 10.1016/j.spc.2021.03.005

Chen, B., Chen, Y., & Liu, X. (2020). Evaluating the effect of urban shelterbelts on dust storm mitigation in northern China. *Environmental Monitoring and Assessment, 192*(5), 302. DOI: 10.1007/s10661-015-4536-0 PMID: 32322989

Chen, B., & Kan, H. (2008). Air pollution and population health: A global challenge. *Environmental Health and Preventive Medicine, 13*(2), 94–101. DOI: 10.1007/s12199-007-0018-5

Chen, M., Liu, W., Lu, D., Chen, H., & Ye, C. (2018). Progress of China's new-type urbanization construction since 2014: A preliminary assessment. *Cities (London, England), 78*, 180–193. DOI: 10.1016/j.cities.2018.02.012

Chen, M., Zhang, H., Liu, W., & Zhang, W. (2014). The global pattern of urbanization and economic growth: Evidence from the last three decades. *PLoS One*, *9*(8), e103799. DOI: 10.1371/journal.pone.0103799 PMID: 25099392

Chen, Z., Wang, J.-N., Ma, G.-X., & Zhang, Y.-S. (2018). China tackles the health effects of air pollution. *Lancet*, *382*(9909), 1959–1960. DOI: 10.1016/S0140-6736(13)62064-4 PMID: 24332069

Cheshmehzangi, A., & Tang, T. (2024). *30 Years of Urban Change in China's 10 Core Cities*. Springer.

Cho, M. H. (2024). Key Factors Affecting Startups' Contribution to SDGs for a Sustainable Future: Integrating a Triple Bottom Line (TBL) Theory. *Journal of Ecohumanism*, *3*(7), 615–631. DOI: 10.62754/joe.v3i7.4232

Christie, A. (2021). Decreases in COVID-19 cases, emergency department visits, hospital admissions, and deaths among older adults following the introduction of COVID-19 vaccine United States, September 6, 2020 May 1, 2021. *MMWR. Morbidity and Mortality Weekly Report*, •••, 70. DOI: 10.15585/mmwr.mm7023e2 PMID: 34111059

Chua, R. Y., Huang, K. G., & Jin, M. (2019). Mapping cultural tightness and its links to innovation, urbanization, and happiness across 31 provinces in China. *Proceedings of the National Academy of Sciences of the United States of America*, *116*(14), 6720–6725. DOI: 10.1073/pnas.1815723116 PMID: 30833399

Chu, F., Zhang, W., & Jiang, Y. (2021). How does policy perception affect green entrepreneurship behavior? An empirical analysis from China. *Discrete Dynamics in Nature and Society*, *2021*, 1–9. DOI: 10.1155/2021/7973046

Church, R. M. (2002). The effective use of secondary data. *Learning and Motivation*, *33*(1), 32–45. DOI: 10.1006/lmot.2001.1098

Coalition, C. C. A. (2019). Air pollution in asia and the pacific: Science-based solutions.

Cojoianu, T. F., Hoepner, A. G., Hu, X., Ramadan, M., Veneri, P., & Wójcik, D. (2024). Are cities venturing green? A global analysis of the impact of green entrepreneurship on city air pollution. *Small Business Economics*, *62*(2), 523–540. DOI: 10.1007/s11187-023-00764-4

Colbeck, I., Nasir, Z. A., & Ali, Z. (2010). The state of ambient air quality in Pakistan—A review. *Environmental Science and Pollution Research International*, *17*(1), 49–63. DOI: 10.1007/s11356-009-0217-2 PMID: 19575252

Coovadia, H., Jewkes, R., Barron, P., Sanders, D., & McIntyre, D. (2009). The health and health system of South Africa: Historical roots of current public health challenges. *Lancet*, *374*(9692), 817–834. DOI: 10.1016/S0140-6736(09)60951-X PMID: 19709728

Coutts, C. (2016). *Green infrastructure and public health*. Routledge. DOI: 10.4324/9781315647623

Criado-Gomis, A., Cervera-Taulet, A., & Iniesta-Bonillo, M.-A. (2017). Sustainable Entrepreneurial Orientation: A Business Strategic Approach for Sustainable Development. *Sustainability (Basel)*, *9*(9), 1667. DOI: 10.3390/su9091667

Cronin, C. J., & Evans, W. N. (2021). Total shutdowns, targeted restrictions, or individual responsibility: How to promote social distancing in the COVID-19 Era? *Journal of Health Economics*, *79*, 102497. DOI: 10.1016/j.jhealeco.2021.102497 PMID: 34325285

Crowe, S. *et al.* (2011) 'The case study approach', *BMC Medical Research Methodology*, 1(1), pp. 02–09.

Cui, Y., Geobey, S., Weber, O., & Lin, H. (2018). The impact of green lending on credit risk in China. *Sustainability (Basel)*, *10*(6), 1–16. DOI: 10.3390/su10062008

Daly, A., & Zannetti, P. (2007). Air Pollution Modeling – An Overview. In P. Zannetti, D. Al-Ajmi, & S. Al-Rashied (Eds.), *AMBIENT AIR POLLUTION*. The Arab School for Science and Technology (ASST). https://home.iitk.ac.in/~anubha/Modeling.pdf

Das, S., Pal, D., & Sarkar, A. (2021). Particulate matter pollution and global agricultural productivity. In *Sustainable Agriculture Reviews 50: Emerging Contaminants in Agriculture* (pp. 79-107). Springer. DOI: 10.1007/978-3-030-63249-6_4

Dawn (2021). Lahore again shows worst air quality. https://shorturl.at/V02d6

Dean, T. J., & McMullen, J. S. (2007). Toward a theory of sustainable entrepreneurship: Reducing environmental degradation through entrepreneurial action. *Journal of Business Venturing*, *22*(1), 50–76. DOI: 10.1016/j.jbusvent.2005.09.003

Dey, S., Sarkar, D., Imam, F., Kumar, A., Mukherjee, A., Purohit, P., Kiesewetter, G., & Klimont, Z. (2023). Pathways for india to reduce air pollution health burden and meet sustainable development goal 3.4 target. *Environmental health perspectives*null, . DOI: 10.1289/isee.2023.MP-052

Diale, C. D., Kanakana-Katumba, G., & Maladzhi, R. W. (2019, December). *IEEE International Conference on Industrial Engineering and Engineering Management (IEEM)* (pp. 384-389). IEEE. Doi: DOI: 10.1109/IEEM44572.2019.8978804

Díaz, C., González, D., & Rodríguez, M. A. (2021). Addressing urban health inequality through integrated social and environmental policies. *International Journal of Environmental Research and Public Health, 18*(6), 3089. DOI: 10.3390/ijerph18063089

Dilotsotlhe, N., & Duh, H. I. (2021). 'Drivers of Middle-Class Consumers' Green Appliance Attitude and Purchase Behavior: A Multi-Theory Application'. *Social Marketing Quarterly, 27*(2), 150–171. DOI: 10.1177/15245004211013737

DiMaggio, P. J., & Powell, W. W. (1983). The iron cage revisited: Institutional isomorphism and collective rationality. *American Sociological Review, 48*(2), 147–160. DOI: 10.2307/2095101

Dinda, S. (2004). Environmental Kuznets Curve Hypothesis: A Survey. *Ecological Economics, 49*(4), 431–455. DOI: 10.1016/j.ecolecon.2004.02.011

Domingo, N. G., Balasubramanian, S., Thakrar, S. K., Clark, M. A., Adams, P. J., Marshall, J. D., Muller, N. Z., Pandis, S. N., Polasky, S., Robinson, A. L., Tessum, C. W., Tilman, D., Tschofen, P., & Hill, J. D. (2021). Air quality–related health damages of food. *Proceedings of the National Academy of Sciences of the United States of America, 118*(20), e2013637118. DOI: 10.1073/pnas.2013637118 PMID: 33972419

Dougherty, S., & Nebreda, A. M. (2022). Going global, locally? Decentralized environmental expenditure and air quality. *Public Sector Economics, 46*(4), 489–503. DOI: 10.3326/pse.46.4.3

Dube, S., & Maroun, W. (2017). Corporate social responsibility reporting by South African mining companies: Evidence of legitimacy theory. *South African Journal of Business Management, 48*(1), 23–34. DOI: 10.4102/sajbm.v48i1.17

Dubihlela, J., & Ngxukumeshe, T. (2016). Customer Attributes and the Repurchase. *The International Business & Economics Research Journal, 15*(4), 163–175. DOI: 10.19030/iber.v15i4.9754

Duran, C. (2024). Bibliometric Analysis of Green Entrepreneurship. *Beykoz Akademi Dergisi, 12*(2), 250–282. DOI: 10.14514/beykozad.1519660

Elkington, J. (1997). *Cannibals with forks: The triple bottom line of 21st-century business*. Capstone.

Elmqvist, T., Andersson, E., Frantzeskaki, N., McPhearson, T., Olsson, P., Gaffney, O., Takeuchi, K., & Folke, C. (2019). Sustainability and resilience for transformation in the urban century. *Nature Sustainability, 2*(4), 267–273. DOI: 10.1038/s41893-019-0250-1

Elsom, D. M. (1996). Smog alert: Managing urban air quality.

EPA. (2023). *Progress Cleaning the Air and Improving People's Health.* United States Environmental Protection Agency. https://www.epa.gov/clean-air-act-overview/progress-cleaning-air-and-improving-peoples-health

Escobedo, F. J., Kroeger, T., & Wagner, J. E. (2011). Urban forests and pollution mitigation: Analyzing ecosystem services and disservices. *Environmental Pollution, 159*(8–9), 2078–2087. DOI: 10.1016/j.envpol.2011.01.010 PMID: 21316130

Esqueda, H. (2019). A profile of the popular entrepreneur: A value-based approach. *Academia (Caracas), 32*(2), 267–281. DOI: 10.1108/ARLA-03-2018-0055

European Environment Agency. (2024). Green bonds in Europe - European Environment Agency, https://www.eea.europa.eu/en/analysis/indicators/green-bonds-8th-eap

Eurostat. (2024). *Green bond issuance and investments in clean technologies (EU-27).* https://ec.europa.eu/eurostat

FAO. 2010. Global forest resources assessment- (2010). FAO forestry paper No. 163. UN Food and Agriculture Organization, Rome. pp 85-146. https://www.fao.org/3/a-i1757e.pdf

FAO. 2023. The State of the Food and Agriculture: Revealing the True Cost of Food to Transform Agrifood Systems. Rome. https://doi.org/.DOI: 10.4060/cc7724en

Farooq, M. S., Feroze, N., Cheng, T., & Feroze, F. (2024). Pakistan's Green Urbanization in the Perspective of Ecological Civilization. *Chinese Journal of Urban and Environmental Studies, 12*(03), 2450017. DOI: 10.1142/S2345748124500179

Farrokhi, F., & Mahmoudi-Hamidabad, A. (2012). Rethinking convenience sampling: Defining quality criteria. *Theory and Practice in Language Studies, 2*(4), 784–792. DOI: 10.4304/tpls.2.4.784-792

Farzaneh, H., Oliveira, J. A. P. D., McLellan, B., & Ohgaki, H. (2019). Towards a low emission transport system: Evaluating the public health and environmental benefits. *Energies, 12*(19), 3747. Advance online publication. DOI: 10.3390/en12193747

Fatkhutdinova, L., Tafeeva, E., Timerbulatova, G., & Zalyalov, R. (2021). Health risks of air pollution with fine particulate matter. *Kazan medical journal, 102*(6), 862-876.

Fazakas, E., Neamtiu, I. A., & Gurzau, E. S. (2024). Health effects of air pollutant mixtures (volatile organic compounds, particulate matter, sulfur and nitrogen oxides)–a review of the literature. *Reviews on Environmental Health, 39*(3), 459–478. DOI: 10.1515/reveh-2022-0252 PMID: 36932657

Fernando, S., & Lawrence, S. (2014). A theoretical framework for CSR practices: Integrating legitimacy theory, stakeholder theory and institutional theory. *Journal of theoretical accounting research, 10*(1), 149-178.

Filippi, F. (2022). A paradigm shift for a transition to sustainable urban transport. *Sustainability (Basel), 14*(5), 2853. DOI: 10.3390/su14052853

Fowler, D., Brimblecombe, P., Burrows, J., Heal, M. R., Grennfelt, P., Stevenson, D. S., Jowett, A., Nemitz, E., Coyle, M., Liu, X., Chang, Y., Fuller, G. W., Sutton, M. A., Klimont, Z., Unsworth, M. H., & Vieno, M. (2020). A chronology of global air quality. *Philosophical Transactions. Series A, Mathematical, Physical, and Engineering Sciences, 378*(2183), 20190314. DOI: 10.1098/rsta.2019.0314

Friedmann, J. (2006). Four theses in the study of China's urbanization. *International Journal of Urban and Regional Research, 30*(2), 440–451. DOI: 10.1111/j.1468-2427.2006.00671.x

Friedman, S. R., Mateu-Gelabert, P., Nikolopoulos, G. K., Cerdá, M., Rossi, D., Jordan, A. E., Townsend, T., Khan, M. R., & Perlman, D. C. (2021). Big Events theory and measures may help explain emerging long-term effects of current crises. *Global Public Health: An International Journal for Research, Policy and Practice, 16*(8-9), 1167–1186. DOI: 10.1080/17441692.2021.1903528 PMID: 33843462

Fu, B., Meadows, M. E., & Zhao, W. (2022). Geography in the Anthropocene: Transforming our world for sustainable development. *Geography and Sustainability, 3*(1), 1–6. DOI: 10.1016/j.geosus.2021.12.004

Fuhrmann, S., Ott, C., Looks, E., & Guenther, T. W. (2017). The contents of assurance statements for sustainability reports and information asymmetry. *Accounting and Business Research, 47*(4), 369–400. DOI: 10.1080/00014788.2016.1263550

Fujikura, R. (2001). Public participation, stakeholder involvement and consensus building in Japan's environmental policy-making. *Environmental Impact Assessment Review, 21*(5), 435–453. DOI: 10.1016/S0195-9255(01)00081-1

Gao, J., Hua, G., AbidAli, R., Huo, B., Mahamane, F., & Li, Z. (2024). The impact of green low-carbon finance on smog reduction and carbon mitigation: Evidence from a spatial panel simultaneous equations model. *The International Journal of Urban Sciences*null, 1-34. DOI: 10.1080/12265934.2024.2438204

Gao, Y., Tsai, S. B. J., Zhang, S., & Li, G. (2018). Green entrepreneurship in transitional economies: Breaking through the constraints of legitimacy. In *Green Production Strategies for Sustainability* (pp. 184–209). IGI Global., DOI: 10.4018/978-1-5225-3537-9.ch011

Gaybullaev, O. (2024). An analysis of the economic impacts of air pollution in the urban areas: Bibliographic review. *Economica, 15*(3-4), 53–68. DOI: 10.47282/economica/2024/15/3-4/15028

Getahun, M. (2014) 'Determinants of capital structure and its impact on the performance of Ethiopian insurance industry'.

Getzmann, A., Lang, S., & Spremann, K. (2014). Target capital structure and adjustment speed in asia. *Asia-Pacific Journal of Financial Studies, 43*(1), 1–30. Advance online publication. DOI: 10.1111/ajfs.12038

GFW. 2024. Global Forest Watch https://www.globalforestwatch.org/dashboards/country/PAK/?category=forest-change&location=WyJjb3VudHJ5IiwiUEFLIl0%3D

Ghani, F., Tsekleves, E., & Thomas, Y. F. (2021). Urbanization and cities as drivers of global health. In *Handbook of Global Health* (pp. 1–28). Springer International Publishing.

Giles-Corti, B., Moudon, A. V., Lowe, M., Cerin, E., Boeing, G., Frumkin, H., Salvo, D., Foster, S., Kleeman, A., Bekessy, S., de Sá, T. H., Nieuwenhuijsen, M., Higgs, C., Hinckson, E., Adlakha, D., Arundel, J., Liu, S., Oyeyemi, A. L., Nitvimol, K., & Sallis, J. F. (2022). What next? Expanding our view of city planning and global health, and implementing and monitoring evidence-informed policy. *The Lancet. Global Health, 10*(6), e919–e926. DOI: 10.1016/S2214-109X(22)00066-3 PMID: 35561726

Gill, A. *et al.* (2009) 'The Determinants of Capital Structure in the Service Industry : Evidence from United States'. .DOI: 10.2174/1874915100902010048

Giusti, L. (2009). A review of waste management practices and their impact on human health. Waste management, 29(8), 2227-2239.

GIZ (Deutsche Gesellschaft für Internationale Zusammenarbeit). (2014). *Green entrepreneurship: A promising path towards a sustainable future in Tunisia and elsewhere.* Retrieved from https://www.greenpolicyplatform.org/sites/default/files/downloads/resource/ GIZ_%20Green%20Entrepreneurship%20in%20Tunisia.pdf

Global trends in motor vehicle pollution control: a 2011 update. (2011). In *COMBUSTION ENGINES* (pp. 106–107).

Global trends in motor vehicle pollution control: a 2011 update. (2011). In *COMBUSTION ENGINES*, 106–107.

Gondhalekar, D., & Ramsauer, T. (2017). Nexus city: Operationalizing the urban water-energy-food nexus for climate change adaptation in Munich, Germany. *Urban Climate*, *19*, 28–40. DOI: 10.1016/j.uclim.2016.11.004

Gong, P., Liang, S., Carlton, E. J., Jiang, Q., Wu, J., Wang, L., & Remais, J. V. (2012). Urbanisation and health in China. *Lancet*, *379*(9818), 843–852. DOI: 10.1016/S0140-6736(11)61878-3 PMID: 22386037

Government of Pakistan. (2021). Pakistan Clean Air Action Plan 2021. Ministry of Climate Change. https://www.mocc.gov.pk

Government of Punjab. (2021). *GOVERNMENT OF THE PUNJAB, ENVIRONMENT PROTECTION DEPARTMENT POLICY NUMBER – No. SOG/EPD/7-26/2013(P-II)*. Government of Punjab.

Greenstone, M., & Fan, C. (2019). *Clean air and its impact on life expectancy in India*. Energy Policy Institute at the University of Chicago (EPIC).

Greyvenstein, R. (2010) *Earnings management on the JSE before and after King II*.

Groenewald, T. (2004). A Phenomenological Research Design Illustrated. *International Journal of Qualitative Methods*, *3*(1), 42–55. DOI: 10.1177/160940690400300104

Gu, C. (2019). Urbanization: Positive and negative effects. *Science Bulletin*, *64*(5), 281–283. DOI: 10.1016/j.scib.2019.01.023 PMID: 36659587

Gulia, S., Nagendra, S. M. S., Khare, M., & Khanna, I. (2021). Urban air quality management—A review. *Atmospheric Environment*, *246*, 118093. DOI: 10.1016/j.atmosenv.2020.118093

Guo, G., Zhang, D., & Wang, Y. (2021). Characteristics of heavy metals in size-fractionated atmospheric particulate matters and associated health risk assessment based on the respiratory deposition. *Environmental Geochemistry and Health*, *43*(1), 285–299. DOI: 10.1007/s10653-020-00706-z PMID: 32895728

Guo, L., Cheng, Z., Tani, M., Cook, S., Zhao, J., & Chen, X. (2025). Air pollution and entrepreneurship. *China Economic Review*, *89*, 102327. DOI: 10.1016/j.chieco.2024.102327

Gurjar, B. R., Jain, A., Sharma, A., Agarwal, A., Gupta, P., Nagpure, A. S., & Lelieveld, J. (2016). Human health risks in megacities due to air pollution. *Atmospheric Environment*, *128*, 233–243. DOI: 10.1016/j.atmosenv.2015.12.039

Gurjar, B. R., Ravindra, K., & Nagpure, A. S. (2016). Air pollution trends over Indian megacities and their local-to-global implications. *Atmospheric Environment*, *142*, 475–495. DOI: 10.1016/j.atmosenv.2016.06.030

Guthrie, J., Petty, R., Yongvanich, K., & Ricceri, F. (2004). Using content analysis as a research method to inquire into intellectual capital reporting. *Journal of Intellectual Capital*, *5*(2), 282–293. DOI: 10.1108/14691930410533704

Guttikunda, S. K., & Gurjar, B. R. (2012). Role of meteorology in seasonality of air pollution in megacity Delhi, India. *Environmental Monitoring and Assessment*, *184*(5), 3199–3211. DOI: 10.1007/s10661-011-2182-8 PMID: 21713474

Gu, W., Pan, H., Hu, Z., & Liu, Z. (2022). The triple bottom line of sustainable entrepreneurship and economic policy uncertainty: An empirical evidence from 22 countries. *International Journal of Environmental Research and Public Health*, *19*(13), 7758. DOI: 10.3390/ijerph19137758

Hajishengallis, G., & Chavakis, T. (2021). Local and systemic mechanisms linking periodontal disease and inflammatory comorbidities. *Nature Reviews. Immunology*, *21*(7), 426–440. DOI: 10.1038/s41577-020-00488-6 PMID: 33510490

Halim, N. D. A., Latif, M. T., & Mohamed, A. F. (2024). Assessing sustainable approaches to urban air quality: A study of malaysia's largest conurbation regions. DOI: 10.5194/egusphere-egu24-12920

Halley, J. M., Van Houtan, K. S., & Mantua, N. (2018). How survival curves affect populations' vulnerability to climate change. *PLoS One*, *13*(9), e0203124. DOI: 10.1371/journal.pone.0203124 PMID: 30188919

Hall, J. K., Daneke, G. A., & Lenox, M. J. (2010). Sustainable development and entrepreneurship: Past contributions and future directions. *Journal of Business Venturing*, *25*(5), 439–448. DOI: 10.1016/j.jbusvent.2010.01.002

Hameed, I., Zaman, U., Waris, I., & Shafique, O. (2021). A serial-mediation model to link entrepreneurship education and green entrepreneurial behavior: Application of resource-based view and flow theory. *International Journal of Environmental Research and Public Health*, *18*(2), 571. DOI: 10.3390/ijerph18020550

Hanif, M. N., & Benitez, I. B. (2025). Soil Pollution in Urban Environments: Sources, Consequences, Potential Mitigation Strategies and the Importance of Sustainable Urban Development. *Water, Air, and Soil Pollution*, *236*(14), 945. DOI: 10.1007/s11270-025-08421-0

Head, B. W. (2022). *Wicked problems in public policy: Understanding and responding to complex challenges*. Springer Nature. DOI: 10.1007/978-3-030-94580-0

Health Effects Institute (HEI). (2020). *State of Global Air 2020.* https://www.stateofglobalair.org/

He, C., Huang, C., Zhang, P., Wu, Q., Wang, Y., Zeng, Z., Zhang, L., Miao, C., Yue, F., Tang, Y., Zhou, J., Zhou, Y., Wang, Y., Yang, L., Hong, S., Peng, S., Gong, X., & Chen, W. (2024). Global pm2.5 exposure inequality hindering progress toward air quality-related sdgs. DOI: 10.21203/rs.3.rs-5072048/v1

Hector, H. (2021). Understanding and development of sustainability investing in the recent history. *Humanum*, *41*(2), 33–39.

He, J., Yang, Y., Liao, Z., Xu, A., & Fang, K. (2022). Linking SDG 7 to assess the renewable energy footprint of nations by 2030. *Applied Energy*, *317*, 119167. DOI: 10.1016/j.apenergy.2022.119167

Hendricks, M. D., & Van Zandt, S. (2021). Unequal protection revisited: Planning for environmental justice, hazard vulnerability, and critical infrastructure in communities of color. *Environmental Justice*, *14*(2), 87–97. DOI: 10.1089/env.2020.0054

Hou, D. (2021). Sustainable remediation in China: Elimination, immobilization, or dilution. *Environmental Science & Technology*, *55*(23), 15572 15574. DOI: 10.1021/acs.est.1c06044 PMID: 34758622

Hox, J. J., & Boeije, H. R. (2004) 'Data Collection, Primary vs. Secondary', *Encyclopedia of Social Measurement*, pp. 593–599. DOI: 10.1016/B0-12-369398-5/00041-4

Hu, Y., Chao, K., Zhu, Z., Yue, J., Qie, X., & Wang, M. (2024). A study on a health impact assessment and healthcare cost calculation of beijing–tianjin–hebei residents under pm2.5 and o3 pollution. *Sustainability*null, . DOI: 10.3390/su16104030

Huang, J., Liu, R., Wang, W., Wang, Z. A., Wang, C., & Jin, Y. J. (2024). Unleashing Fintech's potential: A catalyst for green bonds issuance. *Journal of International Financial Markets, Institutions and Money*, *93*, 102009. DOI: 10.1016/j.intfin.2024.102009

Hui, C., Foxcroft, L. C., Richardson, D. M., & MacFadyen, S. (2011). Defining optimal sampling effort for large-scale monitoring of invasive alien plants: A Bayesian method for estimating abundance and distribution. *Journal of Applied Ecology*, *48*(3), 768–776. DOI: 10.1111/j.1365-2664.2011.01974.x

Hynes, M., Crippen, A., Lemons, K., & Varnell, E. (2024). *Zeroing in on zero-emission buses*.

Imran, M., Khan, S., Nassani, A. A., Haffar, M., Khan, H. U. R., & Zaman, K. (2023). Access to sustainable healthcare infrastructure: A review of industrial emissions, coal fires, and particulate matter. *Environmental Science and Pollution Research International*, *30*(26), 69080–69095. DOI: 10.1007/s11356-023-27218-4 PMID: 37129815

International Energy Agency (IEA). (2023). *World energy outlook 2023*. IEA.

Ionescu, L. (2021). Corporate environmental performance, climate change mitigation, and green innovation behavior in sustainable finance. *Economics, Management, and Financial Markets, 16*(3), 94–106. DOI: 10.22381/emfm16320216

IPCC. (2023). Summary for Policymakers. In *Climate Change 2023: Synthesis Report. Contribution of Working Groups I, II and III to the Sixth Assessment Report of the Intergovernmental Panel on Climate Change* [Core Writing Team, H. Lee and J. Romero (eds.)]. (pp. 1–34). IPCC., DOI: 10.59327/IPCC/AR6-9789291691647

IQAir. (2022). *World air quality report 2022*. IQAir.

IQAir. (2023). *World air quality report 2023*. IQAir.

IQAir. (2024). *2023 World Air Quality Report: Highlighting global air quality performance*. Geneva: IQAir. Retrieved from https://www.greenpeace.org/static/planet4-india-stateless/2024/03/44a856c8-2023_world_air_quality_report.pdf

Iqbal, M., Khan, A., Rauf, A., & Ahmad, Z. (2023). Air pollution and its socio-economic impacts in South Asia: Challenges and strategies. *Journal of Environmental Research and Public Health, 20*(4), 215–228. DOI: 10.3390/ijerph2004215 PMID: 36833854

Iqbal, N., Khan, A., Gill, A. S., & Abbas, Q. (2020). Nexus between sustainable entrepreneurship and environmental pollution: Evidence from developing economy. *Environmental Science and Pollution Research International, 27*(29), 36242–36253. DOI: 10.1007/s11356-020-09642-y

Iram, S., Qaisar, I., Shabbir, R., Pomee, M. S., Schmidt, M., & Hertig, E. (2025). Impact of air pollution and smog on human health in Pakistan: A systematic review. *Environments (Basel, Switzerland), 12*(2), 46. DOI: 10.3390/environments12020046

IRENA – International Renewable Energy Agency. (2023). *Decentralised solar PV: A gender perspective*. https://www.irena.org/publications

Irfan, M., Razzaq, A., Sharif, A., & Yang, X. (2022). Influence mechanism between green finance and green innovation: Exploring regional policy intervention effects in China. *Technological Forecasting and Social Change, 182*, 121882. DOI: 10.1016/j.techfore.2022.121882

Isha, N., Ali, S., Khalid, A., Naseer, I. A., Raza, H., & Chang, Y. (2024). Bioremediation of smog: Current trends and future perspectives. *Processes (Basel, Switzerland), 12*(10), 2266. DOI: 10.3390/pr12102266

Jabeen, N., Farwa, U., & Jadoon, M. (2017). Urbanization in Pakistan: A governance perspective. *Journal of the Research Society of Pakistan, 54*(1), 127–136.

Jacob, D. J. (1999). *Introduction to atmospheric chemistry*. Princeton University Press. DOI: 10.2307/j.ctt7t8hg

Jacob, D. J., & Winner, D. A. (2009). Effect of climate change on air quality. *Atmospheric Environment, 43*(1), 51–63. DOI: 10.1016/j.atmosenv.2008.09.051

Jaiswal, A., Sagar, R., Pandey, A., Yadav, D., Ansari, M., & Rawat, R. (2024). Building Resilient Urban Futures: Adapting Cities to Climate Change Challenges. In *Cities of Tomorrow: Urban Resilience and Climate Change Preparedness* (pp. 51–68). Springer. DOI: 10.1007/978-981-97-9658-8_3

Janardhanan, N. (2017). Energy Transition in India: Role of Climate Change Policies in Energy Security. *International Studies, 54*(1-4), 231–249. DOI: 10.1177/0020881718791851

Jena, M. C., Mishra, S. K., & Moharana, H. S. (2023). Challenges and way forward to maintain air quality standard in urban areas. *Global environmental engineers*null, . DOI: 10.15377/2410-3624.2023.10.4

Jennings, P. D., & Zandbergen, P. A. (1995). Ecologically sustainable organizations: An institutional approach. *Academy of Management Review, 20*(4), 1015–1052. DOI: 10.2307/258964

Johnston, F. H., Borchers-Arriagada, N., Morgan, G. G., Jalaludin, B., Palmer, A. J., Williamson, G. J., & Bowman, D. M. J. S. (2021). Unprecedented health costs of smoke-related PM2.5 from the 2019–2020 Australian bushfires. *Nature Sustainability, 4*(1), 42–47. DOI: 10.1038/s41893-020-00610-5

Johnston, F. H., Henderson, S. B., Chen, Y., Randerson, J. T., Marlier, M., DeFries, R. S., Kinney, P., Bowman, D. M. J. S., & Brauer, M. (2019). Estimated global mortality attributable to smoke from landscape fires. *Environmental Health Perspectives, 120*(5), 695–701. DOI: 10.1289/ehp.1104422 PMID: 22456494

JSE. (2016). *JSE limited listings requirements*. Johannesburg Stock Exchange.

Junejo, Z. U. B. (2025). Pakistan Today. Death in the air demanding national action. https://shorturl.at/zuf77

Jung, J., Herbohn, K., & Clarkson, P. (2018). Carbon Risk, Carbon Risk Awareness and the Cost of Debt Financing. *Journal of Business Ethics, 150*(4), 1151–1171. DOI: 10.1007/s10551-016-3207-6

Kachynska, N., Prakhovnik, N., Zemlyanska, O., Ilchuk, O., & Kovtun, A. (2022). Ecological modernization of enterprises: environmental risk management. *Health Education and Health Promotion, 10*(1), 175-182. https://hehp.modares.ac.ir/article-5-59219-en.html

Kaddoura, Y., & Westerlund, J. (2023). Estimation of Panel Data Models with Random Interactive Effects and Multiple Structural Breaks when T is Fixed. *Journal of Business & Economic Statistics, 41*(3), 778–790. DOI: 10.1080/07350015.2022.2067546

Kamal, A., Malik, R. N., Martellini, T., & Cincinelli, A. (2015). Source, profile, and carcinogenic risk assessment for cohorts occupationally exposed to dust-bound PAHs in Lahore and Rawalpindi cities (Punjab province, Pakistan). *Environmental Science and Pollution Research International, 22*(14), 10580–10591. DOI: 10.1007/s11356-015-4215-2 PMID: 25739839

Kamala, P. N., Wingard, H. C., & Cronje, C. J. (2015). Evolution of corporate environmental reports in South Africa. *Journal of Accounting and Management, 5*(2), 31–34.

Kampa, M., & Castanas, E. (2008). Human health effects of air pollution. *Environmental Pollution, 151*(2), 362–367. DOI: 10.1016/j.envpol.2007.06.012 PMID: 17646040

Kandpal, P. C. (2024). *Combating air pollution: Comparisons between Delhi and Mexico City*. Springer Cham., DOI: 10.1007/978-3-031-68027-4

Karadayı, T., Yazıcı, S., & Akdemir Ömür, G. (2025). Climate Change and Entrepreneurship: A Review and Research Agenda. *Global Business and Organizational Excellence, 44*(4), 16–43. DOI: 10.1002/joe.22281

Kesa, D. D., Wu, M., Abdillah, F., Harjadi, D., & Ningsih, A. (2024). The Impact of Digital Entrepreneurship Courses and Motivation on Fostering a Green Entrepreneurial Spirit in Students' Perceptions. *International Journal of Educational Qualitative Quantitative Research, 3*(2), 46–55. DOI: 10.58418/ijeqqr.v3i2.120

Khan, M., Zhang, Y., & Khan, S. (2025). Echoes of Survival: Climate Change Impact & Typologies of Adaptation among Vulnerable Communities toward Climate-Induced Food Insecurity in Pakistan. *Research on World Agricultural Economy*, 290-318.

Khan, Z. M. (2024). *Unpacking the Smog Crisis in Punjab – Key Areas of Research.* https://shorturl.at/F2iw7

Khanam, Z., Sultana, F. M., & Mushtaq, F. (2023). Environmental pollution control measures and strategies: an overview of recent developments. *Geospatial Analytics for Environmental Pollution Modeling: Analysis, Control and Management*, 385-414.

Khandelwal, H., Thalla, A. K., Kumar, S., & Kumar, R. (2019). Life cycle assessment of municipal solid waste management options: A review. *Journal of Cleaner Production, 209*, 630–653. DOI: 10.1016/j.jclepro.2018.10.233

Khattak, M., & Khan, S. (2024). Smog and Health in Punjab: Analyzing Awareness, Protective Behaviors, and Public Support for Air Quality Policies. *Annals of Human and Social Sciences*, (05), 293–305.

Khosla, A. (2024). Sustainable urban development: Integrating smart city technologies for environmental resilience. *Innovative research thoughts, 10* (3), 138-142. DOI: 10.36676/irt.v10.i4.1472

Kim Oanh, N. T., Upadhyay, N., Zhuang, Y.-H., Hao, Z.-P., Murthy, D. V. S., Lestari, P., Villarin, J. T., Chengchua, K., Co, H. X., & Dung, N. T. (2006). Particulate air pollution in six Asian cities: Spatial and temporal distributions, and associated sources. *Atmospheric Environment, 40*(18), 3367–3380. DOI: 10.1016/j.atmosenv.2006.01.050

Kiron, D. (2012). Get Ready: Mandated Integrated Reporting Is The Future of Corporate Reporting. *MIT Sloan Management Review, 3*(53), 1–3.

Kirwa, K., Szpiro, A. A., Sheppard, L., Sampson, P. D., Wang, M., Keller, J. P., Young, M. T., Kim, S.-Y., Larson, T. V., & Kaufman, J. D. (2021). Fine-scale air pollution models for epidemiologic research: Insights from approaches developed in the multi-ethnic study of atherosclerosis and air pollution (MESA air). *Current Environmental Health Reports, 8*(2), 113–126. DOI: 10.1007/s40572-021-00310-y PMID: 34086258

Kiss, B., Sekulova, F., Hörschelmann, K., Salk, C. F., Takahashi, W., & Wamsler, C. (2022). Citizen participation in the governance of nature-based solutions. *Environmental Policy and Governance, 32*(3), 247–272. DOI: 10.1002/eet.1987

Kitsikopoulos, C., Schwaibold, U., & Taylor, D. (2018). Limited progress in sustainable development: Factors influencing the environmental management and reporting of South African JSE-listed companies. *Business Strategy and the Environment, 27*(8), 1295–1301. DOI: 10.1002/bse.2176

Kling, G., Volz, U., Murinde, V., & Ayas, S. (2021). 'The impact of climate vulnerability on firms' cost of capital and access to finance'. *World Development, 137*(1), 105–131. DOI: 10.1016/j.worlddev.2020.105131

Klusak, P., ... (2021) *Rising temperatures, falling ratings: The effect of climate change on sovereign creditworthiness,* SSRN *Electronic Journal.* DOI: 10.2139/ssrn.3811958

Kordsachia, O., Focke, M., & Velte, P. (2022). 'Do sustainable institutional investors contribute to firms' environmental performance? Empirical evidence from Europe'. *Review ofManagerial Science*, *16*(5), 1409–1436. DOI: 10.1007/s11846-021-00484-7

Kozak, S. (2021) 'Will the reduction of co2 emissions lower the cost of debt financing? The case of eu countries', *Energies*, 14(24), pp. 01–13. .DOI: 10.3390/en14248361

Kroll, C., Warchold, A., & Pradhan, P. (2019). Sustainable Development Goals (SDGs): Are we successful in turning trade-offs into synergies? *Palgrave Communications*, *5*(1), 140. DOI: 10.1057/s41599-019-0335-5

Krutilla, K., & Graham, J. D. (2023). *Benefit-cost analysis of air pollution, energy, and climate regulations*. Cambridge University Press. DOI: 10.1017/9781009189460

Kubiszewski, I., Mulder, K., Jarvis, D., & Costanza, R. (2022). Toward better measurement of sustainable development and wellbeing: A small number of SDG indicators reliably predict life satisfaction. *Sustainable Development (Bradford)*, *30*(1), 139–148. DOI: 10.1002/sd.2234

Kumar, N. M., Goel, S., & Mallick, P. K. (2018). Smart cities in India: Features, policies, current status, and challenges. 2018 Technologies for Smart-City Energy Security and Power (ICSESP), 1-4.

Kuo, L., Kuo, P. W., & Chen, C. C. (2021). Mandatory CSR disclosure, CSR assurance, and the cost of debt capital: Evidence from Taiwan. *Sustainability (Basel)*, *13*(4), 1–19. DOI: 10.3390/su13041768

Kuznets, S. (1955). Economic Growth and Income Inequality. *The American Economic Review*, *XLV*(1), •••.

Kyaw, K., Olugbode, M., & Petracci, B. (2021). 'Stakeholder engagement: Investors' environmental risk aversion and corporate earnings'. *Business Strategy and the Environment*, *1*(1), 1–12. DOI: 10.1002/bse.2951

La Flèche, M., Cuss, C. W., Noernberg, T., Shotyk, W., & Karst, J. (2021). Trace metals as indicators of tree rooting in bituminous soils. *Land Degradation & Development*, *32*(5), 1970–1980. DOI: 10.1002/ldr.3848

Lah, O. (2025). Breaking the silos: Integrated approaches to foster sustainable development and climate action. *Sustainable Earth Reviews*, *8*(1), 1. DOI: 10.1186/s42055-024-00102-w

Lakhani, L., & Herbert, S. L. (2022). Theoretical references applied in integrated reporting and supporting reporting research. *Suid-Afrikaanse Tydskrif vir Ekonomiese en Bestuurswetenskappe*, *25*(1), 1–12. DOI: 10.4102/sajems.v25i1.4427

Lal, R., Kumar, P., & Banerjee, T. (2021). Air pollution exposure and children's health in India: A review. *Environmental Research, 194*, 110700. DOI: 10.1016/j.envres.2020.110700

Lata, L. N., Reddel, T., Head, B. W., & Craven, L. (2024). Advancing collaborative social outcomes through place-based solutions—Aligning policy and funding systems. *Policy and Society*, •••, puae018. DOI: 10.1093/polsoc/puae018

Leedy, P. and Ormorod, J. (2010) *Practical Research: Planning and Design.*

Lelieveld, J., Klingmüller, K., Pozzer, A., Pöschl, U., Fnais, M., Daiber, A., & Münzel, T. (2019). Cardiovascular disease burden from ambient air pollution in Europe reassessed using novel hazard ratio functions. *European Heart Journal, 40*(20), 1590–1596. DOI: 10.1093/eurheartj/ehz135 PMID: 30860255

Lens. *Lens.* Retrieved October, 2025. https://www.lens.org/lens/

Leppo, K., Ollila, E., Pena, S., Wismar, M., & Cook, S. (2013). Health in all policies: seizing opportunities, implementing policies. Sosiaali-ja terveysministeriö.

Liaqut, A., Tariq, S., & Younes, I. (2023). A study on optical properties, classification, and transport of aerosols during the smog period over South Asia using remote sensing. *Environmental Science and Pollution Research International, 30*(26), 69096–69121. DOI: 10.1007/s11356-023-27047-5 PMID: 37129820

Lin, Z. (2023). The economic impact of air pollution in beijing. *Future of Business and Finance*null, 265-282. DOI: 10.1007/978-3-031-29463-1_14

Lin, J., Arnold, D., & Nguyen, M. T. (2023). Welfare in crisis: Labor and social protection in the global south. *Journal of Labor and Society, 27*(3), 297–317. DOI: 10.1163/24714607-bja10128

Li, T., Zhang, Y., Wang, J., Xu, D., & Yin, Z. (2017). All-cause mortality risk associated with long-term exposure to ambient PM2.5 in China: A cohort study. *The Lancet. Public Health, 2*(12), e547–e555. DOI: 10.1016/S2468-2667(17)30241-4 PMID: 30314593

Litman, T. M. (2022). Evaluating transportation equity: Guidance for incorporating distributional impacts in transport planning. Institute of Transportation Engineers. *ITE Journal, 92*(4), 43–49.

Liu, L., & Jensen, M. B. (2018). Green infrastructure for sustainable urban water management: Practices of five forerunner cities. *Cities (London, England), 74*, 126–133. DOI: 10.1016/j.cities.2017.11.013

Liu, L., Zhao, J., & Li, Z. (2024). Sustainable urban development: A comprehensive model for assessing health risks from compounded pollution in xi'an. *Sustainability (Basel), 16*(22), 9799–9799. DOI: 10.3390/su16229799

Liu, Y., Tong, D., Cheng, J., Davis, S. J., Yu, S., Yarlagadda, B., Clarke, L. E., Brauer, M., Cohen, A. J., Kan, H., Xue, T., & Zhang, Q. (2022). Role of climate goals and clean-air policies on reducing future air pollution deaths in China: A modelling study. *The Lancet. Planetary Health, 6*(2), e92–e99. DOI: 10.1016/S2542-5196(21)00326-0 PMID: 35150635

Lluri, L., Lluri, F., & Sefa, S. (2024). Potential strategies to improve air quality in tirana for current and future generations. *Interdisciplinary journal of research and development, 11* (3), 96-96. DOI: 10.56345/ijrdv11n313

Lohr, V. I. (2010). Benefits of nature: What we are learning about why people respond to nature. *Journal of Physiological Anthropology, 29*(3), 111–117. DOI: 10.2114/jpa2.29.111 PMID: 17435348

Lu, S. (2024). Efforts to improve air quality in major regions - costs and benefits, successes and failures. *Highlights in Science Engineering and Technology, 121* null, 625-630. DOI: 10.54097/83bcmw36

Lu, Z. N., Chen, H., Hao, Y., Wang, J., Song, X., & Mok, T. M. (2017). The dynamic relationship between environmental pollution, economic development and public health: Evidence from China. *Journal of Cleaner Production, 166*, 134–147. DOI: 10.1016/j.jclepro.2017.08.010

Lv, Y. (2023). Transitioning to sustainable energy: Opportunities, challenges, and the potential of blockchain technology. *Frontiers in Energy Research, 11*, 1258044. DOI: 10.3389/fenrg.2023.1258044

Madlener, R., & Sunak, Y. (2011). Impacts of urbanization on urban structures and energy demand: What can we learn for urban energy planning and urbanization management? *Sustainable Cities and Society, 1*(1), 45–53. DOI: 10.1016/j.scs.2010.08.006

Mage, D., Ozolins, G., Peterson, P., Webster, A., Orthofer, R., Vandeweerd, V., & Gwynne, M. (1996). Urban air pollution in megacities of the world. *Atmospheric Environment, 30*(5), 681–686. DOI: 10.1016/1352-2310(95)00219-7

Magnanelli, B. S., & Izzo, M. F. (2017). Corporate social performance and cost of debt: The relationship. *Social Responsibility Journal, 13*(2), 250–265. DOI: 10.1108/SRJ-06-2016-0103

Mahdavi, M., Awaafo, A., Jurado, F., Vera, D., & Ramos, R. A. V. (2023). Wind, solar and biogas power generation in water-stressed areas of Morocco considering water and biomass availability constraints and carbon emission limits. *Energy*, *282*, 128756. DOI: 10.1016/j.energy.2023.128756

Majeed, M. M., & Munir, A. (2020). Pakistan: Country report on children's environmental health. *Reviews on Environmental Health*, *35*(1), 57–63. DOI: 10.1515/reveh-2019-0087 PMID: 31926105

Malik, A., Islam, J., Zaib, G., Ashraf, M. H., Zahid, A., Rashid, A. R., Zia, T., & Ali, Q. (2024). ISLAM, J., ZAIB, G., Zahid, A., Zahid, A., Rashid, A., Zia, T., & Ali, Q. (2024). Smog crisis in lahore: Evaluating air quality trends and public health implications. *Bulletin of Biological and Allied Sciences Research*, *2024*(1), 87–87. DOI: 10.54112/bbasr.v2024i1.87

Malik, A., Islam, J., Zaib, G., Zahid, A., Rashid, A. R., & Zia, T. (2024). Interplay of smog and health conditions: A public health perspective on respiratory, hypertensive, and cardiovascular morbidity. *Journal of Physical. Biomedical and Biological Sciences*, *2024*(1), 37–37. DOI: 10.64013/jpbab.v2024i1.37

Mancini, L., Marcheggiani, S., Puccinelli, C., Lacchetti, I., Carere, M., & Bouley, T. (2017). Global environmental changes and the impact on ecosystems and human health. *Energia, ambiente e innovazione, 3*, 98-104.

Mannucci, P. M., & Franchini, M. (2017). Health effects of ambient air pollution in developing countries. *International Journal of Environmental Research and Public Health*, *14*(9), 1048. DOI: 10.3390/ijerph14091048 PMID: 28895888

Ma, R., Abid, N., Yang, S., & Ahmad, F. (2023). From crisis to resilience: Strengthening climate action in OECD countries through environmental policy and energy transition. *Environmental Science and Pollution Research International*, *30*(54), 115480–115495. DOI: 10.1007/s11356-023-29970-z PMID: 37882927

Marlier, M. E., Jina, A. S., Kinney, P. L., & DeFries, R. S. (2016). Extreme air pollution in global megacities. *Current Climate Change Reports*, *2*(1), 15–27. DOI: 10.1007/s40641-016-0032-z

Marshall, G., & Jonker, L. (2011). An introduction to inferential statistics: A review and practical guide. *Radiography*, *17*(1), e1–e6. DOI: 10.1016/j.radi.2009.12.006

Marx, B., & van Dyk, V. (2011). Sustainability reporting and assurance. *Meditari Accountancy Research*, *19*(1/2), 39–55. DOI: 10.1108/10222521111178628

Mazziotta, C., Tognon, M., Martini, F., Torreggiani, E., & Rotondo, J. C. (2023). Probiotics mechanism of action on immune cells and beneficial effects on human health. *Cells*, *12*(1), 184. DOI: 10.3390/cells12010184 PMID: 36611977

Meadows, D. H. (2008). *Thinking in systems: A primer*. Chelsea Green Publishing.

Mehmood, A., Usama, R. M., Mehmood, A., & Rizwan, S. (2024). Evaluating the impact of smog emergency policies on air quality and public perception by using geospatial technologies: A case study of lahore, pakistan. DOI: 10.21203/rs.3.rs-5537502/v1

Mehta, A. K., Chandel, P. S., Shukla, R., & Selvakumar, P. (2024). Opportunities for green entrepreneurship. In Mehta, A. (Ed.), *Green innovation and sustainability practices* (pp. 223–238). IGI Global., DOI: 10.4018/979-8-3693-7046-9.ch009

Meirun, T., Makhloufi, L., & Ghozali Hassan, M. (2020). Environmental outcomes of green entrepreneurship harmonization. *Sustainability (Basel)*, *12*(24), 10615. DOI: 10.3390/su122410615

Meissner, H.. (2011). Best Practices for Mixed Methods Research in the Health Sciences. *Office of Behavioral and Social Sciences Research*, *29*(1), 1–39. DOI: 10.1002/cdq.12009

Melhim, S. H., & Isaifan, R. J. (2025). The Energy-Economy Nexus of Advanced Air Pollution Control Technologies: Pathways to Sustainable Development. *Energies*, *18*(9), 2378. DOI: 10.3390/en18092378

Mencaroni, M., Dal Ferro, N., Furlanetto, J., Longo, M., Lazzaro, B., Sartori, L., Grant, B., Smith, W., & Morari, F. (2021). Identifying N fertilizer management strategies to reduce ammonia volatilization: Towards a site-specific approach. *Journal of Environmental Management*, *277*, 111445. DOI: 10.1016/j.jenvman.2020.111445 PMID: 33070018

Mertler, C. A., & Reinhart, R. V. (2016) 'Advanced and multivariate statistical methods: Practical application and interpretation', in *A guide to multivariate techniques*, pp. 13–22. DOI: 10.4324/9781315266978

Meyer, N., & Auriacombe, C. (2019). Good urban governance and city resilience: An afrocentric approach to sustainable development. *Sustainability (Basel)*, *11*(19), 5514. DOI: 10.3390/su11195514

Ministry of Environment. Forest and Climate Change (MoEFCC). (2019). *National Clean Air Programme (NCAP)*. Government of India.

Miörner, J., Binz, C., & Fuenfschilling, L. (2021). Understanding transformation patterns in different socio-technical systems–A scheme of analysis. *GEIST–Geography of Innovation and Sustainability Transitions, Working Paper Series.*

Mishra, S. (2017). Is smog innocuous? Air pollution and cardiovascular disease. *Indian Heart Journal, 69*(4), 425–429. DOI: 10.1016/j.ihj.2017.07.016

MOEF. (2020). *Korean New Deal: National Strategy for a Great Transformation.* Republic of Korea Ministry of Economy and Finance. https://stip.oecd.org/stip/interactive-dashboards/policy-initiatives/2023%2Fdata%2FpolicyInitiatives%2F99991604

Mohamad Taghvaee, V., Assari Arani, A., Nodehi, M., Khodaparast Shirazi, J., Agheli, L., Neshat Ghojogh, H. M., Salehnia, N., Mirzaee, A., Taheri, S., Mohammadi Saber, R., Faramarzi, H., Alvandi, R., & Ahmadi Rahbarian, H. (2023). Sustainable development goals: Transportation, health and public policy. *Review of Economics and Political Science, 8*(2), 134–161. DOI: 10.1108/REPS-12-2019-0168

Moikwatlhai, K., & Yasseen, Y. (2019). ESG reporting and the institutional shareholder base: A quantitative study of listed companies on the JSE. *Southern African Journal of Accountability and Auditing Research, 21*(1), 31–46.

Mol, A. P., & Sonnenfeld, D. A. (2000). *Ecological modernization around the world: Perspectives and critical debates.* Routledge.

Monaco, S. (2024). SDG 8. Promote sustained, inclusive and sustainable economic growth, full and productive employment, and decent work for all. In *Identity, Territories, and Sustainability: Challenges and Opportunities for Achieving the UN Sustainable Development Goals* (pp. 81-86). Emerald Publishing Limited.

Mondal, S., & Singh, S. (2025). A bibliometric and thematic analysis of green entrepreneurship in business research: Current status and future research directions. *Environment, Development and Sustainability, 27*(1), 67–139. DOI: 10.1007/s10668-023-03905-7

Morgan, D. L. (2007). Paradigms Lost and Pragmatism Regained. *Journal of Mixed Methods Research, 1*(1), 48–76. DOI: 10.1177/2345678906292462

Mosley, S. (2014). Environmental history of air pollution and protection. In *The Oxford Handbook of Environmental History* (pp. 1–27). Oxford University Press. DOI: 10.1007/978-3-319-09180-8_5

Most Environmentally Friendly Countries. (2025). https://worldpopulationreview.com/country-rankings/most-environmentally-friendly-countries

Moulin, O. (2025). India's grassroot innovations are powering global climate resilience – Dipa Bagai, NRDC India, https://etedge-insights.com/sdgs-and-esg/climate-change/indias-grassroot-innovations-are-powering-global-climate-resilience-dipa-bagai-nrdc-india/

Mukherjee, A., Owen, R., Scott, J. M., & Lyon, F. (2024a). Financing green innovation startups: A systematic literature review on early-stage SME funding. *Venture Capital*, •••, 1–27. DOI: 10.1080/13691066.2024.2410730

Mukhtar, F. (2018). The rising menace of smog: Time to act now. *Journal of Ayub Medical College, Abbottabad*, *30*(1), 1–2. PMID: 29504318

Mundorf, N., Redding, C. A., & Paiva, A. L. (2018). Sustainable transportation attitudes and health behavior change: Evaluation of a brief stage-targeted video intervention. *International Journal of Environmental Research and Public Health*, *15*(1), 150. DOI: 10.3390/ijerph15010150

Muo, I., & Azeez, A. A. (2019). Green entrepreneurship: Literature review and agenda for future research. *International Journal of Entrepreneurial Knowledge*, *7*(2). Advance online publication. DOI: 10.37335/ijek.v7i2.90

Muttakin, M. B., Mihret, D., Lemma, T. T., & Khan, A. (2020). Integrated reporting, financial reporting quality and cost of debt. *International Journal of Accounting and Information Management*, *28*(3), 517–534. DOI: 10.1108/IJAIM-10-2019-0124

MWK. (2022). *GreenTech Innovation Strategy*. Federal Ministry for Economic Affairs and Climate Action. https://www.bmwk.de

Nadeem, M., Aziz, A., Al-Rashid, M. A., Tesoriere, G., Asim, M., & Campisi, T. (2021). Scaling the potential of compact city development: The case of Lahore, Pakistan. *Sustainability (Basel)*, *13*(9), 5257. DOI: 10.3390/su13095257

Nadhifa, S. R., Prabarini, A. M., & Wati, I. R. (2024). Green entrepreneurial behavior: A review and research agenda. *Research Horizon*, *4*(3), 295–304. https://journal.lifescifi.com/index.php/RH/article/view/278

Narain, U., & Krupnick, A. (2016). The impact of Delhi's Odd-Even driving policy on air pollution. *Resources for the Future*. https://www.rff.org/publications/working-papers/the-impact-of-delhis-odd-even-driving-policy-on-air-pollution

NASA. (2021). *Monitoring the impact of the Great Green Wall on dust and vegetation using satellite data.* National Aeronautics and Space Administration. https://earthobservatory.nasa.gov/

Nasar-u-Minallah, M., Jabbar, M., Zia, S., & Perveen, N. (2024). Assessing and anticipating environmental challenges in Lahore, Pakistan: future implications of air pollution on sustainable development and environmental governance. *Environmental Monitorin*

Nasim, S., & Sharif, F. (2020). To adopt, or not to adopt,'why' is the question: A case for clean kiln technologies in developing countries. *Journal of Cleaner Production, 257*, 120553. DOI: 10.1016/j.jclepro.2020.120553

Naveed, Z., & Khayyam, U. (2023). Smog and cognitive issues in the school going children of Lahore and Islamabad, Pakistan. *International Journal of Environmental Science and Technology*, *20*(4), 4151–4166. DOI: 10.1007/s13762-022-04264-y

Nelly, V., de Snyder, S., Friel, S., Fotso, J. C., Khadr, Z., Meresman, S., & Patil-deshmukh, A. (2011). Social Conditions and Urban Health Inequities: Realities, Challenges and Opportunities to Transform the Urban Landscape through Research and Action. *Journal of Urban Health*, *88*(6), 1183–1193. DOI: 10.1007/s11524-011-9609-y

Neumann, T. (2022). Impact of green entrepreneurship on sustainable development: An ex-post empirical analysis. *Journal of Cleaner Production*, *377*, 134317. DOI: 10.1016/j.jclepro.2022.134317

News, A. P. (2023). *EV sales boom in Nepal, helping to save on oil imports and alleviate smog*. https://apnews.com/article/nepal-ev-energy-hydropower241d486f228051c73ff2793e043b0cca

Nguyen, P. Y., Astell-Burt, T., Rahimi-Ardabili, H., & Feng, X. (2021). Green space quality and health: A systematic review. *International Journal of Environmental Research and Public Health*, *18*(21), 11028. DOI: 10.3390/ijerph182111028

Nix, E., Paulose, J., Shrubsole, C., Altamirano-Medina, H., Davies, M., Khosla, R., Belesova, K., & Wilkinson, P. (2020). Evaluating housing health hazards: Prevalence, practices and priorities in Delhi's informal settlements. *Journal of Urban Health*, *97*(4), 502–518. DOI: 10.1007/s11524-020-00442-w

None, I., & Kumar Datta, S. (2011). Pro-environmental Concern Influencing Green Buying: A Study on Indian Consumers. *International Journal of Business and Management*, *6*(6), 124–133. DOI: 10.5539/ijbm.v6n6p124

Nowak, D. J., Crane, D. E., & Stevens, J. C. (2014). Air pollution removal by urban trees and shrubs in the United States. *Urban Forestry & Urban Greening*, *4*(3–4), 115–123. DOI: 10.1016/j.ufug.2006.01.007

Nowak, D. J., Hirabayashi, S., Bodine, A., & Hoehn, R. (2018). Tree and forest effects on air quality and human health in the United States. *Environmental Pollution*, *193*, 119–129. DOI: 10.1016/j.envpol.2014.05.028 PMID: 25016465

Nyambuu, U., & Semmler, W. (2023). Fossil fuel resources, environment, and climate change. In *Sustainable macroeconomics, climate risks and energy transitions: Dynamic modeling, empirics, and policies* (pp. 45–58). Springer International Publishing. DOI: 10.1007/978-3-031-27982-9_4

Oberndorfer, E., Lundholm, J., Bass, B., Coffman, R. R., Doshi, H., Dunnett, N., Gaffin, S., Köhler, M., Liu, K. K. Y., & Rowe, B. (2007). Green roofs as urban ecosystems: Ecological structures, functions, and services. *Bioscience*, *57*(10), 823–833. DOI: 10.1641/B571005

Odeyemi, O., Usman, F. O., Mhlongo, N. Z., Elufioye, O. A., & Ike, C. U. (2024). Sustainable entrepreneurship: A review of green business practices and environmental impact. *World Journal of Advanced Research and Reviews*, *21*(2), 346–358. DOI: 10.30574/wjarr.2024.21.2.0461

OECD. (2022a). *Climate-resilient finance and investment*. OECD Environment Working Papers. https://www.oecd.org/en/publications/climate-resilient-finance-and-investment_223ad3b9-en.html

OECD. (2022b). *Policies to Support Green Entrepreneurship*. OECD Report. https://www.oecd.org/en/publications/policies-to-support-green-entrepreneurship_e92b1946-en.html

Ofogbe, N. S., Nnamani, C., Anisiuba, C., & Ezuwore-Obodoekwe, C. (2021a). the Relationship Between the Corporate Social Responsibility and the Cost of the Capital of the Companies Listed on the Nigerian Stock Exchange. *Economic Horizons*, *23*(3), 277–292. DOI: 10.5937/ekonhor2103289O

Ogwu, M. C., & Izah, S. C. (2024). Air pollution and the sustainable development goals. In *Sustainable Strategies for Air Pollution Mitigation: Development, Economics, and Technologies* (pp. 201-233). Springer. DOI: 10.1007/698_2024_1118

Ojha, N., Soni, M., Kumar, M., Gunthe, S. S., Chen, Y., & Ansari, T. U. (2022). Mechanisms and pathways for coordinated control of fine particulate matter and ozone. *Current Pollution Reports*, *8*(4), 594–604. DOI: 10.1007/s40726-022-00229-4 PMID: 35991936

Olivier, J. G., & Berdowski, J. J. (2021). Global emission sources and sinks. In *The climate system* (pp. 33–77). CRC Press. DOI: 10.1201/9781003211266-2

Olson, C., & Lenzmann, F. (2016). The social and economic consequences of the fossil fuel supply chain. *MRS Energy & Sustainability : a Review Journal, 3*(1), E6. DOI: 10.1557/mre.2016.7

Ompad, D. C., Galea, S., Caiaffa, W. T., & Vlahov, D. (2007). Social determinants of the health of urban populations: Methodologic considerations. *Journal of Urban Health, 84*(S1, Suppl 1), 42–53. DOI: 10.1007/s11524-007-9168-4

Onditi, A. (2016) 'Journal of Marketing and Consumer Research www.iiste.org ISSN', *An International Peer-reviewed Journal*, 29, pp. 37–45.

Onnoghen, N. U., Onwuzurike, U., & Anabaraonye, B. (2024). Enhancing climate resilience through green entrepreneurship education in Nigeria. *J Civ Eng Appl, 5*(1), 33–36.

Organisation for Economic Co-operation and Development (OECD). (2021). *The economic consequences of outdoor air pollution.* OECD Publishing., DOI: 10.1787/9789264257474-

Organisation for Economic Co-operation and Development (OECD). *OECD Economic Surveys: Türkiye 2023.* Paris: OECD Publishing. (2023, February). Retrieved from https://www.oecd.org/en/publications/2023/02/oecd-economic-surveys-turkiye-2023_816ce823.html

OriginLab. (2025). *Origin and OriginPro - data analysis and graphing software.* Retrieved October, 2025, from https://www.originlab.com/

Özatac, N., Gökmenoğlu, S., & Taşpınar, N. (2017). Testing the EKC Hypothesis by Considering Trade Openness, Urbanization and Financial Development: The Case of Turkey. *Environmental Science and Pollution Research International, 24*(20), 16690–16701. DOI: 10.1007/s11356-017-9317-6

Page, M. J., McKenzie, J. E., Bossuyt, P. M., Boutron, I., Hoffmann, T. C., Mulrow, C. D., & Moher, D. (2021). The PRISMA 2020 statement: an updated guideline for reporting systematic reviews. *bmj, 372*. Progress, challenges and prospects of urban air pollution and climate in iran. DOI: 10.21203/rs.3.rs-2155113/v1

Pakistan Bureau of Statistics. (2023). *7th Population & Housing Census 2023 – Final Results.* Islamabad: Pakistan Bureau of Statistics. Retrieved from https://www.pbs.gov.pk/wp-content/uploads/2020/07/National-Census-Report-2023.pdf

Pakistan Today. (2024, March 3). *Death toll from pneumonia rises to 456 in Punjab.* https://shorturl.at/C1i6l

Panayotou, T. (1993). *Empirical tests and policy analysis of environmental degradation at different stages of economicdevelopment (No. 992927783402676).* International Labour Organization.

Parikh, N. I., Gonzalez, J. M., Anderson, C. A., Judd, S. E., Rexrode, K. M., Hlatky, M. A., Gunderson, E. P., Stuart, J. J., Vaidya, D., & Epidemiology, A. H. A. C. (2021). Adverse pregnancy outcomes and cardiovascular disease risk: unique opportunities for cardiovascular disease prevention in women: a scientific statement from the American Heart Association. *Circulation*, *143*(18), e902–e916. DOI: 10.1161/CIR.0000000000000961 PMID: 33779213

Patel, J. B., & Raval, Z. (2024). The impacts of urbanization on ecological systems: A comprehensive study of the complex challenges arising from rapid urban growth. *Research Review Journal of Indian Knowledge Systems*, *1*(1), 1–10. DOI: 10.31305/rrjiks.2024.v1.n1.001

Pérez-Vizcaíno, P., Sánchez de la Campa, A. M., Sánchez-Rodas, D., Alastuey, A., Querol, X., & de la Rosa, J. Pm10 Chemical Fingerprints and Source Assessment Guiding Air Quality Improvements by 2030 in Andalusia, Southern Spain. *Southern Spain*.

PLANET, L. (2025). Accelerating Access to Clean Air for a Livable Planet.

Protano, C., Buomprisco, G., Cammalleri, V., Pocino, R. N., Marotta, D., Simonazzi, S., Cardoni, F., Petyx, M., Iavicoli, S., & Vitali, M. (2021). The carcinogenic effects of formaldehyde occupational exposure: A systematic review. *Cancers (Basel)*, *14*(1), 165. DOI: 10.3390/cancers14010165 PMID: 35008329

Puaschunder, J. (2020). *Governance & Climate Justice: Global South & Developing Nations*. Palgrave Macmillan Cham. DOI: 10.1007/978-3-319-63281-0

Qayyum, U., Sabir, S., & Anjum, S. (2021). Urbanization, informal economy, and ecological footprint quality in South Asia. *Environmental Science and Pollution Research International*, *28*(47), 67011–67021. DOI: 10.1007/s11356-021-15111-x PMID: 34244937

Qazi, A., Hussain, F., Rahim, N. A., Hardaker, G., Alghazzawi, D., Shaban, K., & Haruna, K. (2019). Towards sustainable energy: A systematic review of renewable energy sources, technologies, and public opinions. *IEEE Access : Practical Innovations, Open Solutions*, *7*, 63837–63851. DOI: 10.1109/ACCESS.2019.2906402

Rafiq-uz-Zaman, M., Khalid, N., Faridi, M. Z., & Khan, Z. A. (2024). Smog mitigation and climate resilience in pakistan: A policy-centric review of strategies and challenges. DOI: 10.71085/sss.03.04.175

Raimo, N., Caragnano, A., Zito, M., Vitolla, F., & Mariani, M. (2021). Extending the benefits of ESG disclosure: The effect on the cost of debt financing. *Corporate Social Responsibility and Environmental Management*, *28*(4), 1412–1421. DOI: 10.1002/csr.2134

Rani, B., Singh, U., Chuhan, A. K., Sharma, D., & Maheshwari, R. (2011). Photochemical smog pollution and its mitigation measures. In *Journal of Advanced Scientific Research*. http://www.sciensage.info/jasr

Raza, M. H., Abid, M., Faisal, M., Yan, T., Akhtar, S., & Adnan, K. M. (2022). Environmental and health impacts of crop residue burning: Scope of sustainable crop residue management practices. *International Journal of Environmental Research and Public Health*, *19*(8), 4753. DOI: 10.3390/ijerph19084753 PMID: 35457622

Raza, W., Saeed, S., Saulat, H., Gul, H., Sarfraz, M., Sonne, C., Sohn, Z.-H., Brown, R. J. C., & Kim, K. H. (2021). A review on the deteriorating situation of smog and its preventive measures in Pakistan. *Journal of Cleaner Production*, *279*, 123676. DOI: 10.1016/j.jclepro.2020.123676

Razzaq, A., Zafar, M. M., Zahra, L. T., Qadir, F., Qiao, F., Ullah, M. S., Shehzad, S., Rasool, G., & Jiang, X. (2024). Smog: Lahore needs global attention to fix it. *Environmental challenges, 16* null, 100999-100999. DOI: 10.1016/j.envc.2024.100999

Reckien, D., Creutzig, F., Fernandez, B., Lwasa, S., Tovar-Restrepo, M., Mcevoy, D., & Satterthwaite, D. (2017). Climate change, equity and the Sustainable Development Goals: An urban perspective. *Environment and Urbanization*, *29*(1), 159–182. DOI: 10.1177/0956247816677778

Reuters. (2024). IQAir World Air Quality Report 2024 (Pakistan 73.7 μg/m^3 in 2023).

Revesz, R. L. (2022). Air pollution and environmental justice. *Ecology Law Quarterly*, *49*, 187.

Richelle, J. (2025). The EU Legal Framework on Transboundary Air Quality and Pollution: Exploring the Mismatch between Transnationality in Legislation and Its Use in Practice. *Journal for European Environmental & Planning Law*, *22*(1-2), 110–127. DOI: 10.1163/18760104-22010206

Rizzitello, E., Piazza, M., & Perrone, G. (2025). Unlocking green startup investments: How environmental policy pressures drive Venture Capital funding decisions. *Technological Forecasting and Social Change*, *217*, 1–17. DOI: 10.1016/j.techfore.2025.124158

Roy, I. (2024). Technology Push, Demand Pull, Networks, and Public Policy: What Drives and Facilitates Innovation? In *Science, Technology and Innovation Ecosystem: An Indian and Global Perspective* (pp. 133-147). Springer.

Royal College of Physicians. (2016). *Every breath we take: The lifelong impact of air pollution*. RCP.

Sah, S. K. (2024). Public health policy responses to air pollution in india: Challenges and strategies for sustainable development. *Visioner, 16*(2), 163–170. DOI: 10.54783/jv.v16i2.1057

Saleem, Z., Saeed, H., Yousaf, M., Asif, U., Hashmi, F. K., Salman, M., & Hassali, M. A. (2019). Evaluating smog awareness and preventive practices among Pakistani general population: A cross-sectional survey. *International Journal of Health Promotion and Education, 57*(3), 161–173. DOI: 10.1080/14635240.2019.1576535

Salisu, A., Omoke, P., & Fadiya, O. (2022). Climate Policy Uncertainty and Crude Oil Market Volatility. *Energy Research Letters, 3*(1), 1–5. DOI: 10.46557/001c.38781

Sanchez-Triana, E., Enriquez, S., Afzal, J., Nakagawa, A., & Khan, A. S. (Eds.). (2014). *Cleaning Pakistan's air: Policy options to address the cost of outdoor air pollution*. The World Bank., https://shorturl.at/E3QU7 DOI: 10.1596/978-1-4648-0235-5

Santika, I. W., Wardana, I. M., Setiawan, P. Y., & Widagda, I. G. (2022). Entrepreneurship education and green entrepreneurial intention: A conceptual framework. *Linguistics and Culture Review*, 797-810. https://doi.org/DOI: 10.21744/lingcure.v6nS1.2159

Sarfraz. Z. (2020). The Social and Economic Burden of Smog in Pakistan, Editorial;, Pakistan Journal of Surgery & Medicine.

Saunders, M., Lewis, P., & Thornhill, A. (2018). *Research Methods for Business Students* (8th ed.). Pearson Education.

Savitz, A. (2013). *The triple bottom line: How today's best-run companies are achieving economic, social and environmental success*. John Wiley & Sons.

Schaltegger, S., & Wagner, M. (2011). Sustainable entrepreneurship and sustainability innovation: Categories and interactions. *Business Strategy and the Environment, 20*(4), 222–237. DOI: 10.1002/bse.682

Scoones, I., Stirling, A., Abrol, D., Atela, J., Charli-Joseph, L., Eakin, H., Ely, A., Olsson, P., Pereira, L., Priya, R., van Zwanenberg, P., & Yang, L. (2020). Transformations to sustainability: Combining structural, systemic and enabling approaches. *Current Opinion in Environmental Sustainability*, *42*, 65–75. DOI: 10.1016/j.cosust.2019.12.004

SDGS-UN. (n.d.) https://sdgs.un.org/partnerships

Seaward, M., & Richardson, D. (2024). Atmospheric sources of metal pollution and effects on vegetation. In *Heavy metal tolerance in plants* (pp. 75–92). CRC press. DOI: 10.1201/9781003574903-7

Seidman, G. W. (2023). *Manufacturing militance: workers' movements in Brazil and South Africa, 1970-1985*. Univ of California Press. DOI: 10.2307/jj.5973032

Seinfeld, J. H., & Pandis, S. N. (2016). *Atmospheric chemistry and physics: From air pollution to climate change* (3rd ed.). Wiley.

Sekar, A., Sharma, R., & Mahajan, A. (2024). Synergistic impact of air pollution and heat on health and economy in india. DOI: 10.1007/978-3-031-56564-9_2

Serrano-Cinca, C., Gutiérrez-Nieto, B., & Reyes, N. M. (2016). A social and environmental approach to microfinance credit scoring. *Journal of Cleaner Production*, *112*, 3504–3513. DOI: 10.1016/j.jclepro.2015.09.103

Seto, K. C., & Shepherd, J. M. (2009). Global urban land-use trends and climate impacts. *Current Opinion in Environmental Sustainability*, *1*(1), 89–95. DOI: 10.1016/j.cosust.2009.07.012

Shamsi, S. R. A., Ashmore, M. R., Bell, J. N. B., Maggs, R., Kafayat, U., & Wahid, A. (2000). The impacts of air pollution on crops in developing countries—a case study in Pakistan. In *Environmental stress: Indication, mitigation and eco-conservation* (pp. 63–71). Springer Netherlands. DOI: 10.1007/978-94-015-9532-2_6

Shang, Y., Sang, S., Tiwari, A. K., Khan, S., & Zhao, X. (2024). Impacts of renewable energy on climate risk: A global perspective for energy transition in a climate adaptation framework. *Applied Energy*, *362*, 122994. DOI: 10.1016/j.apenergy.2024.122994

Sharma, M., & Dikshit, O. (2016). Comprehensive study on air pollution and greenhouse gases in India. *Indian Journal of Environmental Protection*, *36*(5), 402–411.

Sharma, M., & Kumar, S. (2022). Analysing the spatial patterns and trends of urban growth in Rohtak city, India. *Sustainable Environment*, *8*(1), 2051268. DOI: 10.1080/27658511.2022.2051268

Shen, J., Cai, W., Chen, X., Chen, X., Zhao, Z., Ma, Z., Yang, F., & Zhang, S. (2022). Synergies of carbon neutrality, air pollution control, and health improvement — a case study of china energy interconnection scenario. *Global energy interconnection, 5* (5), 531-542. DOI: 10.1016/j.gloei.2022.10.007

Shen, Y. (2024). Air quality assessment and pollution mitigation strategies in beijing: A comprehensive study on pm2.5 trends, influencing factors, and case studies. *Highlights in Science Engineering and Technology, 88* null, 695-702. DOI: 10.54097/0wzh2456

Shen, Y., & Zhang, X. (2024). Blue sky protection campaign: Assessing the role of digital technology in reducing air pollution. *Systems*null, . DOI: 10.3390/systems12020055

Shepherd, D. A., & Patzelt, H. (2011). The new field of sustainable entrepreneurship: Studying entrepreneurial action linking "what is to be sustained" with "what is to be developed". *Entrepreneurship Theory and Practice, 35*(1), 137–163. DOI: 10.1111/j.1540-6520.2010.00426.x

Sherin, A. (2023). Beyond COP: Pakistan's evolving role in a global climate context. *Khyber Medical University Journal, 15*(4), 208–210.

Siddqiue, M. U., Jamil, M., & Arshad, A. (2023). Digital Solutions and Technological Innovations Driving Public Awareness and Community Initiatives for Smog Reduction in Lahore: A Descriptive Overview. *Pakistan Social Sciences Review, 7*(4), 425–438. DOI: 10.35484/pssr.2023(7-IV)39

Sierra-vargas, M. P., & Teran, L. M. (2012). Air pollution: Impact and prevention. *Respirology (Carlton, Vic.), 17*(7), 1031–1038. DOI: 10.1111/j.1440-1843.2012.02213.x

Singh, S. (2018). *The Great Smog of India*. Penguin Random House India.

Sobko, O., Gavkalova, N., Kurbatska, L., Boichyk, V., & Krysovatyy, I. (2024). Green entrepreneurship strategies: Integrating eco-friendly energy technologies with the digital economy. *Grassroots Journal of Natural Resources, 7*(3), s92–s114. DOI: 10.33002/nr2581.6853.0703ukr05

Solís Arce, J. S., Warren, S. S., Meriggi, N. F., Scacco, A., McMurry, N., Voors, M., Syunyaev, G., Malik, A. A., Aboutajdine, S., Adeojo, O., Anigo, D., Armand, A., Asad, S., Atyera, M., Augsburg, B., Awasthi, M., Ayesiga, G. E., Bancalari, A., Björkman Nyqvist, M., & Omer, S. B. (2021). COVID-19 vaccine acceptance and hesitancy in low-and middle-income countries. *Nature Medicine, 27*(8), 1385–1394. DOI: 10.1038/s41591-021-01454-y PMID: 34272499

Soni, A., Gupta, S. K., Rajamohan, N., & Yusuf, M. (2025). Waste-to-energy technologies: A sustainable pathway for resource recovery and materials management. *Materials Advances*, *6*(14), 4598–4622. Advance online publication. DOI: 10.1039/D5MA00449G

Sourgens, F. G., & Sempertegui, L. (2023). *Principles of international energy transition law: law as an instrument of development in the 21st century*. Oxford University Press. DOI: 10.1093/law/9780198876083.001.0001

Sovacool, B. K., Newell, P., Carley, S., & Fanzo, J. (2022). Equity, technological innovation and sustainable behaviour in a low-carbon future. *Nature Human Behaviour*, *6*(3), 326–337. DOI: 10.1038/s41562-021-01257-8 PMID: 35102347

Sriram, S., & Albadrani, M. (2022). Impoverishing effects of out-of-pocket healthcare expenditures in India. *Journal of Family Medicine and Primary Care*, *11*(11), 7120–7128. DOI: 10.4103/jfmpc.jfmpc_590_22 PMID: 36993034

Srivastava, J., Sampath, A., & Gopalakrishnan, B. (2022). Is ESG the key to unlock debt financing during the COVID-19 pandemic? International evidence. *Finance Research Letters*, *49*(6), 103–125. DOI: 10.1016/j.frl.2022.103125

Steyn, M. (2014). 'Organisational benefits and implementation challenges of mandatory integrated reporting: Perspectives of senior executives at South African listed companies', *Sustainability Accounting. Management and Policy Journal*, *5*(4), 476–503. DOI: 10.1108/SAMPJ-11-2013-0052

Stivanello, P. (2021). Co-benefits of Plus Energy Buildings: more than just energy efficiency to advocate for tomorrow's building technology.

Subhanullah, M., Hassan, N., Rahman, G., Rawan, B., Ullah, W., & Ilyas, M. (2025). Concentration of particulate matter and its impact on public health in different cities in Pakistan-a review. *Environmental Forensics*, *26*(2), 146–162. DOI: 10.1080/15275922.2024.2366794

Sultana, S., Zainal, D., & Zulkifli, N. (2017). The influence of environmental, social and governance (ESG) on investment decisions: The Bangladesh perspective. *Pertanika Journal of Social Science & Humanities*, *25*(11), 155–173.

Tahir, M., Aziz, S., Khan, A., & Shahi, M. (2023). Spreading of smog in Punjab areas o f Pakistan due to violation of environmental laws. *Intl. J. Soc. Sci.: Curr. Fut. Res. Trends*, *18*, 74–89.

Tajammul, K. I. (2023). Critical review and comparative analysis of the Government of Punjab's 'Policy on Controlling Smog, 2017'with Counterpart Strategies in London, Beijing and Los Angeles. *Journal of Development Policy Research & Practice (JoDPRP)*, 181-202.

Tang, A., & Kemp, L. (2021). A fate worse than warming? Stratospheric aerosol injection and global catastrophic risk. *Frontiers in Climate*, *3*, 720312. DOI: 10.3389/fclim.2021.720312

Tan, P. Y., Wang, J., & Sia, A. (2013). Perspectives on five decades of the Singapore Green Plan: From "Garden City" to "City in a Garden". *Cities (London, England)*, *32*, 24–32. DOI: 10.1016/j.cities.2013.02.001

The World Bank. (2022). *Pakistan Country Climate and Development Report*. World Bank.

The World Bank. (2024). World Development Indicators. https://databank.worldbank.org/source/world-development-indicators

Tibrewal, K., & Venkataraman, C. (2021). Climate co-benefits of air quality and clean energy policy in India. *Nature Sustainability*, *4*(4), 305–313. DOI: 10.1038/s41893-020-00666-3

Tien, N. H., Tien, N. V., Mai, N. P., & Duc, L. D. M. (2023). Green entrepreneurship: A game changer in Vietnam business landscape. *International Journal of Entrepreneurship and Small Business*, *48*(4), 408–431. DOI: 10.1504/IJESB.2023.130829

Tirgil, Abdullah, Yasin Acar and Önder Özgür. 2021. "Revisiting the environmental Kuznets curve: evidence from Turkey." *Environment, Development and Sustainability: A Multidisciplinary Approach to the Theory and Practice of Sustainable Development*, Springer, vol. 23(10), pages 14585-14604, October.

Tonne, C., Adair, L., Adlakha, D., Anguelovski, I., Belesova, K., Berger, M., Brelsford, C., Dadvand, P., Dimitrova, A., Giles-Corti, B., Heinz, A., Mehran, N., Nieuwenhuijsen, M., Pelletier, F., Ranzani, O., Rodenstein, M., Rybski, D., Samavati, S., Satterthwaite, D., & Adli, M. (2021). Defining pathways to healthy sustainable urban development. *Environment International*, *146*, 106236. DOI: 10.1016/j.envint.2020.106236 PMID: 33161201

Tranfield, D., Denyer, D., & Smart, P. (2003). Towards a methodology for developing evidence-informed management knowledge by means of systematic review. *British Journal of Management*, *14*(3), 207–222. DOI: 10.1111/1467-8551.00375

Trapp, C. T., & Kanbach, D. K. (2021). Green entrepreneurship and business models: Deriving green technology business model archetypes. *Journal of Cleaner Production, 297*, 126694. DOI: 10.1016/j.jclepro.2021.126694

Turkish Statistical Institute-TÜİK. (2025). "Inflation & Price" statistics (Consumer Price Index). Retrieved from TÜİK database. Retrieved [19.01.2025] from https://data.tuik.gov.tr/Kategori/GetKategori?dil=2&p=enflasyon-ve-fiyat-106

Tzoulas, K., Korpela, K., Venn, S., Yli-Pelkonen, V., Kaźmierczak, A., Niemela, J., & James, P. (2007). Promoting ecosystem and human health in urban areas using green infrastructure: A literature review. *Landscape and Urban Planning, 81*(3), 167–178. DOI: 10.1016/j.landurbplan.2007.02.001

UNDP. 2024. The Sustainable Development Goals Report 2024. United Nations. https://unstats.un.org/sdgs/report/2024/The-Sustainable-Development-Goals-Report-2024.pdf

UNEP. (2020). *Five Cities Tackling Air Pollution.* https://www.unep.org

United Nations Department of Economic and Social Affairs (UN DESA). (2018). *World urbanization prospects: The 2018 revision.* UN.

United Nations Environment Programme (UNEP). (2019). *Air pollution in Asia and the Pacific: Science-based solutions.* Nairobi: UNEP. https://www.unep.org/resources/report/air-pollution-asia-and-pacific-science-based-solutions

United Nations Environment Programme (UNEP). (2019). *Air pollution in Asia and the Pacific: Science-based solutions.* UNEP.

United Nations Environment Programme. (2011). UNEP billion tree campaign reaches 12 billionmilestone.https://www.unep.org/newscentre/Default.aspx?DocumentID=2659&ArticleID=8930

United Nations Environment Programme. (2016). Gender and environment: Air pollution impacts on women. https://www.unep.org

United Nations. (2015). *Transforming our world: The 2030 Agenda for Sustainable Development.* A/70/L.1 https://www.refworld.org/legal/resolution/unga/2015/en/111816

United Nations. (2015). *Transforming our world: The 2030 Agenda for Sustainable Development.* UN Publishing. https://sdgs.un.org/2030agenda

United Nations. (2015). *Transforming our world: the 2030 Agenda for Sustainable Development.* United Nations General Assembly.

United Nations. (2019). World population prospects 2019. *Department of Economic and Social Affairs, 141*.

United States Environmental Protection Agency (US EPA). (2022). *Integrated science assessment (ISA) for oxides of nitrogen, oxides of sulfur, and particulate matter ecological criteria*. EPA.

Urban Unit. (2024). https://urbanunit.gov.pk/

Usman, N. M., Amjad, N. S., & Khan, N. A. (2023). Clearing the Air: Legal Strategies for Combating Smog and Pollution. *Journal of Strategic Policy and Global Affairs*, *04*(01), 15–21. DOI: 10.58669/jspga.v04.i01.02

Van Eck, N. J., & Waltman, L. (2022). *VOSviewer* manual (Version 1.6.18). Leiden University, CWTS. Retrieved from https://www.vosviewer.com/documentation/Manual_VOSviewer_1.6.18.pdf

Vandenberg, L. N., Rayasam, S. D., Axelrad, D. A., Bennett, D. H., Brown, P., Carignan, C. C., Chartres, N., Diamond, M. L., Joglekar, R., Shamasunder, B., Shrader-Frechette, K., Subra, W. A., Zarker, K., & Woodruff, T. J. (2023). Addressing systemic problems with exposure assessments to protect the public's health. *Environmental Health : A Global Access Science Source*, *21*(S1, Suppl 1), 121. DOI: 10.1186/s12940-022-00917-0 PMID: 36635700

Velasco Perez, M., Sotelo Navarro, P. X., Vazquez Morillas, A., Espinosa Valdemar, R. M., & Hermoso Lopez Araiza, J. P. (2021). Waste management and environmental impact of absorbent hygiene products: A review. *Waste Management & Research*, *39*(6), 767–783. DOI: 10.1177/0734242X20954271

Venkataraman, C., Brauer, M., Tibrewal, K., Sadavarte, P., Mahtta, R., Zhang, Q., & Wang, Y. (2018). Source influence on emission pathways and ambient PM2.5 pollution over India (2015–2050). *Atmospheric Chemistry and Physics*, *18*(11), 8017–8039. DOI: 10.5194/acp-18-8017-2018 PMID: 33679902

Ventriglio, A., Bellomo, A., di Gioia, I., Di Sabatino, D., Favale, D., De Berardis, D., & Cianconi, P. (2021). Environmental pollution and mental health: A narrative review of literature. *CNS Spectrums*, *26*(1), 51–61. DOI: 10.1017/S1092852920001303

Wallace, R., Green, S., & Agarwal, G. (2016). Promoting the health benefits of walking and bicycling to work: A qualitative exploration of the role of healthcare providers in addressing barriers to active commuting. *Sports and Exercise Medicine–Open Journal*, *2*(2), 24–32. DOI: 10.17140/SEMOJ-2-135

Wang, F., Lui, J., & Wang, J.-D. (2023). Losses of life expectancy and productivity associated with COVID-19 pandemic in Canada: Policy implication for future communicable disease control. *International Journal of Environmental Research and Public Health, 20*(3), 2419. DOI: 10.3390/ijerph20032419 PMID: 36767785

Wang, G., Xia, W., Xiao, Y., Guan, X. L., & Zhang, X. (2024). The synergy between co2 and air pollution emissions in chinese cities by 2060: An assessment based on the emissions inventory and dynamic projection model. *Sustainability (Basel), 16*(21), 9338–9338. DOI: 10.3390/su16219338

Wang, L., Zhang, F., Pilot, E., Yu, J., Nie, C., Holdaway, J., Yang, L., Li, Y., Wang, W., Vardoulakis, S., & Krafft, T. (2018). Taking action on air pollution control in the beijing-tianjin-hebei (bth) region: Progress, challenges and opportunities. *International Journal of Environmental Research and Public Health, 15*(2), 306. Advance online publication. DOI: 10.3390/ijerph15020306 PMID: 29425189

Wang, M., Liao, G., & Li, Y. (2021). The relationship between environmental regulation, pollution and corporate environmental responsibility. *International Journal of Environmental Research and Public Health, 18*(15), 8018. DOI: 10.3390/ijerph18158018 PMID: 34360311

Wang, W., Cao, Q., Zhuo, C., Mou, Y., Pu, Z., & Zhou, Y. (2021). COVID-19 to green entrepreneurial intention: Role of green entrepreneurial self-efficacy, optimism, ecological values, social responsibility, and green entrepreneurial motivation. *Frontiers in Psychology, 12*, 732904. DOI: 10.3389/fpsyg.2021.732904

Wang, Y. (2020). Urban land and sustainable resource use: Unpacking the countervailing effects of urbanization on water use in China, 1990–2014. *Land Use Policy, 90*, 104307. DOI: 10.1016/j.landusepol.2019.104307

WCED (World Commission on Environment and Development). (1987). *Our Common Future*. Oxford University Press.

Weiland, S., Hickmann, T., Lederer, M., & Schwindenhammer, S. (2021). *The 2030 agenda for sustainable development: transformative change through the sustainable development goals?* (Vol. 9). Cogitatio Press.

Wei, X., Jiang, Y., & Gan, T. (2024). Air pollution and entrepreneurship: Evidence from China. *Applied Economics, 56*(24), 2860–2874. DOI: 10.1080/00036846.2023.2203454

Westlund, H., & Borsekova, K. (2023). Rural problems, policies and possibilities in a post-urban world. *Regional Science Policy & Practice, 15*(4), 717–729. DOI: 10.1111/rsp3.12664

WHO. (2021). Global Air Quality Guidelines. iris.who.int

WHO. (2023). https://www.who.int/emergencies/situation-report

WHO-World Health Organization. (2023). *Air pollution*. https://www.who.int/news-room/fact-sheets/detail/ambient-(outdoor)-air-quality-and-health

Wichmann, H. E., Mueller, W., Allhoff, P., Beckmann, M., Bocter, N., Csicsaky, M. J., Jung, M., Molik, B., & Schoeneberg, G. (1989). Health effects during a smog episode in West Germany in 1985. *Environmental Health Perspectives*, *79*, 89–99. DOI: 10.1289/ehp.897989

Williams, J. J., Freeman, R., Spooner, F., & Newbold, T. (2022). Vertebrate population trends are influenced by interactions between land use, climatic position, habitat loss and climate change. *Global Change Biology*, *28*(3), 797–815. DOI: 10.1111/gcb.15978 PMID: 34837311

Wolf, M. J., Esty, D. C., Kim, H., Bell, M. L., Brigham, S., Nortonsmith, Q., Zaharieva, S., Wendling, Z. A., De Sherbinin, A., & Emerson, J. W. (2022). New insights for tracking global and local trends in exposure to air pollutants. *Environmental Science & Technology*, *56*(7), 3984–3996. DOI: 10.1021/acs.est.1c08080

World Bank, (2025). Pakistan: Country Climate and Development Report (air pollution ≈ 6.5% of GDP).

World Bank. (2019). The cost of air pollution: Strengthening the economic case for action. https://www.worldbank.org/en/topic/environment/publication/the-cost-of-air-pollution

World Bank. (2020). *The cost of air pollution: Strengthening the economic case for action*. World Bank Group.

World Bank. (2020). *The cost of air pollution: Strengthening the economic case for action*. World Bank Publications. https://openknowledge.worldbank.org/handle/10986/25013

World Bank. (2021). *Air pollution and development: Strengthening policy responses in developing countries*. World Bank Group.

World Bank. (2022). *Pakistan country climate and development report*. https://shorturl.at/25GTE

World Bank. (2023). Air Pollution and Public Health in South Asia / airshed analysis._lib.icimod.org

World Health Organization (WHO). (2020). *Health and the environment: A global perspective*. World Health Organization. https://www.who.int/publications/i/item/health-and-environment-a-global-perspective

World Health Organization (WHO). (2021). *WHO global air quality guidelines: Particulate matter (PM2.5 and PM10), ozone, nitrogen dioxide, sulfur dioxide, and carbon monoxide*. WHO.

World Health Organization (WHO). (2022). *Air pollution and health: Summary report*. WHO.

World Health Organization (WHO). (2022). *Ambient (outdoor) air pollution*. https://www.who.int/news-room/fact-sheets/detail/ambient-(outdoor)-air-pollution

World Health Organization (WHO). (2023). *Air pollution*. https://www.who.int/health-topics/air-pollution

World Health Organization. (2018). Air pollution and child health: Prescribing clean air. https://www.who.int/publications/i/item/air-pollution-and-child-health

World Health Organization. (2023). *Air pollution*. https://www.who.int/health-topics/air-pollution

Xie, Y., Liu, X., Li, D., Zhao, M., Weng, Z., Zhang, L., & Xu, M. (2023). Health and economic benefits of reducing air pollution embodied in gba's green and low-carbon development. *urban climate*null, . DOI: 10.1016/j.uclim.2023.101755

Xing, J., Loughlin, D., Ren, L., Klimont, Z., & Woo, J. (2024). Air quality, health, and equity benefits of carbon neutrality and clean air pathways in china.. *Environmental Science & Technology*null, . DOI: 10.1021/acs.est.3c10076

Xu, H., & Lee, H. S. and Ho Eom, T. (2007) 'Introduction to Panel Data Analysis', *Handbook of Research Methods in Public Administration*, 1(1), pp. 571–587. DOI: 10.1201/9781420013276.ch32

Xu, T., Zhao, L., & Zhou, X. (2024). The impact of environmental policies on air quality improvement and public health in beijing. *Studies in social science & humanities, 3* (6), 42-48. DOI: 10.56397/SSSH.2024.06.06

Xu, H., Xu, X., & Yu, J. (2021). The Impact of Mandatory CSR Disclosure on the Cost of Debt Financing: Evidence from China. *Emerging Markets Finance & Trade*, *57*(8), 2191–2205. DOI: 10.1080/1540496X.2019.1657401

Xu, Z., Peng, J., Qiu, S., Liu, Y., Dong, J., & Zhang, H. (2022). Responses of spatial relationships between ecosystem services and the Sustainable Development Goals to urbanization. *The Science of the Total Environment, 850*, 157868. DOI: 10.1016/j.scitotenv.2022.157868 PMID: 35944627

Yang, B., Wang, Q., Zhao, L., & Wang, H. (2024). How does carbon biased technological progress promote carbon haze collaborative governance?evidence from chinese cities. *Environmental research, 257* null, 119312-119312. DOI: 10.1016/j.envres.2024.119312

Yang, G., Zha, D., Cao, D., & Zhang, G. (2024). Time for a change: Rethinking the global renewable energy transition from the Sustainable Development Goals and the Paris Climate Agreement. *Innovation (Cambridge (Mass.)), 5*(2), 100582. DOI: 10.1016/j.xinn.2024.100582 PMID: 38379793

Yasir, N., Babar, M., Mehmood, H. S., Xie, R., & Guo, G. (2023). The environmental values play a role in the development of green entrepreneurship to achieve sustainable entrepreneurial intention. *Sustainability (Basel), 15*(8), 6451. DOI: 10.3390/su15086451

Ye, Y., & Tao, Q. (2023). The dynamic relationship among economic development, air pollution, and health production in china: The dnsbm efficiency model. *Frontiers in Environmental Science*null, . DOI: 10.3389/fenvs.2023.1205712

Yeh, T.-C., Bai, Y.-M., Hsu, J.-W., Huang, K.-L., Tsai, S.-J., Chu, H.-T., Liang, C.-S., & Chen, M.-H. (2021). Bipolar women's antepartum psychotropic exposure and offspring risk of attention-deficit/hyperactivity disorder and autism spectrum disorder. *Journal of Affective Disorders, 295*, 1407–1414. DOI: 10.1016/j.jad.2021.09.016 PMID: 34565590

Yi, G. (2020). From green entrepreneurial intentions to green entrepreneurial behaviors: The role of university entrepreneurial support and external institutional support. *The International Entrepreneurship and Management Journal, 17*(2), 963–979. DOI: 10.1007/s11365-020-00649-y

Yousaf, H. S., Abbas, M., Ghani, N., Chaudhary, H., Fatima, A., Ahmad, Z., & Yasin, S. A. (2021). A comparative assessment of air pollutants of smog in wagah border and other sites in Lahore, Pakistan. *Brazilian Journal of Biology, 84*, e252471. DOI: 10.1590/1519-6984.252471 PMID: 34932634

Yumkella, K., Batchelor, S., Haselip, J. A., & Brown, E. (2021). Solving the clean cooking conundrum in Africa: Technology options in support of SDG7 and the Paris Agreement on Climate Change. In *Scaling up investment in climate technologies: Pathways to realising technology development and transfer in support of the Paris Agreement* (pp. 13-26).

Zeeshan, N., Murtaza, G., Ahmad, H. R., Awan, A. N., Shahbaz, M., & Freer-Smith, P. (2024). Particulate and gaseous air pollutants exceed WHO guideline values and have the potential to damage human health in Faisalabad, Metropolitan, Pakistan. *Environmental Monitoring and Assessment, 196*(7), 659. DOI: 10.1007/s10661-024-12763-3 PMID: 38916809

Zeffer, W. (2024). An Examination of Environmental Health Literacy: Capacities, Structures, and Engagement (Doctoral dissertation, University of Pittsburgh).

Zeng, J., & Ren, J. (2022). How does green entrepreneurship affect environmental improvement? Empirical findings from 293 enterprises. *The International Entrepreneurship and Management Journal, 18*(1), 409–434. DOI: 10.1007/s11365-021-00780-4

Zhang, H., & Dou, W. (2024). Impact of air pollution prevention and control on urban green economy efficiency: Evidence from china. *Dental science reports, 14* (1), . DOI: 10.1038/s41598-024-79467-z

Zhang, W., Xu, H., Yu, X., Li, J., Zhang, Y., Dai, R., Meng, W., Shen, H., Chen, Y., Shen, G., Ma, J., Li, B., Liu, J., Wang, X., & Tao, S. (2023). Rigorous regional air quality standards for substantial health benefits. *Earth's Future*null, . DOI: 10.1029/2023EF003860

Zhang, J., Liu, Y., Cui, L., Liu, S., Yin, X., & Li, H. (2017). Ambient air pollution, smog episodes and mortality in Jinan, China. *Scientific Reports*, *7*(1), 11209. Advance online publication. DOI: 10.1038/s41598-017-11338-2

Zhang, L. (2008). Conceptualizing China's urbanization under reforms. *Habitat International*, *32*(4), 452–470. DOI: 10.1016/j.habitatint.2008.01.001

Zhang, L., Xu, M., Chen, H., Li, Y., & Chen, S. (2022). Globalization, green economy and environmental challenges: State of the art review for practical implications. *Frontiers in Environmental Science*, *10*, 870271. DOI: 10.3389/fenvs.2022.870271

Zhang, Q., Zheng, Y., Tong, D., Shao, M., Wang, S., Zhang, Y., Xu, X., Wang, J., He, H., Liu, W., Ding, Y., Lei, Y., Li, J., Wang, Z., Zhang, X., Wang, Y., Cheng, J., Liu, Y., Shi, Q., & Hao, J. (2019). Drivers of improved PM2.5 air quality in China from 2013 to 2017. *Proceedings of the National Academy of Sciences of the United States of America*, *116*(49), 24463–24469. DOI: 10.1073/pnas.1907956116 PMID: 31740599

Zhang, X., Deng, T., Zhang, Y., & Jin, R. (2025). Combined intervention: Structural-efficiency-intensity indicators can halve nitrogen-related resource-environmental effects by 2050 in Beijing. *Journal of Environmental Management*, *386*, 125759. DOI: 10.1016/j.jenvman.2025.125759 PMID: 40382935

Zhang, X., Jiao, Z., Zhao, C., Qu, Y., Liu, Q., Zhang, H., Tong, Y., Wang, C., Li, S., Guo, J., Zhu, Z., Yin, S., & Cui, L. (2022). Review of land surface albedo: Variance characteristics, climate effect and management strategy. *Remote Sensing (Basel)*, *14*(6), 1382. DOI: 10.3390/rs14061382

Zhang, X., Zhang, X., & Chen, X. (2018). The impact of exposure to air pollution on labor supply: Evidence from China. *Journal of Environmental Economics and Management*, *90*, 293–309. DOI: 10.1016/j.jeem.2018.05.001

Zhao, A., Sun, J., & Liu, Y. (2023). Understanding bacterial biofilms: From definition to treatment strategies. *Frontiers in Cellular and Infection Microbiology*, *13*, 1137947. DOI: 10.3389/fcimb.2023.1137947 PMID: 37091673

Zhao, Z., Pan, Y., Zhu, J., Wu, J., & Zhu, R. (2022). The impact of urbanization on the delivery of public service–related SDGs in China. *Sustainable Cities and Society*, *80*, 103776. DOI: 10.1016/j.scs.2022.103776

Zheng, S., Pozzer, A., Cao, C. X., & Lelieveld, J. (2018). Long-term (2001–2012) PM2.5 exposures and associated mortality burdens in China. *Proceedings of the National Academy of Sciences of the United States of America*, *112*(11), 3527–3532. DOI: 10.1073/pnas.1422939112

Zhou, Y., Zhang, X., Zhang, C., Chen, B., & Gu, B. (2024). Mitigating air pollution benefits multiple sustainable development goals in china.. *Environmental Pollution*-null, 123992-123992. https://doi.org/DOI: 10.1016/j.envpol.2024.123992

Zhou, X., Cao, Y., & Ma, Y. (2020). Phytoremediation of air pollutants: A review. *Environmental Science and Pollution Research International*, *27*(4), 4489–4509. DOI: 10.1007/s11356-019-07053-6 PMID: 31832956

Zia, H., Khan, T., Fatima, H. S., Khurram, M., Harris, N. R., & Khalil, A. (2022). Impacts of urbanization on green spaces of the densely populated City of Karachi, Pakistan-An analysis of 8 years of data for estimating land cover changes.

Zickafoose, A., Ilesanmi, O., Diaz-Manrique, M., Adeyemi, A. E., Walumbe, B., Strong, R., Wingenbach, G., Rodriguez, M. T., & Dooley, K. (2024). Barriers and challenges affecting quality education (Sustainable Development Goal# 4) in sub-Saharan Africa by 2030. *Sustainability (Basel)*, *16*(7), 2657. DOI: 10.3390/su16072657

Zikria Saleem, H. S., Yousaf, M., Asif, U., Khurshid, F., Hashmi, M. S., & Hassali, M. A. (2019). Evaluating smog awareness and preventive practices among Pakistani general population: a cross-sectional survey.

Zulfiqar, A., Ghaffar, M. M., Shahzad, M., Weis, C., Malik, M. I., Shafait, F., & Wehn, N. (2021). AI-Forest Watch: Semantic segmentation based end-to-end framework for forest estimation and change detection using multi-spectral remote sensing imagery. *Journal of Applied Remote Sensing*, *15*(2), 024518. DOI: 10.1117/1.JRS.15.024518

Zusman, E., Unger, C., Borgford-Parnell, N., & Mar, K. A. (2021). One atmosphere: Integrating air pollution and climate policy and governance. *Atmosphere (Basel)*, *12*(12), 1570. DOI: 10.3390/atmos12121570

About the Contributors

Muhammad Ashfaq (Tenured Professor) Professor Dr. Muhammad Ashfaq is currently working as Professor/Dean, Faculty of Management Studies, the University of Faisalabad-Pakistan. He completed his graduation in 1985 and earned MS degree in Agricultural Economics in 1988. He won the scholarship in 1990-91, from USAID for training in Agricultural Marketing and Sales from Colorado State University, USA. Later on he won scholarship from the World Bank for higher studies in Australia and went to University of New England, Armidale Australia in 1995 and successfully completed his PhD degree. There he carried out his doctoral level research on the key issue of agricultural policy in Pakistan. During his stay in Australia, he also actively participated in various socio-cultural activities. In the year 2006, he also won a prestigious Higher Education Commission funded fellowship and went to University of Laval, Canada for his post doctorate studies. Dr. Ashfaq has rendered services as consultant to various national and international organizations. He is also acting as PhD/MS supervisor for ICDD, Kassel University, Germany. He has also won the TWAS Associateship Award in 2011 and Australian Alumni Excellence Award in 2013 for Innovation and Research. He has supervised 12 PhD students and more than 150 MS/M.Phil/MSc students. He is the author/co-author of 5 books; 14 chapters and more than 100 research papers published in peer reviewed journals. He is also working as editor/section editor of different journals. He has completed 32 research projects as PI/Co-PI which were funded by national and international organisations like USAID, ACIAR, ICARDA, IFPRI, ILRI, AgMIP, JBIC, ADB, JICA, TiKA, UK Met Dept. and Govt of Pakistan. Experience: Prof. Ashfaq have more than 32 years of rich teaching and research experience and have interest and worked on various research projects in the following areas; o Irrigated Agriculture o Impact Assessment of Interventions o Supply Chain Management o Climate Change and Poverty Contact Details

***

Muhammad Aamir Aamir Well-organized and motivated person, with an outgoing personality concerned to work innovatively for the enhancement and betterment of the management and economics, artificial intelligence, machine learning, deep learning,, marketing, ICT, Strategic management, federated learning, user design, user interface in smart homes Education and Research.

Maham Akhtar is a passionate researcher in Economics and a Gold Medalist from GCUF, Pakistan. Her research interests span Health Economics, Development Economics, Environmental Economics and Sustainability, and Behavioral Economics. Currently, she is pursuing her Master's internationally, with the aim of contributing to impactful economic research that addresses critical global challenges.

Murat Avci received his PhD in Management from the Kyrgyz-Turkish Manas University. He works at Kyrgyz-Turkish Manas University. His focus is on the fields of entrepreneurship, management, and organisational behaviour.

Susanta Das is a seasoned educator with a proven record of teaching and mentoring diverse students, working at multiple institutions, and providing service to the various initiatives of organizations. He received his Ph.D. and M.A. degrees from Western Michigan University (WMU), USA, and M.Sc. degree from Banaras Hindu University (BHU), India all in Physics. He continued his research as a Marie-Curie post-doctoral fellow at Stockholm University, Stockholm, Sweden on beam diagnostics for the DESIREE (Double ElectroStatic Ion Ring ExpEriment) in the project DITANET (Diagnostic Techniques for particle Accelerators – a Marie-Curie initial training NETwork), at the Indian Institute of Science Education and Research-Kolkata, India on high-pressure physics as a Project Scientist-B, and at the University of Electro-Communications, Tokyo, Japan on ion-surface interactions as a Post-doctoral fellow. Throughout his career, Dr. Das worked and collaborated with many researchers, Ph.D., master, and visiting students from different countries. He further visited many countries to discuss research and an international conference participant (UK, Italy, Germany, Greece, Belgium, Bulgaria, Romania, Brazil etc.). He co-authored several research articles in WoS/Scopus indexed international journals, conference proceedings, and scholarly book chapters published by renowned international publishing houses. Beyond his research endeavors, Dr. Das has a storied history of service to academic and administrative committees, exhibiting his commitment to the growth and development of educational institutions. His experience includes tenure at Central University South Bihar, Sri Sri University, and P.K. University before assuming his current role at Ajeenkya DY Patil University. He received the Marie-Curie post-doctoral fellowship in Sweden, Science Academies' Summer Research Fellowship in India, Gwen Frostic Doctoral

Fellowship, Department Graduate Research and Creative Scholar Award by WMU, and Leo R. Parpart Doctoral Fellowship by Dept. of Physics, WMU, among many others, throughout his academic journey. At present, Dr. Das continues to delve into cutting-edge fields, with a keen interest in nanotechnology, quantum computing, and data science. His multifaceted contributions, spanning teaching, research, and administrative leadership, showcase a dedicated professional who is instrumental in advancing the vision and mission of the institutions he serves.

Lokeshwari D. V. is a research scholar at Dayananda Sagar University, Bengaluru, India. She holds an M Com degree and KSET-qualified, with academic and research expertise in banking and finance, financial markets, ESG performance, risk management,behavioural Finance,Investment Mangement and artificial intelligence in the banking sector. She is currently pursuing her PhD, focusing on the impact of Environmental, Social, and Governance (ESG) scores on the financial performance and risk profile of Indian banks, with particular emphasis on the pre- and post-mandatory ESG disclosure regime in India. She has published research articles in Scopus-indexed, UGC CARE–listed, and peer-reviewed international journals, and has authored three books with ISBN numbers. Her research interests include ESG and sustainable finance, AI-driven banking, corporate governance, bank performance measurement, and digital transformation in financial services.

Manavi Gilotra is an Assistant Professor at the School of Engineering, Ajeenkya DY Patil University, Pune, India. She holds expertise in Inventory Theory and her research interests include operations research, optimization techniques, and supply chain management. Dr. Gilotra's academic work focuses on developing and analyzing efficient inventory models to enhance decision-making in complex industrial systems.

Shruthi MP currently working as Assistant Professor. She is awarded PhD from Visvesvaraya Technological University, Belgavi. She has five years of academic experience and four years research experience. During this time, the faculty has published four research papers at national and international journals. Her major expertise and interest lie in Finance and Investment Management.

Shivansu Sachan is Pursuing Ph.D. in Management, a Master's holder from Integral University, Lucknow, Bachelor's from IHM, Lucknow. Academic and professional accomplishments of 6 years with renowned organizations and institutes. His Teaching Career has been very satisfying and have continuously upgraded the skills and knowledge by attending many courses and training programs, seminars, conferences organized by different organizations and also delivered many lectures as a Hospitality sector. He is working as an Assistant Professor at Chhatrapati Shahu ji Maharaj University Kanpur .

Shreya Singh hails from Shimla, Himachal Pradesh. She earned her B.Sc. (Hons.) in Horticulture from Dr. Yashwant Singh Parmar University of Horticulture and Forestry at Nauni, Solan, Himachal Pradesh. Following that, she pursued her M.Sc. in Agriculture, specializing in Seed Science and Technology, at Chaudhary Sarwan Kumar Himachal Pradesh Krishi Viswavidyalaya at Palampur, Kangra, Himachal Pradesh. Dr. Singh went on to complete her Ph.D. at Chandra Shekhar Azad University of Agriculture and Technology in Kanpur, Uttar Pradesh, as an ICAR candidate (ICAR-SRF 2020). Throughout her academic journey, she has maintained an exemplary record, working extensively on seed longevity, DUS characterization and seed treatment on legume crops (soybean and mungbean). Dr. Singh has also qualified the ICAR ASRB NET in Seed Science and Technology in 2021 and 2023, as well as the UGC-NET in Environmental Science in 2022. She appeared for the ARS Mains as well as Interview in 2021. Dr. Singh has published numerous research papers, articles, and book chapters in internationally recognized journals, and has co-edited a book entitled Textbook on Practices for Agricultural Sustainability. She has actively participated in and presented posters at various national and international conferences, where she has also won awards. Upon completing her Ph.D. in 2024, Dr. Singh began her professional career as an Assistant Professor (Plant Science) at the School of Advanced Agriculture Sciences and Technology, CSJM University, Kanpur, on July 11, 2024.

Aslı Taşbaşı is a faculty member and the Head of the Department of Economics at Işık University in İstanbul, Türkiye. She received her PhD in economics from Marmara University and completed her doctoral thesis research at Lund University's Department of Economics, Sweden, with a TÜBİTAK Scholarship. Professor Taşbaşı has been working in the field of environmental and natural resources economics for over 15 years and teaches undergraduate and graduate courses in this area. Her work in environmental economics has been recognized with a UNESCO scholarship and she has published articles in national and international peer-reviewed journals, edited international books and authored book chapters addressing issues related to renewable energy, water access, clean water and sanitation, climate justice and sustainable foreign trade. Having spoken at various academic meetings on the economic dimensions of the climate crisis, Professor Taşbaşı is also the moderator of the Green Economy Seminars organized by Sustainable Economic Systems Research Lab- SES Lab and the editor of the lab's publications.

Samina Firoz Waglawala is a distinguished scholar in English Literature, holding a Doctor of Philosophy (Ph.D.) degree. Her groundbreaking doctoral thesis, "A Discourse on the Depiction of Female Protagonists by Male American Writers in Selected Novels," explores the nuanced portrayal of women through the lens of male authors, offering critical insights into gender representation in literature. Currently serving as an Assistant Professor at Ajeenkya DY Patil University, Pune, Dr. Waglawala brings a wealth of expertise and a passion for literature to her teaching. Her dynamic approach fosters a deeper understanding of literary theories and cultural narratives among her students. Known for her analytical rigor and exceptional command of English literature, she has also impressed peers and academics, including Dr. Surendra Sinh Gohil, who praised her during her Ph.D. viva defense. Dr. Waglawala continues to inspire and contribute to the literary academic community.

Index

A

B

C

D

E

F

G

H

I

P

R

S

T

U

V